Contact mechanics

Contact mechanics

K. L. JOHNSON

Emeritus Professor of Engineering, University of Cambridge

CAMBRIDGE
UNIVERSITY PRESS

PUBLISHED BY THE PRESS SYNDICATE OF THE UNIVERSITY OF CAMBRIDGE
The Pitt Building, Trumpington Street, Cambridge, United Kingdom

CAMBRIDGE UNIVERSITY PRESS
The Edinburgh Building, Cambridge CB2 2RU, UK
40 West 20th Street, New York, NY 10011–4211, USA
477 Williamstown Road, Port Melbourne, VIC 3207, Australia
Ruiz de Alarcón 13, 28014 Madrid, Spain
Dock House, The Waterfront, Cape Town 8001, South Africa

http://www.cambridge.org

First published 1985
First paperback edition (with corrections) 1987
Ninth printing 2003

Printed in the United Kingdom at the University Press, Cambridge

A catalogue record for this book is available from the British Library

ISBN 0 521 34796 3 paperback

Contents

Preface

The subject of contact mechanics may be said to have started in 1882 with the publication by Heinrich Hertz of his classic paper *On the contact of elastic solids*. At that time Hertz was only 24, and was working as a research assistant to Helmholtz in the University of Berlin. His interest in the problem was aroused by experiments on optical interference between glass lenses. The question arose whether elastic deformation of the lenses under the action of the force holding them in contact could have a significant influence on the pattern of interference fringes. It is easy to imagine how the hypothesis of an elliptical area of contact could have been prompted by observations of interference fringes such as those shown in Fig. 4.1 (p. 86). His knowledge of electrostatic potential theory then enabled him to show, by analogy, that an ellipsoidal – Hertzian – distribution of contact pressure would produce elastic displacements in the two bodies which were compatible with the proposed elliptical area of contact.

Hertz presented his theory to the Berlin Physical Society in January 1881 when members of the audience were quick to perceive its technological importance and persuaded him to publish a second paper in a technical journal. However, developments in the theory did not appear in the literature until the beginning of this century, stimulated by engineering developments on the railways, in marine reduction gears and in the rolling contact bearing industry.

The Hertz theory is restricted to frictionless surfaces and perfectly elastic solids. Progress in contact mechanics in the second half of this century has been associated largely with the removal of these restrictions. A proper treatment of friction at the interface of bodies in contact has enabled the elastic theory to be extended to both slipping and rolling contact in a realistic way. At the same time development of the theories of plasticity and linear viscoelasticity have enabled the stresses and deformations at the contact of inelastic bodies to be examined.

Somewhat surprisingly, in view of the technological importance of the subject, books on contact mechanics have been few. In 1953 the book by L.A.Galin, *Contact Problems in the Theory of Elasticity*, appeared in Russian summarising the pioneering work of Muskhelishvili in elastic contact mechanics. An up-to-date and thorough treatment of the same field by Gladwell, *Contact Problems in the Classical Theory of Elasticity*, was published in 1980. These books exclude rolling contacts and are restricted to perfectly elastic solids. Analyses of the contact of inelastic solids are scattered through the technical journals or are given brief treatment in the books on the Theory of Plasticity. The aim of the present book, however, is to provide an introduction to most aspects of the mechanics of contact between non-conforming surfaces. Bodies whose surfaces are non-conforming touch first at a point or along a line and, even under load, the dimensions of the contact patch are generally small compared with the dimensions of the bodies themselves. In these circumstances the contact stresses comprise a local 'stress concentration' which can be considered independently of the stresses in the bulk of the two bodies. This fact was clearly appreciated by Hertz who wrote: 'We can confine our attention to that part of each body which is very close to the point of contact, since here the stresses are extremely great compared with those occurring elsewhere, and consequently depend only to the smallest extent on the forces applied to other parts of the bodies.' On the other hand, bodies whose surfaces conform to each other are likely to make contact over an area whose size is comparable with the significant dimensions of the two bodies. The contact stresses then become part of the general stress distribution throughout the bodies and cannot be separated from it. We shall not be concerned with conformal contact problems of this sort.

This book is written by an engineer primarily for the use of professional engineers. Where possible the mathematical treatment is tailored to the level of a first Degree in Engineering. The approach which has been followed is to build up stress distributions by the simple superposition of basic 'point force' solutions – the Green's function method. Complex potentials and integral transform methods, which have played an important role in the modern development of elastic contact stress theory, are only mentioned in passing. In this respect the more mathematically sophisticated reader will find Gladwell's book a valuable complement to Chapters 2–5.

This is a user's book rather than a course text-book. The material is grouped according to application: stationary contacts, sliding, rolling and impact, rather than the usual academic division into elastic, plastic and viscoelastic problems. The stresses and deformations in an elastic half-space under the action of surface tractions, which provide the theoretical basis for the solutions of elastic contact problems, have been treated in Chapters 2 and 3. Results derived there are used throughout the book. These chapters may be regarded as appendices which are not necessary for a qualitative understanding of the later chapters.

In my own study of contact mechanics, which has led to this book, I owe a particular debt of gratitude to R.D. Mindlin, whose pioneering work on the influence of tangential forces on elastic contacts stimulated my early interest in the subject, and to D. Tabor whose revealing experiments and physical insight into surface interactions gave rise to many challenging contact problems.

Several chapters of the book have been read and improved by colleagues whose knowledge and experience in those areas greatly exceeds my own: Dr J.R. Barber, Prof. J. Duffy, Prof. G.M. Gladwell, Dr J.A. Greenwood, Prof. J.J. Kalker, Prof. S.R. Reid, Dr W.J. Stronge and Dr T.R. Thomas. The complete manuscript was read by Dr S.L. Grassie who made many valuable suggestions for improvements in presentation. Responsibility for errors, however, is mine alone and I should be very grateful if readers would inform me of any errors which they detect.

The diagrams were carefully drawn by Mr A. Bailey and the manuscript was most efficiently typed by Mrs Rosalie Orriss and Mrs Sarah Cook. Finally my wife assisted in innumerable ways; without her patience and encouragement the book would never have reached completion.

Cambridge   K. L. Johnson
1984

1

Motion and forces at a point of contact

1.1 Frame of reference

This book is concerned with the stresses and deformation which arise when the surfaces of two solid bodies are brought into contact. We distinguish between *conforming* and *non-conforming* contacts. A contact is said to be conforming if the surfaces of the two bodies 'fit' exactly or even closely together without deformation. Flat slider bearings and journal bearings are examples of conforming contact. Bodies which have dissimilar profiles are said to be non-conforming. When brought into contact without deformation they will touch first at a point – 'point contact' – or along a line – 'line contact'. For example, in a ball-bearing the ball makes point contact with the races, whereas in a roller bearing the roller makes line contact. Line contact arises when the profiles of the bodies are conforming in one direction and non-conforming in the perpendicular direction. The contact area between non-conforming bodies is generally small compared with the dimensions of the bodies themselves; the stresses are highly concentrated in the region close to the contact zone and are not greatly influenced by the shape of the bodies at a distance from the contact area. These are the circumstances with which we shall be mainly concerned in this book.

The points of surface contact which are found in engineering practice frequently execute complex motions and are called upon to transmit both forces and moments. For example, the point of contact between a pair of gear teeth itself moves in space, while at that point the two surfaces move relative to each other with a motion which combines both rolling and sliding. In this preliminary chapter we begin by defining a frame of reference in which the motions and forces which arise in any particular circumstances can be generalised. This approach enables the problems of contact mechanics to be formulated and studied independently of technological particularities and, further, it facilitates the application of the results of such studies to the widest variety of engineering problems.

Non-conforming surfaces brought into contact by a negligibly small force touch at a single point. We take this point O as origin of rectangular coordinate axes $Oxyz$. The two bodies, lower and upper as shown in Fig. 1.1, are denoted by suffixes 1 and 2 respectively. The Oz axis is chosen to coincide with the common normal to the two surfaces at O. Thus the x–y plane is the tangent plane to the two surfaces, sometimes called the osculating plane. The directions of the axes Ox and Oy are chosen for convenience to coincide, where possible, with axes of symmetry of the surface profiles.

Line contact, which arises when two cylindrical bodies are brought into contact with their axes parallel, appears to constitute a special case. Their profiles are non-conforming in the plane of cross-section, but they do conform along a line of contact in the plane containing the axes of the cylinders. Nevertheless this important case is covered by the general treatment as follows: we choose the x-axis to lie in the plane of cross-section and the y-axis parallel to the axes of the cylinders.

The undeformed shapes of two surfaces are specified in this frame by the functions:

$$z_1 = f_1(x, y)$$
$$z_2 = f_2(x, y)$$

Thus the separation between them before loading is given by

$$h = z_1 + z_2 = f(x, y) \tag{1.1}$$

Fig. 1.1. Non-conforming surfaces in contact at O.

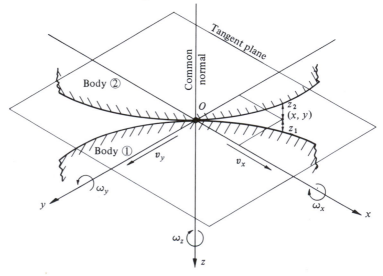

1.2   Relative motion of the surfaces – sliding, rolling and spin

The motion of a body at any instant of time may be defined by the linear velocity vector of an arbitrarily chosen point of reference in the body together with the angular velocity vector of the body. If we now take reference points in each body coincident with the point of contact O at the given instant, body (1) has linear velocity $\mathbf{V}_1$ and angular velocity $\mathbf{\Omega}_1$, and body (2) has linear velocity $\mathbf{V}_2$ and angular velocity $\mathbf{\Omega}_2$. The frame of reference defined above moves with the linear velocity of the contact point $\mathbf{V}_O$ and rotates with angular velocity $\mathbf{\Omega}_O$ in order to maintain its orientation relative to the common normal and tangent plane at the contact point.

Within the frame of reference the two bodies have linear velocities at O:

$$\left. \begin{aligned} \mathbf{v}_1 &= \mathbf{V}_1 - \mathbf{V}_O \\ \mathbf{v}_2 &= \mathbf{V}_2 - \mathbf{V}_O \end{aligned} \right\} \tag{1.2}$$

and angular velocities:

$$\left. \begin{aligned} \boldsymbol{\omega}_1 &= \mathbf{\Omega}_1 - \mathbf{\Omega}_O \\ \boldsymbol{\omega}_2 &= \mathbf{\Omega}_2 - \mathbf{\Omega}_O \end{aligned} \right\} \tag{1.3}$$

We now consider the cartesian components of $\mathbf{v}_1$, $\mathbf{v}_2$, $\boldsymbol{\omega}_1$ and $\boldsymbol{\omega}_2$. If contact is continuous, so the surfaces are neither separating nor overlapping, their velocity components along the common normal must be equal, viz:

$$V_{z1} = V_{z2} = V_{zO}$$

i.e.
$$\left. \vphantom{\begin{aligned}1\\2\\3\end{aligned}} \right\} \tag{1.4}$$

$$v_{z1} = v_{z2} = 0$$

We now define *sliding* as the relative linear velocity between the two surfaces at O and denote it by $\Delta\mathbf{v}$.

$$\Delta\mathbf{v} = \mathbf{v}_1 - \mathbf{v}_2 = \mathbf{V}_1 - \mathbf{V}_2$$

The sliding velocity has components:

$$\Delta v_x = v_{x1} - v_{x2}$$

and
$$\left. \vphantom{\begin{aligned}1\\2\\3\end{aligned}} \right\} \tag{1.5}$$

$$\Delta v_y = v_{y1} - v_{y2}$$

Rolling is defined as a relative angular velocity between the two bodies about an axis lying in the tangent plane. The angular velocity of roll has components:

$$\Delta\omega_x = \omega_{x1} - \omega_{x2} = \Omega_{x1} - \Omega_{x2}$$

and
$$\left. \vphantom{\begin{aligned}1\\2\\3\end{aligned}} \right\} \tag{1.6}$$

$$\Delta\omega_y = \omega_{y1} - \omega_{y2} = \Omega_{y1} - \Omega_{y2}$$

Finally *spin* motion is defined as a relative angular velocity about the common normal, viz.:

$$\Delta\omega_z = \omega_{z1} - \omega_{z2} = \Omega_{z1} - \Omega_{z2} \tag{1.7}$$

Any motion of contacting surfaces must satisfy the condition of continuous contact (1.4) and can be regarded as the combination of sliding, rolling and spin. For example, the wheels of a vehicle normally roll without slide or spin. When it turns a corner spin is introduced; if it skids with the wheels locked, it slides without rolling.

1.3   Forces transmitted at a point of contact

The resultant force transmitted from one surface to another through a point of contact is resolved into a *normal force P* acting along the common normal, which generally must be compressive, and a *tangential force Q* in the tangent plane sustained by friction. The magnitude of **Q** must be less than or, in the limit, equal to the force of limiting friction, i.e.

$$Q \leqslant \mu P \tag{1.8}$$

where μ is the coefficient of limiting friction. **Q** is resolved into components Q_x and Q_y parallel to axes Ox and Oy. In a purely sliding contact the tangential force reaches its limiting value in a direction opposed to the sliding velocity,

Fig. 1.2. Forces and moments acting on contact area *S*.

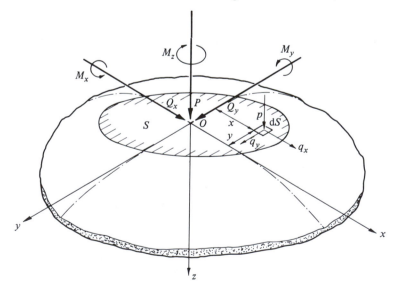

from which:

$$Q_x = -\frac{\Delta v_x}{|\Delta v|}\mu P$$
$$Q_y = -\frac{\Delta v_y}{|\Delta v|}\mu P$$ (1.9)

The force transmitted at a nominal point of contact has the effect of compressing deformable solids so that they make contact over an area of finite size. As a result it becomes possible for the contact to transmit a resultant moment in addition to a force (Fig. 1.2). The components of this moment M_x and M_y are defined as *rolling moments*. They provide the resistance to a rolling motion commonly called 'rolling friction' and in most practical problems are small enough to be ignored.

The third component M_z, acting about the common normal, arises from friction within the contact area and is referred to as the *spin moment*. When spin accompanies rolling the energy dissipated by the spin moment is combined with that dissipated by the rolling moments to make up the overall rolling resistance.

At this point it is appropriate to define *free rolling* ('inertia rolling' in the Russian literature). We shall use this term to describe a rolling motion in which spin is absent and where the *tangential force* **Q** *at the contact point is zero*. This is the condition of the unpowered and unbraked wheels of a vehicle if rolling resistance and bearing friction are neglected; it is in contrast with the driving wheels or braked wheels which transmit sizable tangential forces at their points of contact with the road or rail.

1.4 Surface tractions

The forces and moments which we have just been discussing are transmitted across the contact interface by surface tractions at the interface. The normal traction (pressure) is denoted by p and the tangential traction (due to friction) by q, shown acting positively on the lower surface in Fig. 1.2. While nothing can be said at this stage about the distribution of p and q over the area of contact S, for overall equilibrium:

$$P = \int_S p \, dS$$ (1.10)

$$Q_x = \int_S q_x \, dS, \quad Q_y = \int_S q_y \, dS$$ (1.11)

With non-conforming contacts (including cylinders having parallel axes) the contact area lies approximately in the x-y plane and slight warping is neglected, whence

$$M_x = \int_S py \, dS, \quad M_y = -\int_S px \, dS \tag{1.12}$$

and

$$M_z = \int_S (q_y x - q_x y) \, dS \tag{1.13}$$

When the bodies have closely conforming curved surfaces, as for example in a deep-groove ball-bearing, the contact area is warped appreciably out of the tangent plane and the expressions for M_x and M_y (1.12) have to be modified to include terms involving the shear tractions q_x and q_y. Examples of the treatment of such problems are given later in §8.5.

To illustrate the approach to contact kinematics and statics presented in this chapter, two examples from engineering practice will be considered briefly.

1.5 Examples

Example (1): involute spur gears

The meshing of a pair of involute spur gear teeth is shown in Fig. 1.3(*a*). The wheels rotate about centres C_1 and C_2. The line $I_1 I_2$ is the common tangent

Fig. 1.3. Contact of involute spur gear teeth.

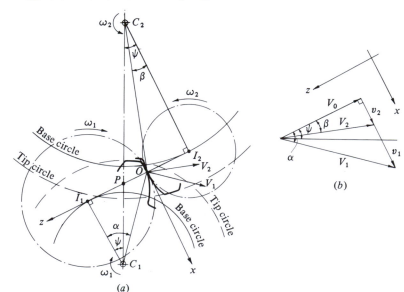

(a)

(b)

to the two base circles from which the involute profiles are generated. P is the pitch point. The teeth are shown in contact at O, which is taken as origin of our coordinate frame of reference. The common normal to the two teeth through O coincides with $I_1 I_2$ and is taken as the z-axis. The x-axis lies in the tangent plane and is taken to be in the plane of rotation as shown.

The point of contact moves along the path $I_1 I_2$ with a velocity V_O: points on the two teeth coincident with O have velocities V_1 and V_2 perpendicular to the radial lines $C_1 O$ and $C_2 O$. Since the path of contact is straight, the frame of reference does not rotate ($\Omega_O = 0$); the wheels rotate with angular velocities $-\omega_1$ and ω_2. (Since the motion lies entirely in the x-z plane, we can omit the suffix y from the angular velocities and the suffix x from the linear velocities.)

Velocities within the frame of reference are shown in Fig. 1.3(b). Applying equation (1.4) for continuity of contact:

$$V_1 \cos \alpha = V_2 \cos \beta = V_O$$

i.e.

$$\omega_1 (C_1 I_1) = \omega_2 (C_2 I_2)$$

therefore

$$\frac{\omega_2}{\omega_1} = \frac{C_1 I_1}{C_2 I_2} = \frac{C_1 P}{C_2 P} \tag{1.14}$$

The angular velocity of rolling about the y-axis is

$$\Delta\omega = -(\omega_1 + \omega_2) \tag{1.15}$$

The velocity of sliding is

$$\begin{aligned}
\Delta v = v_1 - v_2 \\
= V_1 \sin \alpha - V_2 \sin \beta \\
= \omega_1 (OI_1) - \omega_2 (OI_2) \\
= \omega_1 (PI_1 + OP) - \omega_2 (PI_2 - OP)
\end{aligned}$$

i.e.

$$\Delta v = (\omega_1 + \omega_2) OP \tag{1.16}$$

since triangles $C_1 PI_1$ and $C_2 PI_2$ are similar.

Thus the velocity of sliding is equal to the angular velocity of rolling multiplied by the distance of the point of contact from the pitch point. The direction of sliding changes from the arc of approach to the arc of recess and at the pitch point there is pure rolling.

We note that the motion of rolling and sliding at a given instant in the meshing cycle can be reproduced by two circular discs of radii $I_1 O$ and $I_2 O$ rotating with angular velocities $-\omega_1$ and $+\omega_2$ about fixed centres at I_1 and I_2. This is the basis of the *disc machine*, originally developed by Merritt (1935), to simulate the conditions of gear tooth contact in the simple laboratory test. Since the radii of curvature of the involute teeth at O are the same as those of the discs, I_1O and I_2O, the contact stresses under a given contact load are also simulated by the disc machine. The obvious departure from similarity arises from replacing the cyclic behaviour of tooth meshing by a steady motion which reproduces the conditions at only one instant in the meshing cycle.

Example (2): angular contact ball-bearings

An axial cross-section through an angular contact ball-bearing is given in Fig. 1.4, showing a typical ball. The inner and outer races, and the cage (i.e. ball centre C) rotate about the bearing axis with angular velocities Ω_1, Ω_2 and Ω_c respectively. To bring to rest our standard frames of reference, which move with the points of contact between the races and the ball O_i and O_o, we subtract the cage speed from the race speeds, thus

$$\omega_1 = \Omega_1 - \Omega_c, \quad \omega_2 = \Omega_2 - \Omega_c$$

Although the two contact points O_i and O_o are frequently assumed to lie at opposite ends of a ball diameter, they will not do so in general and are deliberately displaced from a diameter in Fig. 1.4. Thus the two sets of axes $O_i x_i y_i z_i$ and $O_o x_o y_o z_o$ will not be in line. If the ball rolls without sliding at the two points of contact, the axis of rotation of the ball (in our prescribed frames of

Fig. 1.4. Angular contact ball bearing, showing contact of the ball (3) with the inner race (1) at O_i and with the outer race (2) at O_o.

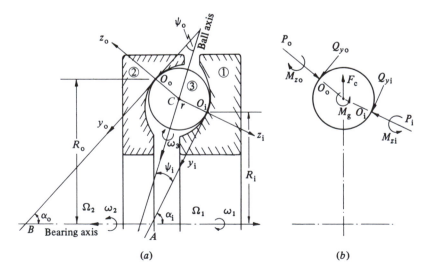

(a)   (b)

reference) must lie in the y-z plane. Its direction in that plane, however, remains to be determined. It is drawn in an arbitrary direction in Fig. 1.4(a), inclined at angle ψ_i, to O_iy_i and ψ_o to O_oy_o. The axes O_iy_i and O_oy_o intersect the bearing axis at points A and B, and at angles α_i and α_o respectively.

For no sliding at O_i:

$$v_{x3} = v_{x1}$$

i.e.

$$\omega_3 r \cos \psi_i = \omega_1 R_i$$

Similarly at O_o

$$\omega_3 r \cos \psi_o = \omega_2 R_o$$

Thus, eliminating ω_3,

$$\frac{\omega_2}{\omega_1} = \frac{R_i \cos \psi_o}{R_o \cos \psi_i} \qquad (1.17)$$

If the points of contact are diametrically opposed, the contact angles α_i and α_o are equal so that $\psi_i = \psi_o$. Only then is the ratio of the race speeds, (1.17), independent of the direction of the axis of rotation of the ball.

We now examine the spin motion at O_i. The angular velocity of spin

$$(\Delta\omega_z)_i = \omega_{z1} - \omega_{z3}$$
$$= \omega_1 \sin \alpha_i - \omega_3 \sin \psi_i$$

i.e.

$$(\Delta\omega_z)_i = \omega_1 \frac{R}{r} \left(\frac{r}{AO_i} - \tan \psi_i \right) \qquad (1.18)$$

From this expression we see that the spin motion at O_i will vanish if the axis of rotation of the ball passes through point A on the axis of the bearing (whereupon $\tan \psi_i = r/(AO_i)$). Similarly, for spin to be absent at O_o, the axis of rotation of the ball must intersect the bearing axis at B. For spin to be absent at both points of contact, either the two tangents O_iy_i and O_oy_o are parallel to the bearing axis, as in a simple radial bearing, or O_i and O_o are so disposed that O_iy_i and O_oy_o intersect the bearing axis at a common point. This latter circumstance is achieved in a taper-roller bearing where the conical races have a common apex on the bearing axis, but never occurs in an angular contact ball-bearing.

We turn now to the forces acting on the ball shown in Fig. 1.4(b). The bearing is assumed to carry a purely axial load so that each ball is identically loaded. Each contact point transmits a normal force $P_{i,o}$ and a tangential force $(Q_y)_{i,o}$. Pressure and friction between the ball and cage pockets introduce small tangential forces in the x-direction at O_i and O_o which are neglected in this example. The rolling friction moments $(M_y)_{i,o}$ will be neglected also, but the spin moments $(M_z)_{i,o}$ play an important role in governing the direction of the axis of rotation of the ball. At high rotational speeds the ball is subjected to an appreciable centrifugal force F_c and a gyroscopic moment M_g.

Consider the equilibrium of the ball; taking moments about the line O_iO_o, it follows that

$$(M_z)_i = (M_z)_o \qquad (1.19)$$

But the positions of the contact points O_i and O_o and the direction of the ball axis ψ_i are not determined by statics alone. In order to proceed further with the analysis it is necessary to know how the tangential forces $(Q_y)_{i,o}$ and the spin moments $M_{zi,o}$ are related to the motions of rolling and spin at O_i and O_o. This question will be considered in Chapter 8, §4.

2

Line loading of an elastic half-space

2.1   The elastic half-space

Non-conforming elastic bodies in contact whose deformation is sufficiently small for the linear small strain theory of elasticity to be applicable
inevitably make contact over an area whose dimensions are small compared
with the radii of curvature of the undeformed surfaces. The contact stresses
are highly concentrated close to the contact region and decrease rapidly in
intensity with distance from the point of contact, so that the region of practical
interest lies close to the contact interface. Thus, provided the dimensions of the
bodies themselves are large compared with the dimensions of the contact area,
the stresses in this region are not critically dependent upon the shape of the
bodies distant from the contact area, nor upon the precise way in which they
are supported. The stresses may be calculated to good approximation by considering each body as a semi-infinite elastic solid bounded by a plane surface:
i.e. an elastic half-space. This idealisation, in which bodies of arbitrary surface
profile are regarded as semi-infinite in extent and having a plane surface, is made
almost universally in elastic contact stress theory. It simplifies the boundary
conditions and makes available the large body of elasticity theory which has
been developed for the elastic half-space.

In this chapter, therefore, we shall study the stresses and deformations in an
elastic half-space loaded one-dimensionally over a narrow strip ('line loading').
In our frame of reference the boundary surface is the x–y plane and the z-axis
is directed into the solid. The loaded strip lies parallel to the y-axis and has
a width $(a + b)$ in the x-direction; it carries normal and tangential tractions
which are a function of x only. We shall assume that a state of plane strain
($\epsilon_y = 0$) is produced in the half-space by the line loading.

For the assumption of plane strain to be justified the thickness of the solid
should be large compared with the width of the loaded region, which is usually

the case. The other extreme of plane stress ($\sigma_y = 0$) would only be realised by the edge loading of a plate whose thickness is small compared with the width of the loaded region, which is a very impractical situation.

The elastic half-space is shown in cross-section in Fig. 2.1. Surface tractions $p(x)$ and $q(x)$ act on the surface over the region from $x = -b$ to $x = a$ while the remainder of the surface is free from traction. It is required to find the stress components σ_x, σ_z and τ_{xz} at all points throughout the solid and the components u_x and u_z of the elastic displacement of any point from its undeformed position. In particular we are interested in the deformed shape of the surface $\bar{u}_z(x)$ (the over bar is used throughout to denote values of the variable at the surface $z = 0$).

The reader is referred to Timoshenko & Goodier: *Theory of Elasticity*, McGraw-Hill, 1951, for a derivation of the elastic equations. For convenience they are summarised below. The stress components must satisfy the equilibrium equations throughout the solid:

$$\left.\begin{aligned}
\frac{\partial \sigma_x}{\partial x} + \frac{\partial \tau_{xz}}{\partial z} &= 0 \\[2mm]
\frac{\partial \sigma_z}{\partial z} + \frac{\partial \tau_{xz}}{\partial x} &= 0
\end{aligned}\right\} \tag{2.1}$$

The corresponding strains ϵ_x, ϵ_z and γ_{xz} must satisfy the compatibility condition:

$$\frac{\partial^2 \epsilon_x}{\partial z^2} + \frac{\partial^2 \epsilon_z}{\partial x^2} = \frac{\partial^2 \gamma_{xz}}{\partial x \partial z} \tag{2.2}$$

Fig. 2.1

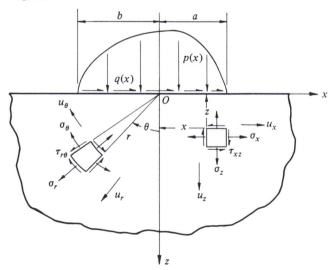

where the strains are related to the displacements by

$$\epsilon_x = \frac{\partial u_x}{\partial x}, \quad \epsilon_z = \frac{\partial u_z}{\partial z}, \quad \gamma_{xz} = \frac{\partial u_x}{\partial z} + \frac{\partial u_z}{\partial x} \tag{2.3}$$

Under conditions of plane strain,

$$\epsilon_y = 0$$
$$\sigma_y = \nu(\sigma_x + \sigma_z) \tag{2.4}$$

whereupon Hooke's law, relating the stresses to the strains, may be written:

$$\left. \begin{aligned} \epsilon_x &= \frac{1}{E} \{(1 - \nu^2)\sigma_x - \nu(1 + \nu)\sigma_z\} \\[2mm] \epsilon_z &= \frac{1}{E} \{(1 - \nu^2)\sigma_z - \nu(1 + \nu)\sigma_x\} \\[2mm] \gamma_{xz} &= \frac{1}{G}\tau_{xz} = \frac{2(1 + \nu)}{E}\tau_{xz} \end{aligned} \right\} \tag{2.5}$$

If a stress function $\phi(x, z)$ is defined by:

$$\sigma_x = \frac{\partial^2 \phi}{\partial z^2}, \quad \sigma_z = \frac{\partial^2 \phi}{\partial x^2}, \quad \tau_{xz} = -\frac{\partial^2 \phi}{\partial x \partial z} \tag{2.6}$$

then the equations of equilibrium (2.1), compatibility (2.2) and Hooke's law (2.5) are satisfied, provided that $\phi(x, z)$ satisfies the biharmonic equation:

$$\left(\frac{\partial^2}{\partial x^2} + \frac{\partial^2}{\partial z^2} \right) \left(\frac{\partial^2 \phi}{\partial x^2} + \frac{\partial^2 \phi}{\partial z^2} \right) = 0 \tag{2.7}$$

In addition the boundary conditions must be satisfied. For the half-space shown in Fig. 2.1 these are as follows. On the boundary $z = 0$, outside the loaded region, the surface is free of stress, i.e.

$$\bar{\sigma}_z = \bar{\tau}_{xz} = 0, \quad x < -b, \quad x > +a \tag{2.8}$$

Within the loaded region

$$\left. \begin{aligned} \bar{\sigma}_z &= -p(x) \\ \bar{\tau}_{xz} &= -q(x) \end{aligned} \right\}, \quad -b \leqslant x \leqslant a \tag{2.9}$$

and the tangential and normal displacements are $\bar{u}_x(x)$ and $\bar{u}_z(x)$. Finally, at a large distance from the loaded region $(x, z \to \infty)$ the stresses must become vanishingly small $(\sigma_x, \sigma_z, \tau_{xz} \to 0)$.

To specify a particular problem for solution two of the four quantities $p(x)$, $q(x)$, $\bar{u}_x(x)$ and $\bar{u}_z(x)$ must be prescribed within the loaded region. Various combinations arise in different contact problems. For example, if a rigid punch is pressed into contact with an elastic half-space the normal displacement $\bar{u}_z(x)$ is prescribed by the known profile of the punch. If the interface is frictionless the second boundary condition is that the shear traction $q(x)$ is zero. Alternatively, if the surface adheres to the punch without slip at the interface, the tangential displacement $\bar{u}_x(x)$ is specified whilst $q(x)$ remains to be found. Special boundary conditions arise if the punch is sliding on the surface of the half-space with a coefficient of friction μ. Only $\bar{u}_z(x)$ is specified but a second boundary condition is provided by the relationship:

$$q(x) = \pm \mu p(x)$$

In some circumstances it is convenient to use cylindrical polar coordinates (r, θ, y). The corresponding equations for the stress components σ_r, σ_θ and $\tau_{r\theta}$, strain components ϵ_r, ϵ_θ and $\gamma_{r\theta}$ and radial and circumferential displacements u_r and u_θ will now be summarised.

The stress function $\phi(r, \theta)$ must satisfy the biharmonic equation:

$$\left(\frac{\partial^2}{\partial r^2} + \frac{1}{r}\frac{\partial}{\partial r} + \frac{1}{r^2}\frac{\partial^2}{\partial \theta^2}\right)\left(\frac{\partial^2 \phi}{\partial r^2} + \frac{1}{r}\frac{\partial \phi}{\partial r} + \frac{1}{r^2}\frac{\partial^2 \phi}{\partial \theta^2}\right) = 0 \qquad (2.10)$$

where

$$
\left.
\begin{aligned}
\sigma_r &= \frac{1}{r}\frac{\partial \phi}{\partial r} + \frac{1}{r^2}\frac{\partial^2 \phi}{\partial \theta^2} \\[2mm]
\sigma_\theta &= \frac{\partial^2 \phi}{\partial r^2} \\[2mm]
\tau_{r\theta} &= -\frac{\partial}{\partial r}\left(\frac{1}{r}\frac{\partial \phi}{\partial \theta}\right)
\end{aligned}
\right\} \qquad (2.11)
$$

The strains are related to the displacements by

$$
\left.
\begin{aligned}
\epsilon_r &= \frac{\partial u_r}{\partial r} \\[2mm]
\epsilon_\theta &= \frac{u_r}{r} + \frac{1}{r}\frac{\partial u_\theta}{\partial \theta} \\[2mm]
\gamma_{r\theta} &= \frac{1}{r}\frac{\partial u_r}{\partial \theta} + \frac{\partial u_\theta}{\partial r} - \frac{u_\theta}{r}
\end{aligned}
\right\} \qquad (2.12)
$$

Equations (2.4) and (2.5) for the stress–strain relationships remain the same with x and z replaced by r and θ.

We shall now proceed to discuss the solutions to particular problems relevant to elastic contact stress theory.

2.2 Concentrated normal force

In this first problem we investigate the stresses produced by a concentrated force of intensity P per unit length distributed along the y-axis and acting in a direction normal to the surface. This loading may be visualised as that produced by a knife-edge pressed into contact with the half-space along the *y*-axis (see Fig. 2.2).

This problem was first solved by Flamant (1892). It is convenient to use polar coordinates in the first instance. The solution is given by the stress function

$$\phi(r, \theta) = Ar\theta \sin \theta \qquad (2.13)$$

where A is an arbitrary constant.

Using equations (2.11), the stress components are

$$\left. \begin{array}{l} \sigma_r = 2A \dfrac{\cos \theta}{r} \\[2mm] \sigma_\theta = \tau_{r\theta} = 0 \end{array} \right\} \qquad (2.14)$$

This system of stresses is referred to as a simple radial distribution directed towards the point of application of the force at *O*. At the surface $\theta = \pm\pi/2$, so that normal stress $\bar{\sigma}_\theta = 0$ except at the origin itself, and the shear stress $\bar{\tau}_{r\theta} = 0$. At a large distance from the point of application of the force ($r \to \infty$) the stresses approach zero, so that all the boundary conditions are satisfied. We note that the stresses decrease in intensity as $1/r$. The theoretically infinite stress at *O* is obviously a consequence of assuming that the load is concentrated along a line. The constant A is found by equating the vertical components of stress acting on a semi-circle of radius r to the applied force P. Thus

$$-P = \int_{-\pi/2}^{\pi/2} \sigma_r \cos \theta \, r \, d\theta = \int_0^{\pi/2} 4A \cos^2 \theta \, d\theta = A\pi$$

Fig. 2.2

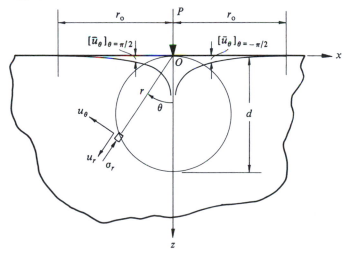

Hence

$$\sigma_r = -\frac{2P}{\pi} \frac{\cos\theta}{r} \tag{2.15}$$

We note that σ_r has a constant magnitude $-2P/\pi d$ on a circle of diameter d which passes through O. Since $\tau_{r\theta} = 0$, σ_r and σ_θ are principal stresses. The principal shear stress τ_1 at (r, θ) has the value $(\sigma_r/2)$ and acts on planes at $45°$ to the radial direction. Hence contours of τ_1 are also a family of circles passing through O. This pattern is clearly demonstrated by the isochromatic fringes in a photo-elastic experiment, as shown in Fig. 4.6(a).

Changing the radial stress distribution of (2.15) into rectangular coordinates we obtain the equivalent stress components

$$\sigma_x = \sigma_r \sin^2\theta = -\frac{2P}{\pi} \frac{x^2 z}{(x^2 + z^2)^2} \tag{2.16a}$$

$$\sigma_z = \sigma_r \cos^2\theta = -\frac{2P}{\pi} \frac{z^3}{(x^2 + z^2)^2} \tag{2.16b}$$

$$\tau_{zx} = \sigma_r \sin\theta \cos\theta = -\frac{2P}{\pi} \frac{xz^2}{(x^2 + z^2)^2} \tag{2.16c}$$

To find the distortion of the solid under the action of the load, we substitute the stresses given by (2.14) and (2.15) into Hooke's law (2.5); this yields the strains, from which we may find the displacements by using equations (2.12) with the result

$$\frac{\partial u_r}{\partial r} = \epsilon_r = -\frac{(1-\nu^2)}{E} \frac{2P}{\pi} \frac{\cos\theta}{r} \tag{2.17a}$$

$$\frac{u_r}{r} + \frac{1}{r} \frac{\partial u_\theta}{\partial\theta} = \epsilon_\theta = \frac{\nu(1+\nu)}{E} \frac{2P}{\pi} \frac{\cos\theta}{r} \tag{2.17b}$$

$$\frac{1}{r} \frac{\partial u_r}{\partial\theta} + \frac{\partial u_\theta}{\partial r} - \frac{u_\theta}{r} = \gamma_{r\theta} = \frac{\tau_{r\theta}}{G} = 0 \tag{2.17c}$$

From these three equations, in the manner demonstrated for plane stress by Timoshenko & Goodier (1951), p. 90, we obtain

$$u_r = -\frac{(1-\nu^2)}{\pi E} 2P\cos\theta \ln r - \frac{(1-2\nu)(1+\nu)}{\pi E} P\theta\sin\theta$$

$$+ C_1 \sin\theta + C_2 \cos\theta \tag{2.18a}$$

and

$$u_\theta = \frac{(1-\nu^2)}{\pi E} 2P\sin\theta \ln r + \frac{\nu(1+\nu)}{\pi E} 2P\sin\theta$$

$$-\frac{(1-2v)(1+v)}{\pi E} P\theta \cos\theta + \frac{(1-2v)(1+v)}{\pi E} P \sin\theta$$

$$+ C_1 \cos\theta - C_2 \sin\theta + C_3 r \qquad (2.18b)$$

If the solid does not tilt, so that points on the z-axis displace only along Oz, then $C_1 = C_3 = 0$. At the surface, where $\theta = \pm\pi/2$,

$$[\bar{u}_r]_{\theta=\frac{\pi}{2}} = [\bar{u}_r]_{\theta=-\frac{\pi}{2}} = -\frac{(1-2v)(1+v)P}{2E} \qquad (2.19a)$$

$$[\bar{u}_\theta]_{\theta=\frac{\pi}{2}} = -[\bar{u}_\theta]_{\theta=-\frac{\pi}{2}} = \frac{(1-v^2)}{\pi E} 2P \ln r + C \qquad (2.19b)$$

where the constant C is determined by choosing a point on the surface at a distance r_0, say (or alternatively on the z-axis below the surface), as a datum for normal displacements. Then

$$[\bar{u}_\theta]_{\theta=\frac{\pi}{2}} = -[\bar{u}_\theta]_{\theta=-\frac{\pi}{2}} = -\frac{(1-v^2)}{\pi E} 2P \ln (r_0/r)$$

The deformed shape of the surface is shown in Fig. 2.2. The infinite displacement at O is to be expected in view of the singularity in stress at that point. Choice of an appropriate value of r_0 presents some difficulty in view of the logarithmic variation of $\bar{u}_\theta$ with r. This is an inevitable feature of two-dimensional deformation of an elastic half-space. To surmount the difficulty it is necessary to consider the actual shape and size of the body and its means of support. This question is discussed further in §5.6.

2.3 Concentrated tangential force

A concentrated force Q per unit length of the y-axis, which acts tangentially to the surface at O as shown in Fig. 2.3, produces a radial stress field similar to that due to a normal force but rotated through $90°$. If we measure θ from the line of action of the force, in this case the Ox direction, the expressions for the stresses are the same as for a normal force, viz.:

$$\left.\begin{array}{l} \sigma_r = -\dfrac{2Q}{\pi}\dfrac{\cos\theta}{r} \\[2mm] \sigma_\theta = \tau_{r\theta} = 0 \end{array}\right\} \qquad (2.20)$$

Contours of constant stress are now semi-circular through O, as shown in Fig. 2.3. Ahead of the force, in the quadrant of positive x, σ_r is compressive, whilst behind the force σ_r is tensile, as we might expect. The expressions for the stress

Fig. 2.3

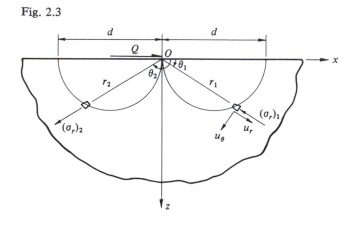

in x–z coordinates may be obtained as before:

$$\sigma_x = -\frac{2Q}{\pi}\frac{x^3}{(x^2+z^2)^2} \tag{2.21a}$$

$$\sigma_z = -\frac{2Q}{\pi}\frac{xz^2}{(x^2+z^2)^2} \tag{2.21b}$$

$$\tau_{xz} = -\frac{2Q}{\pi}\frac{x^2z}{(x^2+z^2)^2} \tag{2.21c}$$

With the appropriate change in the definition of θ, equations (2.18) for the displacement still apply. If there is no rigid body rotation of the solid, nor vertical displacement of points on the z-axis, the surface displacements turn out to be:

$$-[\bar{u}_r]_{\theta=\pi} = [\bar{u}_r]_{\theta=0} = -\frac{(1-\nu^2)}{\pi E}2Q\ln r + C \tag{2.22a}$$

$$[\bar{u}_\theta]_{\theta=\pi} = [\bar{u}_\theta]_{\theta=0} = \frac{(1-2\nu)(1+\nu)}{2E}Q \tag{2.22b}$$

which compare with (2.19) due to normal force. Equation (2.22b) snows that the whole surface ahead of the force $(x>0)$ is depressed by an amount proportional to Q whilst the surface behind Q $(x<0)$ rises by an equal amount. Once again the tangential displacement of the surface varies logarithmically with the distance from O and the datum chosen for this displacement determines the value of the constant C.

2.4   Distributed normal and tangential tractions

In general, a contact surface transmits tangential tractions due to friction in addition to normal pressure. An elastic half-space loaded over the strip $(-b < x < a)$ by a normal pressure $p(x)$ and tangential traction $q(x)$ distributed in any arbitrary manner is shown in Fig. 2.4. We wish to find the stress components due to $p(x)$ and $q(x)$ at any point A in the body of the solid and the displacement of any point C on the surface of the solid.

The tractions acting on the surface at B, distance s from O, on an elemental area of width ds can be regarded as concentrated forces of magnitude $p\,ds$ acting normal to the surface and $q\,ds$ tangential to the surface. The stresses at A due to these forces are given by equations (2.16) and (2.21) in which x is replaced by $(x - s)$. Integrating over the loaded region gives the stress components at A due to the complete distribution of $p(x)$ and $q(x)$. Thus:

$$\sigma_x = -\frac{2z}{\pi}\int_{-b}^{a}\frac{p(s)(x-s)^2\,ds}{\{(x-s)^2+z^2\}^2} - \frac{2}{\pi}\int_{-b}^{a}\frac{q(s)(x-s)^3\,ds}{\{(x-s)^2+z^2\}^2} \qquad (2.23a)$$

$$\sigma_z = -\frac{2z^3}{\pi}\int_{-b}^{a}\frac{p(s)\,ds}{\{(x-s)^2+z^2\}^2} - \frac{2z^2}{\pi}\int_{-b}^{a}\frac{q(s)(x-s)\,ds}{\{(x-s)^2+z^2\}^2} \qquad (2.23b)$$

$$\tau_{xz} = -\frac{2z^2}{\pi}\int_{-b}^{a}\frac{p(s)(x-s)\,ds}{\{(x-s)^2+z^2\}^2} - \frac{2z}{\pi}\int_{-b}^{a}\frac{q(s)(x-s)^2\,ds}{\{(x-s)^2+z^2\}^2} \qquad (2.23c)$$

If the distributions of $p(x)$ and $q(x)$ are known then the stresses can be evaluated although the integration in closed form may be difficult.

The elastic displacements on the surface are deduced in the same way by summation of the displacements due to concentrated forces given in equations (2.19) and (2.22). Denoting the tangential and normal displacement of point C due to the combined action of $p(x)$ and $q(x)$ by $\bar{u}_x$ and $\bar{u}_z$ respectively,

Fig. 2.4

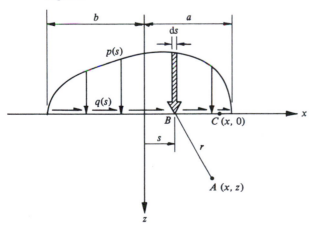

we find

$$\bar{u}_x = -\frac{(1-2v)(1+v)}{2E}\left\{\int_{-b}^{x} p(s)\,ds - \int_{x}^{a} p(s)\,ds\right\}$$

$$-\frac{2(1-v^2)}{\pi E}\int_{-b}^{a} q(s)\ln|x-s|\,ds + C_1 \qquad (2.24a)$$

$$\bar{u}_z = -\frac{2(1-v^2)}{\pi E}\int_{-b}^{a} p(s)\ln|x-s|\,ds$$

$$+\frac{(1-2v)(1+v)}{2E}\left\{\int_{-b}^{x} q(s)\,ds - \int_{x}^{a} q(s)\,ds\right\} + C_2 \qquad (2.24b)$$

The step changes in displacement at the origin which occur in equations (2.19a) and (2.22b) lead to the necessity of splitting the range of integration in the terms in curly brackets in equations (2.24). These equations take on a much neater form, and also a form which is more useful for calculation if we choose to specify the *displacement gradients* at the surface $\partial\bar{u}_x/\partial x$ and $\partial\bar{u}_z/\partial x$ rather than the absolute values of $\bar{u}_x$ and $\bar{u}_z$. The artifice also removes the ambiguity about a datum for displacements inherent in the constants C_1 and C_2. The terms in curly brackets can be differentiated with respect to the limit x, and the other integrals can be differentiated within the integral signs to give

$$\frac{\partial\bar{u}_x}{\partial x} = -\frac{(1-2v)(1+v)}{E}p(x) - \frac{2(1-v^2)}{\pi E}\int_{-b}^{a}\frac{q(s)}{x-s}\,ds \qquad (2.25a)$$

$$\frac{\partial\bar{u}_z}{\partial x} = -\frac{2(1-v^2)}{\pi E}\int_{-b}^{a}\frac{p(s)}{x-s}\,ds + \frac{(1-2v)(1+v)}{E}q(x) \qquad (2.25b)$$

The gradient $\partial\bar{u}_x/\partial x$ will be recognised as the tangential component of strain $\bar{e}_x$ at the surface and the gradient $\partial\bar{u}_z/\partial x$ is the actual slope of the deformed surface.

An important result follows directly from (2.25). Due to the normal pressure $p(x)$ alone $(q(x)=0)$

$$\bar{e}_x = \frac{\partial\bar{u}_x}{\partial x} = -\frac{(1-2v)(1+v)}{E}p(x)$$

But from Hooke's law in plane strain (the first of (2.5)), at the boundary

$$\bar{e}_x = \frac{1}{E}\{(1-v^2)\bar{\sigma}_x - v(1+v)\bar{\sigma}_z\}$$

Equating the two expressions for $\bar{e}_x$ and remembering that $\bar{\sigma}_z = -p(x)$ gives

$$\bar{\sigma}_x = \bar{\sigma}_z = -p(x) \qquad (2.26)$$

Thus under any distribution of surface pressure the tangential and normal direct stresses at the surface are compressive and equal. This state of affairs restricts the tendency of the surface layer to yield plastically under a normal contact pressure.

2.5   Uniform distributions of traction

(a) Normal pressure

The simplest example of a distributed traction arises when the pressure is uniform over the strip $(-a \leqslant x \leqslant a)$ and the shear traction is absent. In equations (2.23) the constant pressure p can be taken outside the integral sign and $q(s)$ is everywhere zero. Performing the integrations and using the notation of Fig. 2.5, we find

$$\sigma_x = -\frac{p}{2\pi}\{2(\theta_1 - \theta_2) + (\sin 2\theta_1 - \sin 2\theta_2)\} \tag{2.27a}$$

$$\sigma_z = -\frac{p}{2\pi}\{2(\theta_1 - \theta_2) - (\sin 2\theta_1 - \sin 2\theta_2)\} \tag{2.27b}$$

$$\tau_{xz} = \frac{p}{2\pi}(\cos 2\theta_1 - \cos 2\theta_2) \tag{2.27c}$$

where

$$\tan \theta_{1,2} = z/(x \mp a)$$

If the angle $(\theta_1 - \theta_2)$ is denoted by α, the principal stresses shown by Mohr's circle in Fig. 2.6 are given by:

$$\sigma_{1,2} = -\frac{p}{\pi}(\alpha \mp \sin \alpha) \tag{2.28}$$

Fig. 2.5

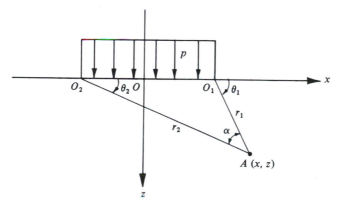

at an angle $(\theta_1 + \theta_2)/2$ to the surface. The principal shear stress has the value

$$\tau_1 = \frac{p}{\pi}\sin\alpha \qquad (2.29)$$

Expressed in this form it is apparent that contours of constant principal stress and constant τ_1 are a family of circles passing through the points O_1 and O_2 as shown in Fig. 2.7(*a*) and by the photoelastic fringes in Fig. 4.6(*b*). The principal shear stress reaches a uniform maximum value p/π along the semi-circle $\alpha = \pi/2$. The trajectories of principal stress are a family of confocal ellipses and hyperbolae with foci O_1 and O_2 as shown in Fig. 2.7(*b*). Finally we note that the stress system we have just been discussing approaches that due to a concentrated normal force at O (§2.2) when r_1 and r_2 become large compared with *a*.

To find the displacements on the surface we use equation (2.25). For a point lying inside the loaded region $(-a \leqslant x \leqslant a)$

$$\frac{\partial \bar{u}_x}{\partial x} = -\frac{(1-2\nu)(1+\nu)}{E}p$$

Then, assuming that the origin does not displace laterally,

$$\bar{u}_x = -\frac{(1-2\nu)(1+\nu)}{E}px \qquad (2.30a)$$

Now

$$\frac{\partial \bar{u}_z}{\partial x} = -\frac{2(1-\nu^2)}{\pi E}\int_{-a}^{a}\frac{ds}{x-s}$$

Fig. 2.6. Mohr's circle for stress due to loading of Fig. 2.5.

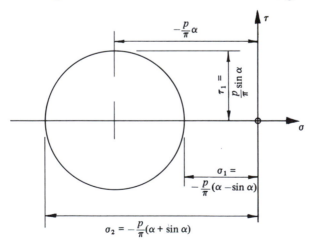

This integral calls for comment: the integrand has a singularity at $s = x$ and changes sign. The integration must be carried out in two parts, from $s = -a$ to $x - \epsilon$ and from $s = x + \epsilon$ to a, where ϵ can be made vanishingly small. The result is then known as the Cauchy Principal Value of the integral, i.e.

$$\int_{-a}^{a} \frac{ds}{x-s} = \int_{-a}^{x-\epsilon} \frac{ds}{x-s} - \int_{x+\epsilon}^{a} \frac{ds}{s-x}$$

$$= [\ln(x-s)]_{-a}^{x-\epsilon} - [\ln(s-x)]_{x+\epsilon}^{a}$$

$$= \ln(a+x) - \ln(a-x)$$

Fig. 2.7. Stresses due to loading of Fig. 2.5: (*a*) Contours of principal stresses σ_1, σ_2 and τ_1; (*b*) Trajectories of principal stress directions.

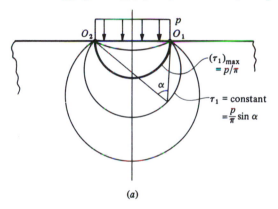

(*a*)

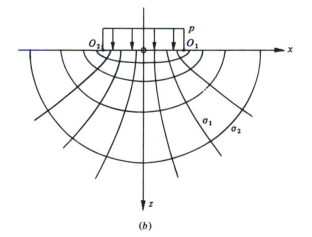

(*b*)

Fig. 2.8

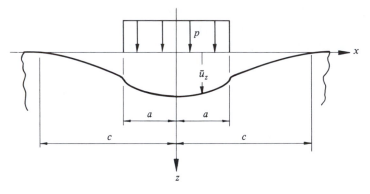

Hence

$$\frac{\partial \bar{u}_z}{\partial x} = -\frac{2(1-\nu^2)}{\pi E} p \{\ln(a+x) - \ln(a-x)\}$$

$$\bar{u}_z = -\frac{(1-\nu^2)}{\pi E} p \left\{(a+x)\ln\left(\frac{a+x}{a}\right)^2 + (a-x)\ln\left(\frac{a-x}{a}\right)^2\right\} + C$$

(2.30b)

For a point outside the loaded region ($|x| > a$)

$$\bar{u}_x = \begin{cases} +\dfrac{(1-2\nu)(1+\nu)}{E} pa, & x < -a \\[2mm] -\dfrac{(1-2\nu)(1+\nu)}{E} pa, & x > a \end{cases}$$

(2.30c)

In this case the integrand in (2.25b) is continuous so that we find

$$\bar{u}_z = -\frac{(1-\nu^2)p}{\pi E} \left\{(x+a)\ln\left(\frac{x+a}{a}\right)^2\right.$$

$$\left. -(x-a)\ln\left(\frac{x-a}{a}\right)^2\right\} + C \qquad (2.30d)$$

which is identical with equation (2.30b). The constant C in equations (2.30b and d) is the same and is fixed by the datum chosen for normal displacements. In Fig. 2.8 the normal displacement is illustrated on the assumption that $\bar{u}_z = 0$ when $x = \pm c$.

(b) Tangential traction

The stresses and surface displacements due to a uniform distribution of tangential traction acting on the strip ($-a \leqslant x \leqslant a$) can be found in the same way. From equations (2.23) putting $p(x) = 0$, we obtain

$$\sigma_x = \frac{q}{2\pi} \{4 \ln (r_1/r_2) - (\cos 2\theta_1 - \cos 2\theta_2)\} \qquad (2.31a)$$

$$\sigma_z = \frac{q}{2\pi} (\cos 2\theta_1 - \cos 2\theta_2) \qquad (2.31b)$$

$$\tau_{xz} = -\frac{q}{2\pi} \{2(\theta_1 - \theta_2) + (\sin 2\theta_1 - \sin 2\theta_2)\} \qquad (2.31c)$$

where $r_{1,2} = \{(x \mp a)^2 + z^2\}^{1/2}$.

Examination of the equations (2.24) for general surface displacements reveals that the surface displacements in the present problem may be obtained directly from those given in equations (2.30) due to uniform normal pressure. Using suffixes p and q to denote displacements due to similar distributions of normal and tangential tractions respectively, we see that

$$(\bar{u}_x)_q = (\bar{u}_z)_p \qquad (2.32a)$$

and

$$(\bar{u}_z)_q = -(\bar{u}_x)_p \qquad (2.32b)$$

provided that the same point is taken as a datum in each case.

The stress distributions in an elastic half-space due to uniformly distributed normal and tangential tractions p and q, given in equations (2.27) and (2.31), have been found by summing the stress components due to concentrated normal or tangential forces (equations (2.16) and (2.21)). An alternative approach is by superposition of appropriate Airy stress functions and subsequent derivation of the stresses by equations (2.6) or (2.11). This method has been applied to the problem of uniform loading of a half-space by Timoshenko & Goodier (1951). Although calculating the stresses by this method is simpler, there is no direct way of arriving at the appropriate stress functions other than by experience and intuition.

It is instructive at this juncture to examine the influence of the discontinuities in p and q at the edges of a uniformly loaded region upon the stresses and displacements at those points. Taking the case of a normal load first, we see from equations (2.27) that the stresses are everywhere finite, but at O_1 and O_2 there is a jump in σ_x from zero outside the region to $-p$ inside it. There is also a jump in τ_{xz} from zero at the surface to p/π just beneath. The surface displacements given by (2.30b) are also finite everywhere (taking a finite value for C) but the *slope* of the surface becomes theoretically infinite at O_1 and O_2. The discontinuity in q at the edge of a region which is loaded tangentially has a strikingly different effect. In equation (2.31a) the logarithmic term leads to an infinite value of σ_x, compressive at O_1 and tensile at O_2, as shown in Fig. 2.9. The

Fig. 2.9

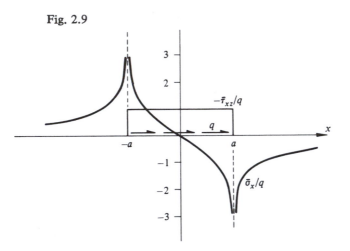

normal displacements of the surface given by equations (2.32) together with (2.30*a* and *c*) are continuous but there is a discontinuity in *slope* at O_1 and O_2. The concentrations of stress implied by the singularities at O_1 and O_2 undoubtedly play a part in the fatigue failure of surfaces subjected to oscillating friction forces – the phenomenon known as fretting fatigue.

2.6   Triangular distributions of traction

Another simple example of distributed loading will be considered. The tractions, normal and tangential, increase uniformly from zero at the surface points O_1 and O_2, situated at $x = \pm a$, to maximum values p_0 and q_0 at O ($x = 0$),

Fig. 2.10

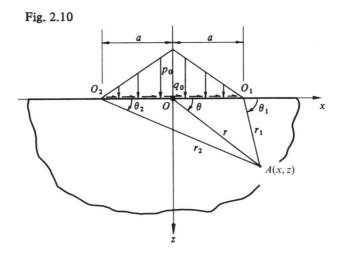

as shown in Fig. 2.10, i.e.

$$p(x) = \frac{p_0}{a}(a - |x|), \quad |x| \leqslant a \tag{2.33}$$

and

$$q(x) = \frac{q_0}{a}(a - |x|), \quad |x| \leqslant a \tag{2.34}$$

These triangular distributions of traction provide the basis for the numerical procedure for two-dimensional contact stress analysis described in §5.9.

When these expressions are substituted into equations (2.23) the integrations are straightforward so that the stresses at any point $A(x, z)$ in the solid may be found. Due to the normal pressure:

$$\sigma_x = \frac{p_0}{\pi a}\{(x - a)\theta_1 + (x + a)\theta_2 - 2x\theta + 2z \ln (r_1 r_2/r^2)\} \tag{2.35a}$$

$$\sigma_z = \frac{p_0}{\pi a}\{(x - a)\theta_1 + (x + a)\theta_2^* - 2x\theta\} \tag{2.35b}$$

$$\tau_{xz} = -\frac{p_0 z}{\pi a}(\theta_1 + \theta_2 - 2\theta) \tag{2.35c}$$

and due to the tangential traction:

$$\sigma_x = \frac{q_0}{\pi a}\{2x \ln (r_1 r_2/r^2) + 2a \ln (r_2/r_1) - 3z(\theta_1 + \theta_2 - 2\theta)\} \tag{2.36a}$$

$$\sigma_z = -\frac{q_0 z}{\pi a}(\theta_1 + \theta_2 - 2\theta) \tag{2.36b}$$

$$\tau_{xz} = \frac{q_0}{\pi a}\{(x - a)\theta_1 + (x + a)\theta_2 - 2x\theta + 2z \ln (r_1 r_2/r^2)\} \tag{2.36c}$$

where $r_1^2 = (x - a)^2 + z^2, r_2^2 = (x + a)^2 + z^2, r^2 = x^2 + z^2$ and $\tan \theta_1 = z/(x - a)$, $\tan \theta_2 = z/(x + a)$, $\tan \theta = z/x$.

The surface displacements are found from equations (2.25). Due to the normal pressure $p(x)$ acting alone, at a point within the loaded region:

$$\frac{\partial \bar{u}_x}{\partial x} = -\frac{(1 - 2v)(1 + v)}{E}\frac{p_0}{a}(a - |x|) \tag{2.37a}$$

i.e.

$$\bar{u}_x = -\frac{(1 - 2v)(1 + v)}{E}\frac{p_0}{a}x(a - \tfrac{1}{2}|x|), \quad |x| \leqslant a \tag{2.37b}$$

relative to a datum at the origin. At a point outside the loaded region:

$$\bar{u}_x = \mp \frac{(1 - 2v)(1 + v)}{E}\frac{p_0 a}{2} \quad \text{for } x \gtrless 0$$

The normal displacement throughout the surface is given by

$$\frac{\partial \bar{u}_z}{\partial x} = -\frac{(1-\nu^2)}{\pi E}\frac{p_0}{a}\left\{(x+a)\ln\left(\frac{x+a}{x}\right)^2 + (x-a)\ln\left(\frac{x-a}{x}\right)^2\right\}$$

i.e.

$$\bar{u}_z = -\frac{(1-\nu^2)}{2\pi E}\frac{p_0}{a}\left\{(x+a)^2\ln\left(\frac{x+a}{a}\right)^2 + (x-a)^2\ln\left(\frac{x-a}{a}\right)^2\right.$$

$$\left. -2x^2\ln(x/a)^2\right\} + C \tag{2.37c}$$

The surface displacements due to a triangular distribution of shear stress are similar and follow from the analogy expressed in equations (2.32).

Examining the stress distributions in equations (2.35) and (2.36) we see that the stress components are all finite and continuous. Equations (2.37) show that the slope of the deformed surface is also finite everywhere. This state of affairs contrasts with that discussed in the last section where there was a discontinuity in traction at the edge of the loaded region.

2.7   Displacements specified in the loaded region

So far we have discussed the stresses and deformations of an elastic half-space to which specified distributions of surface tractions are applied in the loaded region. Since the surface tractions are zero outside the loaded region, the boundary conditions in these cases amount to specifying the distribution of traction over the complete boundary of the half-space. In most contact problems, however, it is the displacements, or a combination of displacements and surface tractions, which are specified within the contact region, whilst outside the contact the surface tractions are specifically zero. It is to these 'mixed boundary-value problems' that we shall turn our attention in this section.

It will be useful to classify the different combinations of boundary conditions with which we have to deal. In all cases the surface of the half-space is considered to be free from traction outside the loaded region and, within the solid, the stresses should decrease as $(1/r)$ at a large distance r from the centre of the loaded region. There are four classes of boundary conditions within the contact region:

Class I: Both tractions, $p(x)$ and $q(x)$, specified. These are the conditions we have discussed in the previous sections. The stresses and surface displacements may be calculated by equations (2.23) and (2.24) respectively.

Class II: Normal displacements $\bar{u}_z(x)$ and tangential traction $q(x)$ specified *or* tangential displacements $\bar{u}_x(x)$ and normal pressure $p(x)$ specified.

The first alternative in this class arises most commonly in the contact of frictionless surfaces, where $q(x)$ is zero everywhere, and the displacements $\bar{u}_z(x)$ are specified by the profile of the two contacting surfaces before deformation. The second alternative arises where the frictional traction $q(x)$ is sought between surfaces which do not slip over all or part of the contact interface, and where the normal traction $p(x)$ is known.

Class III: Normal and tangential displacements $\bar{u}_z(x)$ and $\bar{u}_x(x)$ specified. These boundary conditions arise when surfaces of known profile make contact without interfacial slip. The distributions of both normal and tangential traction are sought.

Class IV: The normal displacement $\bar{u}_z(x)$ is specified, while the tractions are related by $q(x) = \pm\mu p(x)$, where μ is a constant coefficient of friction. This class of boundary conditions clearly arises with solids in sliding contact; $\bar{u}_z(x)$ is specified by their known profiles.

It should be noted that the boundary conditions on different sectors of the loaded region may fall into different classes. For example, two bodies in contact may slip over some portions of the interface, to which the boundary conditions of class IV apply, while not slipping over the remaining portion of the interface where the boundary conditions are of class III.

To formulate two-dimensional problems of an elastic half-space in which displacements are specified over the interval $(-b \leqslant x \leqslant a)$ we use equations (2.25). Using a prime to denote $\partial/\partial x$, we may write these equations:

$$\int_{-b}^{a} \frac{q(s)}{x-s}\, ds = -\frac{\pi(1-2\nu)}{2(1-\nu)}\, p(x) - \frac{\pi E}{2(1-\nu^2)}\, \bar{u}'_x(x) \tag{2.38a}$$

$$\int_{-b}^{a} \frac{p(s)}{x-s}\, ds = \frac{\pi(1-2\nu)}{2(1-\nu)}\, q(x) - \frac{\pi E}{2(1-\nu^2)}\, \bar{u}'_z(x) \tag{2.38b}$$

With known displacements, (2.38) are coupled integral equations for the unknown tractions $p(x)$ and $q(x)$. Within the limits of integration there is a point of singularity when $s = x$, which has led to their being known as 'singular integral equations'. Their application to the theory of elasticity has been advanced notably by Muskhelishvili (1946, 1949) and the Soviet school: Mikhlin (1948) and Galin (1953). The development of this branch of the subject is beyond the scope of this book and only the immediately relevant results will be quoted.

When the boundary conditions are in the form of class II, e.g. $\bar{u}'_z(x)$ and $q(x)$ prescribed, then equations (2.38a) and (2.38b) become uncoupled. Each equation takes the form

$$\int_{-b}^{a} \frac{F(s)}{x-s}\, ds = g(x) \tag{2.39}$$

where $g(x)$ is a known function, made up from a combination of the known component of traction and the known component of displacement gradient, and $F(x)$ is the unknown component of traction. This is a singular integral equation of the first kind; it provides the basis for the solution of most of the two-dimensional elastic contact problems discussed in this book. It has a general solution of the form (see Söhngen, 1954; or Mikhlin, 1948)

$$F(x) = \frac{1}{\pi^2 \{(x+b)(a-x)\}^{1/2}} \int_{-b}^{a} \frac{\{(s+b)(a-s)\}^{1/2} g(s) \, ds}{x-s}$$
$$+ \frac{C}{\pi^2 \{(x+b)(a-x)\}^{1/2}} \qquad (2.40)$$

If the origin is taken at the centre of the loaded region the solution simplifies to

$$F(x) = \frac{1}{\pi^2 (a^2-x^2)^{1/2}} \int_{-a}^{a} \frac{(a^2-s^2)^{1/2} g(s) \, ds}{x-s} + \frac{C}{\pi^2 (a^2-x^2)^{1/2}} \quad (2.41)$$

The constant C is determined by the total load, normal or tangential, from the relationship

$$C = \pi \int_{-a}^{a} F(x) \, dx \qquad (2.42)$$

The integrals in equations (2.40) and (2.41) have a singularity at $s = x$. The *principal value* of these integrals is required, as defined by:

$$\text{P.V.} \int_{-b}^{a} \frac{f(s) \, ds}{x-s} \equiv \underset{\epsilon \to 0}{\text{Limit}} \left[\int_{-b}^{x-\epsilon} \frac{f(s) \, ds}{x-s} + \int_{x+\epsilon}^{a} \frac{f(s) \, ds}{x-s} \right] \qquad (2.43)$$

The principal values of a number of integrals which arise in contact problems are listed in Appendix 1.

The integral equation in which $g(x)$ is of polynomial form:

$$g(x) = Ax^n \qquad (2.44)$$

is of technical importance. An obvious example arises when a rigid frictionless punch or stamp is pressed into contact with an elastic half-space as shown in Fig. 2.11. If the profile of the stamp is of polynomial form

$$z = Bx^{n+1}$$

the normal displacements of the surface are given by

$$\bar{u}_z(x) = \bar{u}_z(0) - Bx^{n+1}$$

thus

$$\bar{u}_z'(x) = -(n+1)Bx^n$$

If the punch is frictionless $q(x) = 0$, so that substituting in equation (2.38b) gives

$$\int_{-b}^{a} \frac{p(s)}{x - s}\, ds = \frac{\pi E}{2(1 - v^2)}(n + 1)Bx^n \tag{2.45}$$

This is an integral equation of the type (2.39) for the pressure $p(x)$, where $g(x)$ is of the form Ax^n. If the contact region is symmetrical about the origin $b = a$, equation (2.45) has a solution of the form expressed in (2.41). The principal value of the following integral is required:

$$I_n \equiv \text{P.V.} \int_{-1}^{+1} \frac{S^n(1 - S^2)^{1/2}\, dS}{X - S} \tag{2.46}$$

where $X = x/a$, $S = s/a$. From the table in Appendix 1

$$I_0 = \text{P.V.} \int_{-1}^{+1} \frac{(1 - S^2)^{1/2}}{X - S}\, dS = \pi X$$

A series for I_n may be developed by writing

$$I_n = X \int_{-1}^{+1} \frac{S^{n-1}(1 - S^2)^{1/2}\, dS}{X - S} - \int_{-1}^{+1} S^{n-1}(1 - S^2)^{1/2}\, dS$$

$$= X I_{n-1} - J_{n-1}$$

$$= X^n I_0 - X^{n-1} J_0 - X^{n-2} J_1 - \cdots - X J_{n-2} - J_{n-1}$$

Fig. 2.11

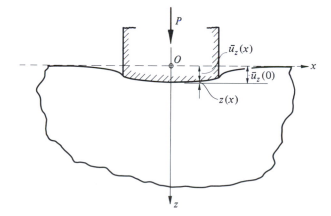

where

$$J_m = \int_{-1}^{+1} S^m (1 - S^2)^{1/2} \, dS$$

$$= \begin{cases} \pi/2 & \text{for } m = 0 \\[2mm] \dfrac{1 \cdot 3 \cdot 5 \ldots (m-1)}{2 \cdot 4 \ldots m(m+2)} \pi & \text{for } m \text{ even} \\[2mm] 0 & \text{for } m \text{ odd} \end{cases}$$

Hence

$$I_n = \begin{cases} \pi \left\{ X^{n+1} - \tfrac{1}{2} X^{n-1} - \tfrac{1}{8} X^{n-3} - \cdots - \dfrac{1 \cdot 3 \cdot 5 \ldots (n-3)}{2 \cdot 4 \ldots n} X \right\} \\[2mm] \qquad \text{for } n \text{ even} \hfill (2.47a) \\[4mm] \pi \left\{ X^{n+1} - \tfrac{1}{2} X^{n-1} - \tfrac{1}{8} X^{n-3} - \cdots - \dfrac{1 \cdot 3 \ldots (n-2)}{2 \cdot 4 \ldots (n+1)} \right\} \\[2mm] \qquad \text{for } n \text{ odd} \hfill (2.47b) \end{cases}$$

If P is the total load on the punch, then by equation (2.42)

$$C = \pi P$$

The pressure distribution under the face of the punch is then given by equation (2.41), i.e.

$$p(x) = -\frac{E(n+1)Ba^{n+1}}{2(1-\nu^2)\pi} \frac{I_n}{(a^2 - x^2)^{1/2}} + \frac{P}{\pi(a^2 - x^2)^{1/2}} \tag{2.48}$$

In this example it is assumed that the load on the punch is sufficient to maintain contact through a positive value of pressure over the whole face of the punch. If n is odd the profile of the punch and the pressure distribution given by equation (2.48) are symmetrical about the centre-line. On the other hand, if n is even, the punch profile is anti-symmetrical and the line of action of the compressive load will be eccentric giving rise to a moment

$$M = \int_{-a}^{a} xp(x) \, dx \tag{2.49}$$

Finally it is apparent from the expression for the pressure given in (2.48) that, in general, the pressure at the edges of the punch rises to a theoretically infinite value.

We turn now to boundary conditions in classes III and IV. When both components of boundary displacement are specified (class III) the integral equations (2.38) can be combined by expressing the required surface tractions as a single complex function:

$$F(x) = p(x) + iq(x) \tag{2.50}$$

Then by adding (2.38a) and (2.38b), we get

$$F(x) - i\,\frac{2(1-v)}{\pi(1-2v)}\int_{-b}^{a}\frac{F(s)\,\mathrm{d}s}{x-s} = -\frac{E}{(1-2v)(1+v)}$$
$$\times\{\bar{u}_x'(x) - i\bar{u}_z'(x)\} \tag{2.51}$$

In the case of sliding motion, where $\bar{u}_z(x)$ is given together with $q(x) = \mu p(x)$ (boundary conditions of class IV), equation (2.38b) becomes

$$p(x) - \frac{2(1-v)}{\pi\mu(1-2v)}\int_{-b}^{a}\frac{p(s)}{x-s}\,\mathrm{d}s = \frac{E}{\mu(1-2v)(1+v)}\,\bar{u}_z'(x) \tag{2.52}$$

To simplify equations (2.51) and (2.52) we shift the origin to the mid-point of the contact region (i.e. put $b = a$), and put $X = x/a$, $S = s/a$. Equations (2.51) and (2.52) are both integral equations of the second kind having the form

$$F(X) + \frac{\lambda}{\pi}\int_{-1}^{+1}\frac{F(S)\,\mathrm{d}S}{X-S} = G(X) \tag{2.53}$$

where $G(X)$, $F(X)$ and λ can be real or complex. The function $G(X)$ is known and it is required to find the function $F(X)$. λ is a parameter whose value depends upon the particular problem. The solution to (2.53) is given by Söhngen (1954) in the form

$$F(X) = F_1(X) + F_0(X) \tag{2.54}$$

where $F_0(X)$ is the solution of the homogeneous equation, i.e. equation (2.53) with the right-hand side put equal to zero. He gives

$$F_1(X) = \frac{1}{1+\lambda^2}\,G(X) - \frac{\lambda}{1+\lambda^2}\,\frac{1}{\pi(1-X^2)^{1/2}}\left(\frac{1+X}{1-X}\right)^{\gamma}$$
$$\times\int_{-1}^{1}(1-S^2)^{1/2}\left(\frac{1-S}{1+S}\right)^{\gamma}\frac{G(S)}{X-S}\,\mathrm{d}S \tag{2.55}$$

where γ is a complex constant related to λ by $\cot(\pi\gamma) = \lambda$, i.e. $e^{2\pi i\gamma} = (i\lambda - 1)/(i\lambda + 1)$, restricted so that its real part $\mathrm{Re}(\gamma)$ lies within the interval $-\frac{1}{2}$ to $+\frac{1}{2}$, and

$$F_0(X) = -\frac{\lambda}{(1+\lambda^2)^{1/2}}\,\frac{1}{\pi(1-X^2)^{1/2}}\left(\frac{1+X}{1-X}\right)^{\gamma}C \tag{2.56}$$

where the constant

$$C = \int_{-1}^{+1}F(X)\,\mathrm{d}X = \frac{1}{a}(P + iQ)$$

If λ is imaginary, so that $\lambda = i\lambda_1$, then λ_1 must lie outside the interval -1 to $+1$. This condition is met in the problems considered here.

We will take first the case of both boundary displacements specified (class III), where we require the solution of equation (2.51). Comparing the general solution given in (2.55) and (2.56) with equation (2.51) we see that λ is imaginary ($\lambda = i\lambda_1$) where

$$\lambda_1 = -\frac{2 - 2\nu}{1 - 2\nu} \tag{2.57}$$

Since ν lies between 0 and $\frac{1}{2}$, $\lambda_1 < -2$ which makes the solution given by (2.56) valid. Thus γ is also imaginary, so that, putting $\gamma = i\eta$, we have

$$e^{-2\pi\eta} = \frac{-\lambda_1 - 1}{-\lambda_1 + 1} = \frac{1}{3 - 4\nu}$$

giving

$$\eta = \frac{1}{2\pi} \ln (3 - 4\nu) \tag{2.58}$$

Substituting for λ and γ from equations (2.57) and (2.58) in (2.56) the required solution is

$$p(X) + iq(X) = F(X) = F_1(X) + F_0(X)$$

where

$$F_1(X) = \frac{(1 - 2\nu)E}{(3 - 4\nu)(1 + \nu)} \{\bar{u}_x'(X) - i\bar{u}_z'(X)\} + i\,\frac{2(1 - \nu)E}{(3 - 4\nu)(1 + \nu)}$$

$$\times \frac{1}{\pi(1 - X^2)^{1/2}} \left(\frac{1 + X}{1 - X}\right)^{i\eta} \int_{-1}^{+1} (1 - S^2)^{1/2} \left(\frac{1 - S}{1 + S}\right)^{i\eta}$$

$$\times \left\{\frac{\bar{u}_x'(S) - i\bar{u}_z'(S)}{X - S}\right\} \mathrm{d}S \tag{2.59a}$$

and

$$F_0(X) = \frac{2(1 - \nu)}{(3 - 4\nu)^{1/2}} \frac{P + iQ}{\pi a(1 - X^2)^{1/2}} \left(\frac{1 + X}{1 - X}\right)^{i\eta} \tag{2.59b}$$

To obtain expressions for the surface tractions $p(X)$ and $q(X)$ requires the evaluation of the integral in (2.59a). So far only a few problems, in which the distributions of displacement $\bar{u}_x'(X)$ and $\bar{u}_z'(X)$ are particularly simple, have been solved in closed form. For an incompressible material ($\nu = 0.5$) the basic integral equations (2.38) become uncoupled. In this case we see from (2.58) that $\eta = 0$, whereupon the general solution to the coupled equations given by equation (2.59), when real and imaginary parts are separated, reduces to the solution of two uncoupled equations of the form of (2.41).

When the boundary conditions are of class IV, by comparing equations (2.52) with the general form (2.53), we see that λ is real: i.e.

$$\lambda = -\frac{2(1-\nu)}{\mu(1-2\nu)} = \cot \pi\gamma \qquad (2.60)$$

Substituting for λ in the general solution (2.55) and (2.56) gives

$$p(X) = -\frac{E \sin \pi\gamma \cos \pi\gamma}{2(1-\nu^2)} \bar{u}_z'(X) + \frac{E \cos^2 \pi\gamma}{2(1-\nu^2)} \frac{1}{\pi(1-X^2)^{1/2}} \left(\frac{1+X}{1-X}\right)^\gamma$$

$$\times \int_{-1}^{+1} (1-S^2)^{1/2} \left(\frac{1-S}{1+S}\right)^\gamma \frac{\bar{u}_z'(S)}{X-S} \, dS + \frac{P \cos \pi\gamma}{\pi a(1-X^2)^{1/2}}$$

$$\times \left(\frac{1+X}{1-X}\right)^\gamma \qquad (2.61)$$

and

$$q(X) = \mu p(X).$$

Once again, for an incompressible material, or when the coefficient of friction approaches zero, γ approaches zero and the integral equations become uncoupled. Equation (2.61) then degenerates into the uncoupled solution (2.41).

Having found the surface tractions $p(X)$ and $q(X)$ to satisfy the displacement boundary conditions, we may find the internal stresses in the solid, in principle at least, by the expressions for stress given in equations (2.23).

An example in the application of the results presented in this section is provided by the indentation of an elastic half-space by a rigid two-dimensional punch which has a flat base. This example will be discussed in the next section.

2.8 Indentation by a rigid flat punch

In this section we consider the stresses produced in an elastic half-space by the action of a rigid punch pressed into the surface as shown in Fig. 2.12. The punch has a flat base of width $2a$ and has sharp square corners; it is long in the y-direction so that plane-strain conditions can be assumed. Since the punch is rigid the surface of the elastic solid must remain flat where it is in contact with the punch. We shall restrict our discussion to indentations in which the punch does not tilt, so that the interface, as well as being flat, remains parallel to the undeformed surface of the solid. Thus our first boundary condition within the contact region is one of specified normal displacement:

$$\bar{u}_z(x) = \text{constant} = \delta_z \qquad (2.62)$$

The second boundary condition in the loaded region depends upon the frictional conditions at the interface. We shall consider four cases:

(a) that the surface of the punch is frictionless, so that $q(x) = 0$;

(b) that friction at the interface is sufficient to prevent any slip between the punch and the surface of the solid so that $\bar{u}_x(x) = \text{constant} = \delta_x$;

(c) that partial slip occurs to limit the tangential traction $|q(x)| \leqslant \mu p(x)$; and

(d) that the punch is sliding along the surface of the half-space from right to left, so that $q(x) = \mu p(x)$ at all points on the interface, where μ is a constant coefficient of sliding friction.

No real punch, of course, can be perfectly rigid, although this condition will be approached closely when a solid of low elastic modulus such as a polymer or rubber is indented by a metal punch. Difficulties arise in allowing for the elasticity of the punch, since the deformation of a square-cornered punch cannot be calculated by the methods appropriate to a half-space. However the results of this section are of importance in circumstances other than that of a punch indentation. We shall use the stresses arising from constant displacements δ_x and δ_z in the solution of other problems (see §§5.5 & 7.2).

(a) Frictionless punch

The boundary conditions:

$$\bar{u}_z(x) = \text{constant}, \quad q(x) = 0 \tag{2.63}$$

are of class II as defined in the last section so that the pressure distribution is given by the integral equation (2.38b) which has the general solution (2.41) in which

$$g(s) = -\frac{\pi E}{2(1 - \nu^2)} \bar{u}_z'(x) = 0$$

Fig. 2.12

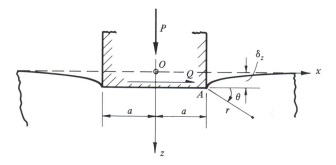

In this case the result reduces to the homogeneous solution while $C = \pi P$:

$$p(x) = \frac{P}{\pi(a^2 - x^2)^{1/2}} \qquad (2.64)$$

This pressure distribution is plotted in Fig. 2.13(a) (curve A). The pressure reaches a theoretically infinite value at the edges of the punch ($x = \pm a$). The stresses within the solid in the vicinity of the corners of the punch have been found by Nadai (1963). The sum of the principal stresses is given by

$$\sigma_1 + \sigma_2 \approx -\frac{2P}{\pi(2ar)^{1/2}} \sin(\theta/2) \qquad (2.65a)$$

and the principal shear stress

$$\tau_1 \approx -\frac{P}{2\pi(2ar)^{1/2}} \sin\theta \qquad (2.65b)$$

Fig. 2.13. (a) Tractions on the face of flat punch shown in Fig. 2.12: curve A – Frictionless, eq. (2.64) for $p(x)$; curve B – No slip, exact eq. (2.69) for $p(x)$; curve C – No slip, exact eq. (2.69) for $q(x)$; curve D – No slip, approx. eq. (2.72) for $q(x)$; curve E – Partial slip, $p(x)$; curve F – Partial slip, $q(x)$ (curves E & F from Spence, 1973). (b) Ratio of tangential traction $q(x)$ to normal traction $p(x)$: curve G – No slip, approx. eqs (2.72) and (2.64); curve H – Partial slip, from Spence (1973) ($\nu = 0.3, \mu = 0.237$ giving $c = 0.5$).

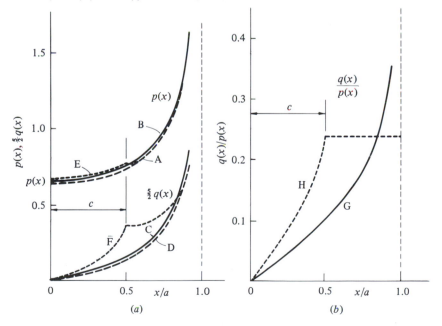

where r and θ are polar coordinates from an origin at $x = \pm a$ and $r \ll a$ (see the photoelastic fringe pattern in Fig. 4.6(c)). At the surface of the solid $\sigma_1 = \sigma_2 = \sigma_x = \sigma_z$. From (2.65b) the principal shear stress is seen to reach a theoretically infinite value as $r \to 0$ so that we would expect a real material to yield plastically close to the corners of the punch even at the lightest load. The displacement of the surface outside the punch can be calculated from (2.24b) with the result

$$\bar{u}_z(x) = \delta_z - \frac{2(1-\nu^2)P}{\pi E} \ln\left\{\frac{x}{a} + \left(\frac{x^2}{a^2} - 1\right)^{1/2}\right\} \tag{2.66}$$

where, as usual, δ_z can only be determined relative to an arbitrarily chosen datum. We note that the surface gradient $\bar{u}_z'$ is infinite at $x = \pm a$. From (2.24a) we find the tangential displacements under the punch to be

$$\bar{u}_x(x) = -\frac{(1-2\nu)(1+\nu)P}{\pi E} \sin^{-1}(x/a) \tag{2.67}$$

For a compressible material ($\nu < 0.5$), this expression shows that points on the surface move towards the centre of the punch. In practice this motion would be opposed by friction, and, if the coefficient of friction were sufficiently high, it might be prevented altogether. We shall now examine this possibility.

(b) No slip
If the surface of the solid adheres completely to the punch during indentation then the boundary conditions are

$$\bar{u}_x(x) = \delta_x \quad \text{and} \quad \bar{u}_z(x) = \delta_z \tag{2.68}$$

where δ_x and δ_z are the (constant) displacements of the punch. These boundary conditions, in which both displacements are specified, are of class III. The integral equations (2.38) for the tractions at the surface of the punch are now coupled and their general solution is given by equation (2.59). Since the displacements are constant, $\bar{u}_x'(x) = \bar{u}_z'(x) = 0$, so that only the solution to the homogeneous equation (2.59b) remains, viz.:

$$p(x) + iq(x) = \frac{2(1-\nu)}{(3-4\nu)^{1/2}} \frac{P+iQ}{\pi(a^2-x^2)^{1/2}} \left(\frac{a+x}{a-x}\right)^{i\eta}$$

$$= \frac{2(1-\nu)}{(3-4\nu)^{1/2}} \frac{P+iQ}{\pi(a^2-x^2)^{1/2}}$$

$$\times \left[\cos\left\{\eta \ln\left(\frac{a+x}{a-x}\right)\right\} + i \sin\left\{\eta \ln\left(\frac{a+x}{a-x}\right)\right\}\right] \tag{2.69}$$

where $\eta = (1/2\pi) \ln(3 - 4\nu)$.

The tractions $p(x)$ and $q(x)$ under the action of a purely normal load ($Q = 0$) have been computed for $\nu = 0.3$ and are plotted in Fig. 2.13(a) (curves B and C).

The nature of the singularities at $x = \pm a$ is startling. From the expression (2.69) it appears that the tractions fluctuate in sign an infinite number of times as $x \to a$! However the maximum value of η is $(\ln 3)/2\pi$ which results in the pressure first becoming negative when $x = \pm a \tanh (\pi^2/2 \ln 3)$ i.e. when $x = \pm 0.9997a$, which is very close to the edge of the punch. We conclude that this anomalous result arises from the inadequacy of the linear theory of elasticity to handle the high strain gradients in the region of the singularity. Away from those points, we might expect equation (2.69) to provide an accurate measure of the stresses.

If a tangential force Q acts on the punch in addition to a normal force, additional shear and normal tractions arise at the interface. With complete adhesion, these are also given by equation (2.69) such that, due to unit loads

$$[q(x)]_Q = [p(x)]_P$$

and

$$(2.70)$$

$$[p(x)]_Q = -[q(x)]_P$$

Since $[q(x)]_P$ is an odd function of x, the influence of the tangential force is to reduce the pressure on the face of the punch where x is positive and to increase it where x is negative. A moment is then required to keep the punch face square. Close to $x = +a$ the pressure would tend to become negative unless the punch were permitted to tilt to maintain contact over the whole face. Problems of a tilted punch have been solved by Muskhelishvili (1949) and are discussed by Gladwell (1980).

At this juncture it is instructive to compare the pressure distribution in the presence of friction computed from equation (2.69) with that in the absence of friction from equation (2.64) (see Fig. 2.13(a)). The difference is not large, showing that the influence of the tangential traction on the normal pressure is small for $\nu = 0.3$. Larger values of ν will make the difference even smaller. Therefore, in more difficult problems than the present one, the integral equations can be uncoupled by assuming that the pressure distribution in the presence of friction is the same as that without friction. Thus we put $q(x) = 0$ in equation (2.38b) and solve it to find $p(x)$ without friction. This solution for $p(x)$ is then substituted in equation (2.38a) to find an approximate solution for $q(x)$. Each integral equation is then of the first kind having a solution of the form (2.41).

If this expedient is used in the present example, the pressure given by (2.64) is substituted in equation (2.38a) to give

$$\int_{-a}^{a} \frac{q(s)}{x-s} \, ds = -\frac{(1-2\nu)}{2(1-\nu)} \frac{P}{(a^2 - x^2)^{1/2}} \qquad (2.71)$$

Using the general solution to this equation given by (2.41), we get

$$q(x) = -\frac{(1-2\nu)}{2\pi^2(1-\nu)}\frac{P}{(a^2-x^2)^{1/2}}\int_{-a}^{a}\frac{ds}{x-s} + \frac{Q}{\pi(a^2-x^2)^{1/2}}$$

$$= -\frac{(1-2\nu)}{2\pi^2(1-\nu)}\frac{P}{(a^2-x^2)^{1/2}}\ln\left(\frac{a+x}{a-x}\right) + \frac{Q}{\pi(a^2-x^2)^{1/2}} \qquad (2.72)$$

This approximate distribution of tangential traction is also plotted in Fig. 2.13(*a*) for $Q = 0$ (curve D). It is almost indistinguishable from the exact solution given by (2.69).

(c) Partial slip

In case (*b*) above it was assumed that friction was capable of preventing slip entirely between the punch and the half-space. The physical possibility of this state of affairs under the action of a purely normal load P can be examined by considering the ratio of tangential to normal traction $q(x)/p(x)$. This ratio is plotted in Fig. 2.13(*b*) (curve G) using the approximate expressions for $q(x)$ and $p(x)$, i.e. equations (2.72) and (2.64) respectively, from which it is apparent that theoretically infinite values are approached at the edges of the contact. (The same conclusion would be reached if the exact expressions for $q(x)$ and $p(x)$ were used.) This means that, in practice, some slip must take place at the edges on the contact.

The problem of partial slip was studied first by Galin (1945) and more completely by Spence (1973). Under a purely normal load the contact is symmetrical about the centre-line so that the no-slip region will be centrally placed from $x = -c$ to $x = +c$, say. The boundary condition $\bar{u}_z(x) = \delta_z = $ constant still applies for $-a < x < a$, but the condition $\bar{u}_x(x) = \delta_x = 0$ is restricted to the no-slip zone $-c < x < c$. In the slip zones $c \leqslant |x| \leqslant a$

$$q(x) = \pm\mu p(x)$$

The extent of the no-slip zone is governed by the values of Poisson's ratio ν and the coefficient of friction μ. The problem is considerably simplified if the integral equations are uncoupled by neglecting the influence of tangential traction on normal pressure. With this approximation Spence has shown that c is given by the relationship

$$\mathbf{K}'(c/a)/\mathbf{K}(c/a) = (1-2\nu)/2(1-\nu)\mu \qquad (2.73)$$

where $\mathbf{K}(c/a)$ is the complete elliptic integral of the second kind and $\mathbf{K}'(c/a) = \mathbf{K}(1 - c^2/a^2)^{1/2}$. The values of $p(x)$ and $q(x)$ and the ratio $q(x)/p(x)$ with partial slip have been calculated for $\nu = 0.3$, $\mu = 0.237$ and are plotted in curves E, F and H in Fig. 2.13(*a* and *b*).

(d) Sliding punch

A punch which is sliding over the surface of a half-space at a speed much less than the velocity of elastic waves, so that inertia forces can be neglected, has the boundary conditions:

$$\bar{u}_z(x) = \text{constant} = \delta_z, \quad q(x) = \mu p(x) \tag{2.74}$$

These are boundary conditions in class IV, so that the coupled integral equations for the surface tractions combine to give equation (2.52) having as its general solution equation (2.61). In the case of the flat punch, $\bar{u}'_z(x) = 0$, so that once again we only require the solution to the homogeneous equation, viz.:

$$p(x) = \frac{P \cos \pi\gamma}{\pi(a^2 - x^2)^{1/2}} \left(\frac{a+x}{a-x}\right)^{\gamma} \tag{2.75}$$

where

$$\cot \pi\gamma = -\frac{2(1-\nu)}{\mu(1-2\nu)}$$

This pressure distribution is plotted for $\nu = 0.3$ and $\mu = 0.5$ in Fig. 2.14, where it is compared with the pressure distribution in the absence of friction.

Fig. 2.14. Pressure on the face of a sliding punch: curve A frictionless; curve B with friction ($\nu = 0.3, \mu = 0.5$).

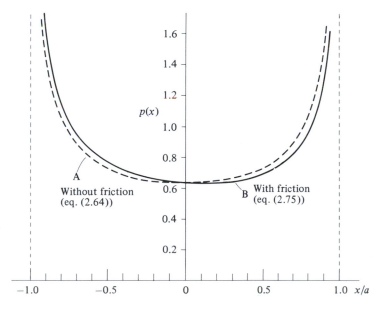

The punch is moving from right to left so that the effect of friction is to reduce the pressure in the front half of the punch and to increase it on the rear. In this case also it is apparent that the influence of frictional traction upon the normal pressure is relatively small.

2.9 Traction parallel to y-axis

A different form of two-dimensional deformation occurs when tangential tractions, whose magnitude and distribution are independent of the y-coordinate, act on the surface of the half-space in the y-direction. Clearly x–z cross-sections of the solid will not remain plane but will be warped by the action of the surface traction. Since all cross-sections will deform alike, however, the stress field and resulting deformations will be independent of y.

We will consider first a concentrated tangential force of magnitude Q_y per unit length acting on the surface along the y-axis, as shown in Fig. 2.15. This force produces a simple shear-stress distribution which can be derived easily as follows.

We think first of two half-spaces with their surfaces glued together to make a complete infinite solid. A force $2Q_y$ per unit length acts along the y-axis. In cylindrical coordinates r, θ, y the stress system in the complete solid must be axially symmetrical and independent of θ and y. By considering the equilibrium of a cylinder of radius r we find

$$2\pi r \tau_{ry} = -2Q_y$$

i.e.

$$\tau_{ry} = -\frac{Q_y}{\pi r} \tag{2.76a}$$

Fig. 2.15

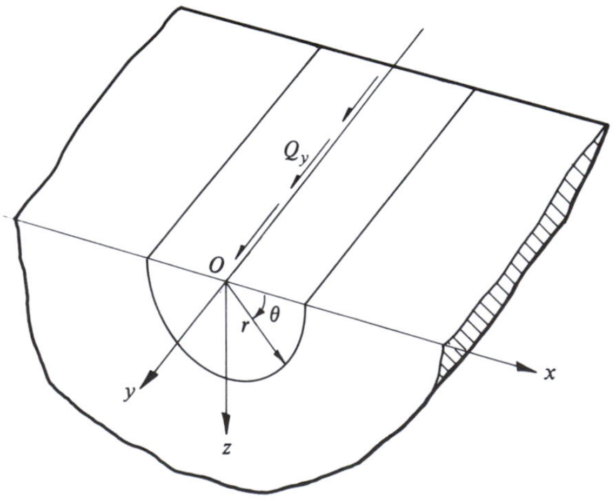

The other stress components must vanish, i.e.

$$\sigma_r = \sigma_\theta = \sigma_y = 0 \tag{2.76b}$$

The strains associated with these stresses are compatible and result in an identical warping of each cross-section of the solid. Now no stresses act across the interface where the two half-spaces are glued together, so that they can be separated without changing the stress distribution. Thus equations (2.76) describe the required stress distribution. In Cartesian coordinates

$$\tau_{xy} = \tau_{ry} \cos \theta = -\frac{Q_y x}{\pi r^2} \tag{2.77a}$$

$$\tau_{yz} = \tau_{ry} \sin \theta = -\frac{Q_y z}{\pi r^2} \tag{2.77b}$$

The deformations are found from the only strain component

$$\frac{\partial u_y}{\partial r} + \frac{\partial u_r}{\partial y} = \gamma_{ry} = \tau_{ry}/G$$

Since the deformation is independent of y, $\partial u_r/\partial y = 0$, so

$$\frac{\partial u_y}{\partial r} = \tau_{ry}/G = -\frac{Q_y}{\pi G r} \tag{2.78}$$

At the surface $z = 0$, this becomes

$$\frac{\partial \bar{u}_y}{\partial x} = -\frac{Q_y}{\pi G x} \tag{2.79}$$

or

$$\bar{u}_y = -\frac{Q_y}{\pi G} \ln |x| + C \tag{2.80}$$

and, neglecting rigid body motions, $\bar{u}_x = \bar{u}_z = 0$. Comparing this result with equations (2.19) and (2.22), we note that the displacement in the direction of the force is similar in form to that produced by a concentrated normal force, or by a tangential force in the x-direction. However in this case the displacement perpendicular to the force is absent.

The stresses and displacements produced by a tangential traction, distributed over the strip $-a \leqslant x \leqslant a$, are found as before by summing the effects of a concentrated load $Q_y = q_y(s)\, ds$ acting on an elemental strip of width ds. As an example we shall consider the tangential traction

$$q_y(x) = q_0(1 - x^2/a^2)^{1/2} \tag{2.81}$$

acting on the strip $-a \leqslant x \leqslant a$. From equations (2.77)

$$\tau_{xy} = -\frac{q_0}{\pi a} \int_{-a}^{a} \frac{(a^2 - s^2)^{1/2}(x - s)\, ds}{(x - s)^2 + z^2} \tag{2.82a}$$

and

$$\tau_{yz} = -\frac{q_0 z}{\pi a} \int_{-a}^{a} \frac{(a^2 - s^2)^{1/2} \, ds}{(x - s)^2 + z^2} \qquad (2.82b)$$

Evaluation of the integrals is straightforward. On the surface ($z = 0$) it is found that

$$\bar{\tau}_{xy} = \begin{cases} -q_0 x/a, & -a \leqslant x \leqslant a \\ -q_0 [(x/a) \mp \{(x/a)^2 - 1\}^{1/2}], & |x| > a \end{cases} \qquad (2.83a)$$

and on the axis of symmetry ($x = 0$)

$$\tau_{yz} = -q_0 [\{1 + (z/a)^2\}^{1/2} - z/a] \qquad (2.83b)$$

The surface displacements follow from equation (2.79)

$$\frac{\partial \bar{u}_y}{\partial x} = -\frac{q_0}{\pi G} \int_{-a}^{a} \frac{(a^2 - s^2)^{1/2}}{(x - s)} \, ds$$

From the list of principal values of integrals of this type given in Appendix 1 it follows that

$$\frac{\partial \bar{u}_y}{\partial x} = -\frac{q_0}{G} \frac{x}{a}$$

Thus

$$\bar{u}_y = -\frac{q_0 x^2}{2Ga} + C \qquad (2.84)$$

The results obtained in this example are used later (§8.3) when studying the contact stresses in rolling cylinders which transmit a tangential force parallel to the axes of the cylinders.

3

Point loading of an elastic half-space

3.1 Potential functions of Boussinesq and Cerruti

In this chapter we consider the stresses and deformations produced in an elastic half-space, bounded by the plane surface $z = 0$, under the action of normal and tangential tractions applied to a closed area S of the surface in the neighbourhood of the origin. Outside the loaded area both normal and tangential tractions are zero. Thus the problem in elasticity is one in which the tractions are specified throughout the whole surface $z = 0$. In view of the restricted area to which the loads are applied, it follows that all the components of stress fall to zero at a long distance from the origin. The loading is two-dimensional: the normal pressure $p(x, y)$ and the tangential tractions $q_x(x,y)$ and $q_y(x,y)$, in general, vary in both x and y directions. The stress system is three-dimensional therefore; in general all six components of stress, σ_x, σ_y, σ_z, τ_{xy}, τ_{yz}, τ_{zx}, will appear.

A special case arises when the loading is axi-symmetric about the z-axis. In cylindrical polar coordinates (r, θ, z) the pressure $p(r)$ and the tangential traction $q(r)$ are independent of θ and $q(r)$, if it is present, acts in a radial direction. The stress components $\tau_{r\theta}$ and $\tau_{\theta z}$ vanish and the other stress components are independent of θ.

The classical approach to finding the stresses and displacements in an elastic half-space due to surface tractions is due to Boussinesq (1885) and Cerruti (1882) who made use of the theory of potential. This approach is presented by Love (1952): only selected results will be quoted here.

The half-space is shown in Fig. 3.1. We take $C(\xi, \eta)$ to be a general surface point within the loaded area S, whilst $A(x, y, z)$ is a general point within the body of the solid. The distance

$$CA \equiv \rho = \{(\xi - x)^2 + (\eta - y)^2 + z^2\}^{1/2} \tag{3.1}$$

Distributions of traction $p(\xi, \eta), q_x(\xi, \eta)$ and $q_y(\xi, \eta)$ act on the area S. The following potential functions, each satisfying Laplace's equation, are defined:

$$F_1 = \int_S \int q_x(\xi, \eta)\Omega \, d\xi \, d\eta$$

$$G_1 = \int_S \int q_y(\xi, \eta)\Omega \, d\xi \, d\eta \tag{3.2}$$

$$H_1 = \int_S \int p(\xi, \eta)\Omega \, d\xi \, d\eta$$

where

$$\Omega = z \ln (\rho + z) - \rho \tag{3.3}$$

In addition we define the potential functions

$$F = \frac{\partial F_1}{\partial z} = \int_S \int q_x(\xi, \eta) \ln (\rho + z) \, d\xi \, d\eta$$

$$G = \frac{\partial G_1}{\partial z} = \int_S \int q_y(\xi, \eta) \ln (\rho + z) \, d\xi \, d\eta \tag{3.4}$$

$$H = \frac{\partial H_1}{\partial z} = \int_S \int p(\xi, \eta) \ln (\rho + z) \, d\xi \, d\eta$$

We now write

$$\psi_1 = \frac{\partial F_1}{\partial x} + \frac{\partial G_1}{\partial y} + \frac{\partial H_1}{\partial z} \tag{3.5}$$

Fig. 3.1

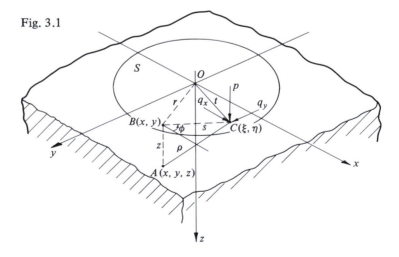

and

$$\psi = \frac{\partial \psi_1}{\partial z} = \frac{\partial F}{\partial x} + \frac{\partial G}{\partial y} + \frac{\partial H}{\partial z} \tag{3.6}$$

Love (1952) shows that the components of elastic displacement u_x, u_y and u_z at any point $A(x, y, z)$ in the solid can be expressed in terms of the above functions as follows:

$$u_x = \frac{1}{4\pi G}\left\{2\frac{\partial F}{\partial z} - \frac{\partial H}{\partial x} + 2v\frac{\partial \psi_1}{\partial x} - z\frac{\partial \psi}{\partial x}\right\} \tag{3.7a}$$

$$u_y = \frac{1}{4\pi G}\left\{2\frac{\partial G}{\partial z} - \frac{\partial H}{\partial y} + 2v\frac{\partial \psi_1}{\partial y} - z\frac{\partial \psi}{\partial y}\right\} \tag{3.7b}$$

$$u_z = \frac{1}{4\pi G}\left\{\frac{\partial H}{\partial z} + (1 - 2v)\psi - z\frac{\partial \psi}{\partial z}\right\} \tag{3.7c}$$

These expressions decrease as $(1/\rho)$ at large distances from the loaded region. They represent, therefore, the elastic displacements of points close to the loaded region relative to the points in the solid at a large distance from the loaded region $(\rho \to \infty)$ where the half-space may be looked upon as fixed. This behaviour in two-dimensional loading, where a datum for displacements can be taken at infinity, compares favourably with one-dimensional loading, considered in the previous chapter, where a variation of displacements as $\ln \rho$ precludes taking a datum at infinity and imposes an arbitrary choice of datum.

The displacements having been found, the stresses are calculated from the corresponding strains by Hooke's law:

$$\sigma_x = \frac{2vG}{1 - 2v}\left(\frac{\partial u_x}{\partial x} + \frac{\partial u_y}{\partial y} + \frac{\partial u_z}{\partial z}\right) + 2G\frac{\partial u_x}{\partial x} \tag{3.8a}$$

$$\sigma_y = \frac{2vG}{1 - 2v}\left(\frac{\partial u_x}{\partial x} + \frac{\partial u_y}{\partial y} + \frac{\partial u_z}{\partial z}\right) + 2G\frac{\partial u_y}{\partial y} \tag{3.8b}$$

$$\sigma_z = \frac{2vG}{1 - 2v}\left(\frac{\partial u_x}{\partial x} + \frac{\partial u_y}{\partial y} + \frac{\partial u_z}{\partial z}\right) + 2G\frac{\partial u_z}{\partial z} \tag{3.8c}$$

$$\tau_{xy} = G\left(\frac{\partial u_x}{\partial y} + \frac{\partial u_y}{\partial x}\right) \tag{3.8d}$$

$$\tau_{yz} = G\left(\frac{\partial u_y}{\partial z} + \frac{\partial u_z}{\partial y}\right) \tag{3.8e}$$

$$\tau_{zx} = G\left(\frac{\partial u_z}{\partial x} + \frac{\partial u_x}{\partial z}\right) \tag{3.8f}$$

Under the action of a purely normal pressure $p(\xi, \eta)$, which would occur in a frictionless contact, the above equations may be simplified. Here

$$F = F_1 = G = G_1 = 0$$

whence

$$\psi_1 = \frac{\partial H_1}{\partial z} = H = \int_S \int p(\xi, \eta) \ln (\rho + z) \, d\xi \, d\eta \tag{3.9}$$

$$\psi = \frac{\partial H}{\partial z} = \frac{\partial \psi_1}{\partial z} = \int_S \int p(\xi, \eta) \, \frac{1}{\rho} \, d\xi \, d\eta \tag{3.10}$$

$$\iota_x = -\frac{1}{4\pi G} \left\{ (1 - 2\nu) \frac{\partial \psi_1}{\partial x} + z \frac{\partial \psi}{\partial x} \right\} \tag{3.11a}$$

$$u_y = -\frac{1}{4\pi G} \left\{ (1 - 2\nu) \frac{\partial \psi_1}{\partial y} + z \frac{\partial \psi}{\partial y} \right\} \tag{3.11b}$$

$$u_z = \frac{1}{4\pi G} \left\{ 2(1 - \nu)\psi - z \frac{\partial \psi}{\partial z} \right\} \tag{3.11c}$$

Remembering that ψ and ψ_1 are harmonic functions of x, y and z, i.e. they both satisfy Laplace's equation,

$$\nabla^2 \psi = 0, \quad \nabla^2 \psi_1 = 0$$

the dilatation Δ is given by

$$\Delta \equiv \frac{\partial u_x}{\partial x} + \frac{\partial u_y}{\partial y} + \frac{\partial u_z}{\partial z} = \frac{1 - 2\nu}{2\pi G} \frac{\partial \psi}{\partial z} \tag{3.12}$$

Substitution of equations (3.11) and (3.12) into equations (3.8) gives expressions for the components of stress at any point in the solid. These are:

$$\sigma_x = \frac{1}{2\pi} \left\{ 2\nu \frac{\partial \psi}{\partial z} - z \frac{\partial^2 \psi}{\partial x^2} - (1 - 2\nu) \frac{\partial^2 \psi_1}{\partial x^2} \right\} \tag{3.13a}$$

$$\sigma_y = \frac{1}{2\pi} \left\{ 2\nu \frac{\partial \psi}{\partial z} - z \frac{\partial^2 \psi}{\partial y^2} - (1 - 2\nu) \frac{\partial^2 \psi_1}{\partial y^2} \right\} \tag{3.13b}$$

$$\sigma_z = \frac{1}{2\pi} \left\{ \frac{\partial \psi}{\partial z} - z \frac{\partial^2 \psi}{\partial z^2} \right\} \tag{3.13c}$$

$$\tau_{xy} = -\frac{1}{2\pi} \left\{ (1 - 2\nu) \frac{\partial^2 \psi_1}{\partial x \partial y} + 2 \frac{\partial^2 \psi}{\partial x \partial y} \right\} \tag{3.13d}$$

$$\tau_{yz} = -\frac{1}{2\pi} z \frac{\partial^2 \psi}{\partial y \partial z} \tag{3.13e}$$

$$\tau_{zx} = -\frac{1}{2\pi} z \frac{\partial^2 \psi}{\partial x \partial z} \tag{3.13f}$$

We note that the stress components σ_z, τ_{yz} and τ_{zx} depend upon the function ψ only. The stress components σ_x and σ_y depend upon the function ψ_1 but their sum does not, thus

$$\sigma_x + \sigma_y = \frac{1}{2\pi}\left\{(1 + 2\nu)\frac{\partial \psi}{\partial z} + z\frac{\partial^2 \psi}{\partial z^2}\right\} \tag{3.14}$$

At the surface of the solid the normal stress

$$\bar{\sigma}_z = \frac{1}{2\pi}\left(\frac{\partial \psi}{\partial z}\right)_{z=0} = \left\{\begin{matrix} -p(\xi, \eta) & \text{inside } S \\ 0 & \text{outside } S \end{matrix}\right\} \tag{3.15}$$

and the surface displacements are

$$\bar{u}_x = -\frac{1 - 2\nu}{4\pi G}\left(\frac{\partial \psi_1}{\partial x}\right)_{z=0} \tag{3.16a}$$

$$\bar{u}_y = -\frac{1 - 2\nu}{4\pi G}\left(\frac{\partial \psi_1}{\partial y}\right)_{z=0} \tag{3.16b}$$

$$\bar{u}_z = \frac{1 - \nu}{2\pi G}\left(\frac{\partial \psi_1}{\partial z}\right)_{z=0} = \frac{1 - \nu}{2\pi G}(\psi)_{z=0} \tag{3.16c}$$

Equations (3.15) and (3.16c) show that the normal pressure and normal displacement within the loaded area depend only on the potential function ψ.

The equations quoted above provide a formal solution to the problem of stresses and deformations in an elastic half-space with prescribed tractions acting on the surface. If the distributions of traction within the area S are known explicitly then, in principle, the stresses and displacements at any point in the solid can be found. In practice, obtaining expressions in closed form for the stresses in any but the simplest problems presents difficulties. In particular circumstances more sophisticated analytical techniques have been developed to overcome some of the difficulties of the classical approach. A change from rectangular to ellipsoidal coordinates enables problems in which the loaded area is bounded by an ellipse to be handled more conveniently (Lur'e, 1964; Galin, 1953; de Pater, 1964). For circular contact areas, the use of a special complex stress function suggested by Rostovtzev (1953) (see also Green & Zerna, 1954) enables the stresses to be found when the displacements are specified within the loaded area.

For the case of axial symmetry Sneddon (1951) has put forward integral transform methods which have been developed by Noble & Spence (1971) (see Gladwell, 1980, for a full discussion of this approach).

An alternative approach, which is the one generally followed in this book, is to start from the stresses and displacements produced by concentrated normal and tangential forces. The stress distribution and deformation resulting from any distributed loading can then be found by superposition. This approach has the merit that it lends itself to numerical analysis and makes possible the solution of problems in which the geometry makes analytical methods impossible.

3.2   Concentrated normal force

The stresses and displacements produced by a concentrated point force P acting normally to the surface at the origin (Fig. 3.2) can be found directly in several ways. Using the results of the previous section, the area S over which the normal traction acts is made to approach zero, thus

$$\rho = (x^2 + y^2 + z^2)^{1/2}$$

and

$$\int_S \int p(\xi, \eta) \, d\xi \, d\eta = P \tag{3.17}$$

The Boussinesq potential functions ψ_1 and ψ, defined in equations (3.9) and (3.10), in this case reduce to

$$\psi_1 = \frac{\partial H_1}{\partial z} = H = P \ln (\rho + z)$$

$$\psi = \frac{\partial H}{\partial z} = P/\rho$$

Substituting (3.11) for the elastic displacements at any point in the solid gives

$$u_x = \frac{P}{4\pi G} \left\{ \frac{xz}{\rho^3} - (1 - 2\nu) \frac{x}{\rho(\rho + z)} \right\} \tag{3.18a}$$

$$u_y = \frac{P}{4\pi G} \left\{ \frac{yz}{\rho^3} - (1 - 2\nu) \frac{y}{\rho(\rho + z)} \right\} \tag{3.18b}$$

Fig. 3.2

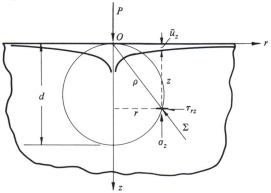

$$u_z = \frac{P}{4\pi G}\left\{\frac{z^2}{\rho^3} + \frac{2(1-v)}{\rho}\right\} \tag{3.18c}$$

where ρ is given by (3.17).

The stress components are then given by equations (3.13) with the results:

$$\sigma_x = \frac{P}{2\pi}\left[\frac{(1-2v)}{r^2}\left\{\left(1-\frac{z}{\rho}\right)\frac{x^2-y^2}{r^2} + \frac{zy^2}{\rho^3}\right\} - \frac{3zx^2}{\rho^5}\right] \tag{3.19a}$$

$$\sigma_y = \frac{P}{2\pi}\left[\frac{(1-2v)}{r^2}\left\{\left(1-\frac{z}{\rho}\right)\frac{y^2-x^2}{r^2} + \frac{zx^2}{\rho^3}\right\} - \frac{3zy^2}{\rho^5}\right] \tag{3.19b}$$

$$\sigma_z = -\frac{3P}{2\pi}\frac{z^3}{\rho^5} \tag{3.19c}$$

$$\tau_{xy} = \frac{P}{2\pi}\left[\frac{(1-2v)}{r^2}\left\{\left(1-\frac{z}{\rho}\right)\frac{xy}{r^2} - \frac{xyz}{\rho^3}\right\} - \frac{3xyz}{\rho^5}\right] \tag{3.19d}$$

$$\tau_{xz} = -\frac{3P}{2\pi}\frac{xz^2}{\rho^5} \tag{3.19e}$$

$$\tau_{yz} = -\frac{3P}{2\pi}\frac{yz^2}{\rho^5} \tag{3.19f}$$

where $r^2 = x^2 + y^2$. The equations in this form are useful for finding, by direct superposition, the stresses due to a distributed normal load.

Alternatively, we may recognise from the outset that the system is axi-symmetric and use polar coordinates. Timoshenko & Goodier (1951) start by introducing a suitable stress function for axi-symmetric problems and use it to deduce the stresses produced by a concentrated normal force acting on the surface of an elastic half-space with the results:

$$\sigma_r = \frac{P}{2\pi}\left\{(1-2v)\left(\frac{1}{r^2} - \frac{z}{\rho r^2}\right) - \frac{3zr^2}{\rho^5}\right\} \tag{3.20a}$$

$$\sigma_\theta = -\frac{P}{2\pi}(1-2v)\left(\frac{1}{r^2} - \frac{z}{\rho r^2} - \frac{z}{\rho^3}\right) \tag{3.20b}$$

$$\sigma_z = -\frac{3P}{2\pi}\frac{z^3}{\rho^5} \tag{3.20c}$$

$$\tau_{rz} = -\frac{3P}{2\pi}\frac{rz^2}{\rho^5} \tag{3.20d}$$

It is easy to see that equations (3.19) and (3.20) are identical by putting $x = r$, $y = 0$, $\sigma_x = \sigma_r$ and $\sigma_y = \sigma_\theta$ in equations (3.19). Note also that

$$\sigma_r + \sigma_\theta + \sigma_z = -\frac{P}{\pi}\frac{(1+v)z}{\rho^3} \tag{3.20e}$$

The direct and shear stresses σ_z and τ_{rz} which act on planes within the solid parallel to the free surface are independent of Poisson's ratio. Consider the resultant stress Σ acting on elements of such parallel planes. Where those planes intersect the spherical surface of diameter d which is tangential to the surface of the half-space at O, as shown in Fig. 3.2, Σ is given by

$$\Sigma = (\sigma_z^2 + \tau_{rz}^2)^{1/2} = \frac{3P}{2\pi} \frac{z^2}{\rho^4} = \frac{3P}{2\pi d^2} = \text{constant}$$

and the direction of the resultant stress acts towards O. There is some analogy in this result with that for a concentrated line load discussed in the last chapter (§2.2 and Fig. 2.2). In the three-dimensional case, however, it should be noted that Σ is not a principal stress and that the principal stress does not act in a radial direction, nor are the surfaces of constant principal shear stress spherical.

Timoshenko & Goodier (1951) derive the strains from the stresses and integrate to obtain the displacements as was done in §2.2 with the results:

$$u_r = \frac{P}{4\pi G} \left\{ \frac{rz}{\rho^3} - (1 - 2v) \frac{\rho - z}{\rho r} \right\} \tag{3.21a}$$

$$u_z = \frac{P}{4\pi G} \left\{ \frac{z^2}{\rho^3} + \frac{2(1 - v)}{\rho} \right\} \tag{3.21b}$$

The results are consistent with equations (3.18). On the surface of the solid $(z = 0)$

$$\bar{u}_r = - \frac{(1 - 2v)}{4\pi G} \frac{P}{r} \tag{3.22a}$$

$$\bar{u}_z = \frac{(1 - v)}{2\pi G} \frac{P}{r} \tag{3.22b}$$

From equation (3.22b) we note that the profile of the deformed surface is a rectangular hyperboloid, which is asymptotic to the undeformed surface at a large distance from O and exhibits a theoretically infinite deflexion at O, as shown in Fig. 3.2.

The stresses and deflexions produced by a normal pressure distributed over an area S of the surface can now be found by superposition using results of the last section for a concentrated force. Referring to Fig. 3.1, we require the surface depression $\bar{u}_z$ at a general surface point $B(x, y)$ and the stress components at an interior point $A(x, y, z)$ due to a distributed pressure $p(\xi, \eta)$ acting on the surface area S. We change to polar coordinates (s, ϕ) with origin at B such that the pressure $p(s, \phi)$ acting on a surface element at C is equivalent to a force of magnitude $ps \, ds \, d\phi$. The displacement of the surface at B due to this force can be written down from equation (3.22b) in which $r = BC = s$. The displacement at B due to the pressure distributed over the whole of S is thus:

$$\bar{u}_z = \frac{1 - \nu^2}{\pi E} \int_S \int p(s, \phi) \, ds \, d\phi \tag{3.23}$$

The stress components at A can be found by integrating the stress components for a concentrated force given by equation (3.19).

3.3 Pressure applied to a polygonal region

(a) Uniform pressure

We shall consider in this section a uniform pressure p applied to a region of the surface consisting of a straight-sided polygon, as shown in Fig. 3.3(*a*). It is required to find the depression $\bar{u}_z$ at a general point $B(x, y)$ on the surface and the stress components at a subsurface point $A(x, y, z)$. BH_1, BH_2, etc. are perpendiculars of length h_1, h_2, etc. from B onto the sides of the polygon DE, EF respectively. The loaded polygon is then made up of the algebraic addition of

Fig. 3.3

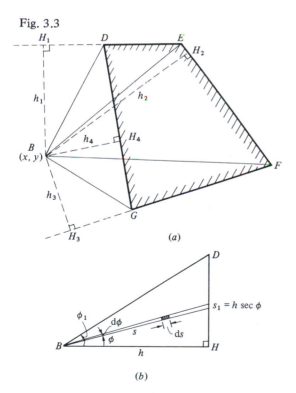

eight right-angle triangles:

$$DEFG = [BEH_1 + BEH_2 + BFH_2 + BFH_3]$$
$$- [BDH_1 + BDH_4 + BGH_3 + BGH_4]$$

A similar breakdown into rectangular triangles would have been possible if *B* had lain inside the polygon. A typical triangular area is shown in Fig. 3.3(*b*).

If the pressure is uniformly distributed, equation (3.23) becomes

$$(\bar{u}_z)_B = \frac{1-v^2}{\pi E} p \int_0^{\phi_1} d\phi \int_0^{s_1} ds$$

$$= \frac{1-v^2}{\pi E} p \int_0^{\phi_1} h \sec \phi \, d\phi$$

$$= \frac{1-v^2}{\pi E} p \frac{h}{2} \ln \left\{ \frac{1 + \sin \phi_1}{1 - \sin \phi_1} \right\} \tag{3.24}$$

The total displacement at *B* due to a uniform pressure on the polygonal region *DEFG* can then be found by combining the results of equation (3.24) for the eight constitutive triangles. The stress components at an interior point $A(x, y, z)$ below *B* can be found by integration of the stress components due to a point force given by equations (3.19), but the procedure is tedious.

The effect of a uniform pressure acting on a rectangular area $2a \times 2b$ has been analysed in detail by Love (1929). The deflexion of a general point (x, y) on the surface is given by

$$\frac{\pi E}{1 - v^2} \frac{\bar{u}_z}{p} = (x + a) \ln \left[\frac{(y + b) + \{(y + b)^2 + (x + a)^2\}^{1/2}}{(y - b) + \{(y - b)^2 + (x + a)^2\}^{1/2}} \right]$$

$$+ (y + b) \ln \left[\frac{(x + a) + \{(y + b)^2 + (x + a)^2\}^{1/2}}{(x - a) + \{(y + b)^2 + (x - a)^2\}^{1/2}} \right]$$

$$+ (x - a) \ln \left[\frac{(y - b) + \{(y - b)^2 + (x - a)^2\}^{1/2}}{(y + b) + \{(y + b)^2 + (x - a)^2\}^{1/2}} \right]$$

$$+ (y - b) \ln \left[\frac{(x - a) + \{(y - b)^2 + (x - a)^2\}^{1/2}}{(x + a) + \{(y - b)^2 + (x + a)^2\}^{1/2}} \right] \tag{3.25}$$

Expressions have been found by Love (1929) from which the stress components at a general point in the solid can be found. Love comments on the fact that the component of shear stress τ_{xy} has a theoretically infinite value at the corner of the rectangle. Elsewhere all the stress components are finite. On the surface, at the centre of the rectangle:

$$[\sigma_x]_0 = -p\{2v + (2/\pi)(1 - 2v) \tan^{-1}(b/a)\} \tag{3.26a}$$

$$[\sigma_y]_0 = -p\{2\nu + (2/\pi)(1 - 2\nu)\tan^{-1}(a/b)\} \tag{3.26b}$$
$$[\sigma_z]_0 = -p \tag{3.26c}$$

These results are useful when a uniformly loaded rectangle is used as a 'boundary element' in the numerical solution of more general contact problems (see §5.9).

(b) Non-uniform pressure

Any general variation in pressure over a polygonal region can only be analysed numerically. In this connection it is useful to consider a triangular area *DEF* on which the pressure varies linearly from p_d at *D* to p_e at *E* to p_f at *F*, as shown in Fig. 3.4, i.e. the pressure distribution forms a plane facet *def*. Any general polygonal region can be divided into triangular elements such as *DEF*. In this way a continuous distribution of pressure over the whole polygonal surface may be approximated by linear facets acting on triangular elements such as the one shown in Fig. 3.4. This representation of a continuous pressure distribution may be regarded as an improved approximation upon a series of boundary elements in each of which the pressure is taken to be uniform, since discontinuities in pressure along the sides of the elements have been eliminated and replaced by discontinuities in pressure gradient.

The element shown in Fig. 3.4 can be further simplified by splitting it into three tetragonal pressure elements: the first having pressure p_d at *D* which falls linearly to zero along the side of the triangle *EF*; the second having pressure p_e

Fig. 3.4

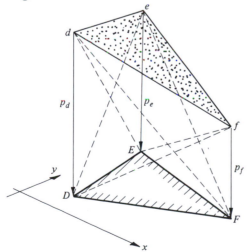

at E which falls linearly to zero along FD; and the third having pressure p_f at F which falls to zero along DE.

Influence coefficients for the normal deflexion of the surface $\bar{u}_z(x, y)$ due to the pressure distribution shown in Fig. 3.4 have been calculated by Kalker & van Randen (1972) in connection with the numerical solution of contact problems. Similarly Johnson & Bentall (1977) considered the deflexion of a surface under the action of a pyramidal distribution of pressure acting on a uniform hexagonal area of an elastic half-space. The maximum pressure p_0 acts at the centre O of the hexagon and falls to zero along the edges. The deflexion at the centre $(\bar{u}_z)_0$ was found to be $3\sqrt{3} \ln 3(1-v^2)p_0c/2\pi E$ and that at an apex to be $(\bar{u}_z)_0/3$, where c is the length of the side of the hexagon. Explicit results for a polynomial distribution of pressure acting on a triangular area have been found by Svec & Gladwell (1971).

The stresses and deformations produced by a pressure distribution of the form $p_0(1-x^2/a^2)^{1/2}$ acting on the rectangle $x = \pm a, y = \pm b$ have been calculated by Kunert (1961).

3.4 Pressure applied to a circular region

A circular region of radius a is shown in Fig. 3.5. It is required to find the displacement at a surface point B and the stresses at an internal point A due to pressure distributed over the circular region. Solutions in closed form can be found for axi-symmetrical pressure distributions of the form:

$$p = p_0(1 - r^2/a^2)^n \tag{3.27}$$

We will consider in detail some particular values of n.

(a) Uniform pressure $(n = 0)$

Regarding the pressure p at C, acting on a surface element of area $s\, ds\, d\phi$, as a concentrated force, the normal displacement is given by equation (3.23), i.e.

$$\bar{u}_z = \frac{1-v^2}{\pi E}\, p \int_s \int d\phi\, ds$$

We will consider first the case where B lies inside the circle (Fig. 3.5(a)). The limits on s are

$$s_{1,2} = -r\cos\phi \pm \{r^2\cos^2\phi + (a^2 - r^2)\}^{1/2} \tag{3.28}$$

Thus

$$\bar{u}_z = \frac{1-v^2}{\pi E}\, p \int_0^\pi 2\{r^2\cos^2\phi + (a^2 - r^2)\}^{1/2}\, d\phi$$

$$= \frac{4(1-v^2)pa}{\pi E} \int_0^{\pi/2} \{1 - (r^2/a^2)\sin^2\phi\}^{1/2}\, d\phi$$

i.e.

$$\bar{u}_z = \frac{4(1-v^2)pa}{\pi E} \; \mathbf{E}(r/a), \quad (r < a) \tag{3.29a}$$

where $\mathbf{E}(r/a)$ is the complete elliptic integral of the second kind with modulus (r/a). At the centre of the circle: $r = 0$, $\mathbf{E}(0) = \pi/2$, thus

$$(\bar{u}_z)_0 = 2(1-v^2)pa/E.$$

At the edge of the circle: $r/a = 1$, $\mathbf{E}(1) = 1$, then

$$(\bar{u}_z)_a = 4(1-v^2)pa/\pi E$$

The mean displacement of the loaded circle is $16(1-v^2)pa/3\pi E$.

In view of the axial symmetry, the displacement tangential to the surface must be radial. By equation (3.22a) the tangential displacement at B due to an element of load at C is

$$\frac{(1-2v)(1+v)}{2\pi E} \; ps \; \mathrm{d}s \; \mathrm{d}\phi}{s}$$

Fig. 3.5. Pressure applied to a circular region. Displacement: (a) at an internal point B; (b) at an external point B.

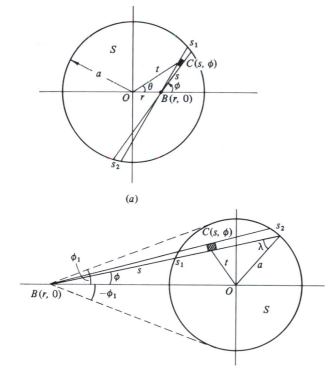

(a)

(b)

in the direction from B towards C. Thus the radial component of this displacement is:

$$\bar{u}_r = \frac{(1 - 2\nu)(1 + \nu)}{2\pi E} p \cos \phi \, ds \, d\phi$$

The total displacement due to the whole load is then

$$\bar{u}_r = \frac{(1 - 2\nu)(1 + \nu)}{2\pi E} p \int_0^{2\pi} \{-r \cos \phi + (r^2 \cos^2 \phi + a^2 - r^2)^{1/2}\}$$
$$\times \cos \phi \, d\phi$$

The second term in the integral vanishes when integrated over the limits 0 to 2π, hence

$$\bar{u}_r = -(1 - 2\nu)(1 + \nu)pr/2E, \quad r \leqslant a \tag{3.29b}$$

We turn now to a point B on the surface outside the circle (Fig. 3.5(b)). In this case the limits on ϕ are $\pm\phi_1$, so that

$$\bar{u}_z = \frac{2(1 - \nu^2)p}{\pi E} \int_0^{\phi_1} (a^2 - r^2 \sin^2 \phi)^{1/2} \, d\phi$$

We change the variable to the angle λ, shown in Fig. 3.5(b), which is related to ϕ by

$$a \sin \lambda = r \sin \phi$$

whereupon the expression for $\bar{u}_z$ becomes:

$$\bar{u}_z = \frac{4(1 - \nu^2)}{\pi E} p \int_0^{\pi/2} \frac{a^2 \cos^2 \lambda \, d\lambda}{r\{1 - (a^2/r^2) \sin^2 \lambda\}}$$
$$= \frac{4(1 - \nu^2)}{\pi E} pr\{E(a/r) - (1 - a^2/r^2)K(a/r)\}, \quad r > a \tag{3.30a}$$

where $K(a/r)$ is the complete elliptic integral of the first kind with modulus (a/r).

The tangential displacement at B is radial and is given by

$$\bar{u}_r = -\frac{2(1 - 2\nu)(1 + \nu)}{\pi E} p \int_0^{\phi_1} \cos \phi \, (a^2 - r^2 \sin^2 \phi)^{1/2} \, d\phi$$

Changing the variable to λ, as before, gives

$$\bar{u}_r = -\frac{2(1 - 2\nu)(1 + \nu)}{\pi E} p \frac{a^2}{r} \int_0^{\pi/2} \cos^2 \lambda \, d\lambda$$
$$= -\frac{(1 - 2\nu)(1 + \nu)}{2E} p \frac{a^2}{r}, \quad r > a \tag{3.30b}$$

Since the $p\pi a^2$ is equal to the total load P acting on the whole area, we note that the tangential displacement outside the loaded region, given by (3.30b), is the same as though the whole load were concentrated at the centre of the circle (see equation (3.22*a*)). It follows by superposition that this conclusion is true for any axially symmetrical distribution of pressure acting in the circle.

The stresses at the surface within the circle may now be found from equations (3.29). Thus:

$$\bar{\epsilon}_r = \frac{\partial \bar{u}_r}{\partial r} = \bar{\epsilon}_\theta = \frac{\bar{u}_r}{r} = -\frac{(1-2\nu)(1+\nu)}{2E} p \tag{3.31}$$

from which, by Hooke's law, we get

$$\bar{\sigma}_r = \bar{\sigma}_\theta = -\tfrac{1}{2}(1+2\nu)p, \quad \bar{\sigma}_z = -p \tag{3.32}$$

To find the stresses within the half-space along the z-axis we make use of equations (3.20) for the stresses due to a concentrated force. Consider an annular element of area $2\pi r \, dr$ at radius r. The load on the annulus is $2\pi r p \, dr$, so substituting in (3.20*c*) and integrating over the circle gives

$$\sigma_z = -3p \int_0^a \frac{rz^3}{(r^2+z^2)^{5/2}} \, dr$$
$$= -p\{1 - z^3/(a^2+z^2)^{3/2}\} \tag{3.33a}$$

Along Oz, $\sigma_r = \sigma_\theta$, hence applying equation (3.20*e*) to an annulus of pressure

$$\sigma_r + \sigma_\theta + \sigma_z = -\frac{(1+\nu)}{\pi} p \int_0^a \frac{2\pi rz \, dr}{(r^2+z^2)^{3/2}}$$
$$= 2(1+\nu)p\{z(a^2+z^2)^{-1/2} - 1\}$$

so that

$$\sigma_r = \sigma_\theta = -p \left\{ \frac{1+2\nu}{2} - \frac{(1+\nu)z}{(a^2+z^2)^{1/2}} + \frac{z^3}{2(a^2+z^2)^{3/2}} \right\} \tag{3.33b}$$

The stress components at other points throughout the half-space have been investigated by Love (1929).

(b) Uniform normal displacement ($n = -\tfrac{1}{2}$)

We shall proceed to show that a pressure distribution of the form

$$p = p_0(1 - r^2/a^2)^{-1/2} \tag{3.34}$$

gives rise to a uniform normal displacement of the loaded circle. This is the pressure, therefore, which would arise on the face of a flat-ended, frictionless cylindrical punch pressed squarely against an elastic half-space. It is the axi-symmetrical analogue of the two-dimensional problem discussed in §2.8.

Referring to Fig. 3.5(*a*):

$$t^2 = r^2 + s^2 + 2rs \cos \phi$$

so that

$$p(s, \phi) = p_0 a(\alpha^2 - 2\beta s - s^2)^{-1/2} \tag{3.35}$$

where $\alpha^2 = a^2 - r^2$ and $\beta = r \cos \phi$. The displacement within the loaded circle, using equation (3.23), is

$$\bar{u}_z(r) = \frac{1 - \nu^2}{\pi E} p_0 a \int_0^{2\pi} d\phi \int_0^{s_1} (\alpha^2 - 2\beta s - s^2)^{-1/2} ds$$

where the limit s_1 is the positive root of

$$\alpha^2 - 2\beta s - s^2 = 0$$

Now

$$\int_0^{s_1} (\alpha^2 - 2\beta s - s^2)^{-1/2} ds = \frac{\pi}{2} - \tan^{-1}(\beta/\alpha)$$

and

$$\tan^{-1}\{\beta(\phi)/\alpha\} = -\tan^{-1}\{\beta(\phi + \pi)/\alpha\}$$

so that the integral of $\tan^{-1}(\beta/\alpha)$ vanishes as ϕ varies from 0 to 2π, whereupon

$$\bar{u}_z = \frac{1 - \nu^2}{\pi E} p_0 a \int_0^{2\pi} \left\{ \frac{\pi}{2} - \tan^{-1}(\beta/\alpha) \right\} d\phi = \pi(1 - \nu^2) p_0 a/E \tag{3.36}$$

which is constant and independent of r. The total force

$$P = \int_0^a 2\pi r p_0 (1 - r^2/a^2)^{-1/2} dr = 2\pi a^2 p_0 \tag{3.37}$$

When B lies outside the loaded circle (Fig. 3.5(b))

$$p(s, \phi) = p_0 a(\alpha^2 + 2\beta s - s^2)^{-1/2}$$

and the limits $s_{1,2}$ are the root of $\alpha^2 + 2\beta s - s^2 = 0$, whereupon

$$\int_{s_1}^{s_2} (\alpha^2 + 2\beta s - s^2)^{-1/2} ds = \pi$$

The limits on ϕ are $\phi_{1,2} = \sin^{-1}(a/r)$, so that

$$\bar{u}_z(r) = \frac{2(1 - \nu^2)}{E} p_0 a \sin^{-1}(a/r) \tag{3.38}$$

Like the two-dimensional punch, the pressure is theoretically infinite at the edge of the punch and the surface has an infinite gradient just outside the edge. Stresses within the half-space have been found by Sneddon (1946).

(c) Hertz pressure ($n = \frac{1}{2}$)

The pressure given by the Hertz theory (see Chapter 4), which is exerted between two frictionless elastic solids of revolution in contact, is given by

$$p(r) = p_0(a^2 - r^2)^{1/2}/a \tag{3.39}$$

from which the total load $P = 2\pi p_0 a^2/3$. The method of finding the deflexions is identical to that in the previous problem (§3.4b) and uses the same notation Thus, within the loaded circle the normal displacement is given by

$$\bar{u}_z(r) = \frac{1-\nu^2}{\pi E} \frac{p_0}{a} \int_0^{2\pi} d\phi \int_0^{s_1} (\alpha^2 - 2\beta s - s^2)^{1/2} \, ds \qquad (3.40)$$

$$\int_0^{s_1} (\alpha^2 - 2\beta s - s^2)^{1/2} \, ds = -\tfrac{1}{2}\alpha\beta + \tfrac{1}{2}(\alpha^2 + \beta^2)\{(\pi/2) - \tan^{-1}(\beta/\alpha)\}$$

The terms $\beta\alpha$ and $\tan^{-1}(\beta/\alpha)$ vanish when integrated with respect to ϕ between the limits 0 and 2π, whereupon

$$\bar{u}_z(r) = \frac{1-\nu^2}{\pi E} \frac{p_0}{a} \int_0^{2\pi} \frac{\pi}{4} (a^2 - r^2 + r^2 \cos^2 \phi) \, d\phi$$

$$= \frac{1-\nu^2}{E} \frac{\pi p_0}{4a} (2a^2 - r^2), \quad r \leqslant a \qquad (3.41a)$$

To find the tangential displacement at B, which by symmetry must be radial, we make use of equation (3.22a). The element of pressure at C causes a tangential displacement at B:

$$\frac{(1-2\nu)(1+\nu)}{2\pi E} p \, ds \, d\phi$$

directed from B towards C. The radial component of this displacement is

$$\frac{(1-2\nu)(1+\nu)}{2\pi E} \cos\phi \, p \, ds \, d\phi$$

so that the resultant tangential displacement at B due to the whole pressure distribution is

$$\bar{u}_r(r) = \frac{(1-2\nu)(1+\nu)}{2\pi E} \frac{p_0}{a} \int_0^{2\pi} \cos\phi \, d\phi \int_0^{s_1} (\alpha^2 - 2\beta s - s^2)^{1/2} \, ds$$

The integration with respect to s is the same as before; integrating with respect to ϕ gives

$$\bar{u}_r(r) = -\frac{(1-2\nu)(1+\nu)}{3E} \frac{a^2}{r} p_0 \{1 - (1 - r^2/a^2)^{3/2}\}, \quad r \leqslant a \quad (3.41b)$$

When the point B lies outside the loaded circle, proceeding in the same way as in the previous case, we find

$$\bar{u}_z = \frac{(1-\nu^2)}{E} \frac{p_0}{2a}$$

$$\times \{(2a^2 - r^2)\sin^{-1}(a/r) + r^2 (a/r)(1 - a^2/r^2)^{1/2}\}, \quad r > a \quad (3.42a)$$

The tangential displacement outside the loaded circle is the same as if the load were concentrated at the centre, so that, by equation (3.22a)

$$\bar{u}_r = -\frac{(1-2\nu)(1+\nu)}{3E}\, p_0\frac{a^2}{r}, \quad r > a \tag{3.42b}$$

The surface strain components: $\bar{\varepsilon}_r = \partial \bar{u}_r/\partial r$ and $\bar{\varepsilon}_\theta = \bar{u}_r/r$ can be found from equations (3.41) and (3.42) which, together with the pressure, determine the stresses in the surface $z = 0$ with the result that, inside the loaded circle,

$$\bar{\sigma}_r/p_0 = \frac{1-2\nu}{3}\,(a^2/r^2)\{1-(1-r^2/a^2)^{3/2}\}-(1-r^2/a^2)^{1/2} \tag{3.43a}$$

$$\bar{\sigma}_\theta/p_0 = -\frac{1-2\nu}{3}\,(a^2/r^2)\{1-(1-r^2/a^2)^{3/2}\}-2\nu(1-r^2/a^2)^{1/2}$$

$$\tag{3.43b}$$

$$\bar{\sigma}_z/p_0 = -(1-r^2/a^2)^{1/2} \tag{3.43c}$$

and outside the circle

$$\bar{\sigma}_r/p_0 = -\bar{\sigma}_\theta/p_0 = (1-2\nu)a^2/3r^2 \tag{3.44}$$

The radial stress is therefore tensile outside the loaded circle. It reaches its maximum value at the edge of the circle at $r = a$. This is the maximum tensile stress occurring anywhere. The stresses along the z-axis may be calculated without difficulty by considering a ring of concentrated force at radius r. They are:

$$\frac{\sigma_r}{p_0} = \frac{\sigma_\theta}{p_0} = -(1+\nu)\{1-(z/a)\tan^{-1}(a/z)\}+\tfrac{1}{2}(1+z^2/a^2)^{-1} \tag{3.45a}$$

$$\frac{\sigma_z}{p_0} = -(1+z^2/a^2)^{-1} \tag{3.45b}$$

The stresses at other points throughout the solid have been calculated by Huber (1904) and Morton & Close (1922). This stress distribution is illustrated in Fig. 4.3 where it is compared with the stresses produced by a uniform pressure acting on a circular area given by equation (3.33). Along the z-axis σ_r, σ_θ and σ_z are principal stresses. The principal shear stress, $\tau_1 = \frac{1}{2}|\sigma_z - \sigma_\theta|$, is also plotted in Fig. 4.3. It has a maximum value which lies below the surface. For the Hertz pressure distribution

$$(\tau_1)_{\max} = 0.31p_0 = 0.47P/\pi a^2 \tag{3.46}$$

at $z = 0.57a$.

For the uniform pressure distribution

$$(\tau_1)_{\max} = 0.33p = 0.33P/\pi a^2 \tag{3.47}$$

at $z = 0.64a$. Both the values above are computed for $\nu = 0.30$.

(d) General pressure ($n = m - \frac{1}{2}$)

A pressure distribution of the form

$$p_m = p_0 a^{1-2m}(a^2 - r^2)^{m-1/2} \qquad (3.48)$$

where m is an integer, will produce a normal displacement within the loaded region given in the notation of equation (3.35) by:

$$\bar{u}_z = \frac{1-\nu^2}{\pi E} \frac{p_0}{a^{2m-1}} \int_0^\pi \mathrm{d}\phi \int_0^{s_1} (\alpha^2 - 2\beta s - s^2)^{m-1/2} \, \mathrm{d}s \qquad (3.49)$$

Reduction formulae for the integral with respect to s enable solutions to be found in closed form (see Lur'e, 1964). The resulting expressions for $\bar{u}_z$ are polynomials in r of order $2m$. Thus, in the examples considered above: $m = 0$ results in a constant displacement; $m = 1$ results in a displacement quadratic in r.

Alternative methods, which link polynomial variation of displacement with pressure distribution of the form $p_m = p_0 a^{1-2m} r^{2m}(a^2 - r^2)^{-1/2}$, have been developed by Popov (1962) using Legendre polynomials and by Steuermann (1939). In this way, if the displacement within the loaded circle can be represented by an even polynomial in r, then the corresponding pressure distributions can be expressed in the appropriate sum of distributions having the form p_m defined above (see §5.3).

3.5 Pressure applied to an elliptical region

It is shown in Chapter 4 that two non-conforming bodies loaded together make contact over an area which is elliptical in shape, so that the stresses and deformation due to pressure and traction on an elliptical region are of practical importance. A circle is a particular case of an ellipse, so that we might expect results for an elliptical region to be qualitatively similar to those derived for a circular region in the previous section. This is indeed the case, so that we are led to consider pressure distributions of the form

$$p(x, y) = p_0 \{1 - (x/a)^2 - (y/b)^2\}^n \qquad (3.50)$$

which act over the region bounded by the ellipse

$$(x/a)^2 + (y/b)^2 - 1 = 0$$

The classical approach, using the potential functions of Boussinesq, is usually followed. Thus, by equation (3.10),

$$\psi(x, y, z) = \int_S \int \{1 - (\xi/a)^2 - (\eta/b)^2\}^n \rho^{-1} \, \mathrm{d}\xi \, \mathrm{d}\eta \qquad (3.51)$$

where $\rho^2 = (\xi - x)^2 + (\eta - y)^2 + z^2$. The normal displacement of the surface is then given by equation (3.16c), viz.:

$$\bar{u}_z(x, y) = \frac{1-\nu}{2\pi G} (\psi)_{z=0}$$

It then follows from potential theory (see, for example, Routh, 1908, p. 129) that for a general point in the solid

$$\psi(x, y, z) = \frac{\Gamma(n + 1)\Gamma(\tfrac{1}{2})}{\Gamma(n + 3/2)} p_0 ab \int_{\lambda_1}^{\infty} \left(1 - \frac{x^2}{a^2 + w} - \frac{y^2}{b^2 + w} - \frac{z^2}{w}\right)^{n+1/2}$$

$$\times \frac{dw}{\{(a^2 + w)(b^2 + w)w\}^{1/2}} \tag{3.52}$$

where Γ denotes a gamma function and where λ_1 is the positive root of the equation

$$\frac{x^2}{a^2 + \lambda} + \frac{y^2}{b^2 + \lambda} + \frac{z^2}{\lambda} = 1 \tag{3.53}$$

λ_1 may be interpreted geometrically as the parameter of an ellipsoid, confocal with the given elliptical pressure region, whose surface passes through the point in question (x, y, z). To find $\psi(x, y, 0)$ at a surface point *within* the loaded region the lower limit of the integral in (3.52) is taken to be zero.

A few cases in which n takes different values will be discussed.

(a) Uniform displacement ($n = -\tfrac{1}{2}$)

Putting $n = -\tfrac{1}{2}$ in equation (3.52) gives:

$$\psi(x, y, z) = \pi p_0 ab \int_{\lambda_1}^{\infty} \frac{dw}{\{(a^2 + w)(b^2 + w)w\}^{1/2}} \tag{3.54}$$

and on the surface $z = 0$, within the loaded region,

$$\psi(x, y, 0) = \pi p_0 ab \int_{0}^{\infty} \frac{dw}{\{(a^2 + w)(b^2 + w)w\}^{1/2}} \tag{3.55}$$

This integral is a constant independent of x and y; it is an elliptic integral which, when put into standard form and substituted into equation (3.16c), quoted above, gives

$$\bar{u}_z = \frac{1 - \nu^2}{E} 2p_0 bK(e) \tag{3.56}$$

where $e = (1 - b^2/a^2)^{1/2}$ is the eccentricity of the ellipse, and $a \geqslant b$. This is the uniform displacement of a half-space under that action of a rigid frictionless punch of elliptical plan-form. The pressure on the face of the punch is

$$p(x, y) = p_0 \{1 - (x/a)^2 - (y/b)^2\}^{-1/2} \tag{3.57}$$

where the total load $P = 2\pi abp_0$. The displacement under a cylindrical punch given by equation (3.36) could be obtained by putting $a = b$ or $e = 1$ in equation (3.56).

(b) Hertz pressure ($n = \frac{1}{2}$)

In this important case the pressure is

$$p = p_0\{1 - (x/a)^2 - (y/b)^2\}^{1/2} \tag{3.58}$$

Thus

$$\psi(x, y, z) = \frac{1}{2}\pi a b p_0 \int_{\lambda_1}^{\infty} \left(1 - \frac{x^2}{a^2 + w} - \frac{y^2}{b^2 + w} - \frac{z^2}{w}\right)$$

$$\times \frac{dw}{\{(a^2 + w)(b^2 + w)w\}^{1/2}} \tag{3.59}$$

and on the surface, within the loaded region,

$$\psi(x, y, 0) = \frac{1}{2}\pi a b p_0 \int_0^{\infty} \left(1 - \frac{x^2}{a^2 + w} - \frac{y^2}{b^2 + w}\right)$$

$$\times \frac{dw}{\{(a^2 + w)(b^2 + w)w\}^{1/2}} \tag{3.60}$$

The surface displacement within the loaded region may then be written:

$$\bar{u}_z = \frac{1 - \nu^2}{\pi E} (L - Mx^2 - Ny^2) \tag{3.61}$$

where

$$M = \frac{\pi p_0 a b}{2} \int_0^{\infty} \frac{dw}{\{(a^2 + w)^3(b^2 + w)w\}^{1/2}} = \frac{\pi p_0 b}{e^2 a^2} \{K(e) - E(e)\} \tag{3.62a}$$

$$N = \frac{\pi p_0 a b}{2} \int_0^{\infty} \frac{dw}{\{(a^2 + w)(b^2 + w)^3 w\}^{1/2}} = \frac{\pi p_0 b}{e^2 a^2} \left\{\frac{a^2}{b^2} E(e) - K(e)\right\} \tag{3.62b}$$

and

$$L = \frac{\pi p_0 a b}{2} \int_0^{\infty} \frac{dw}{\{(a^2 + w)(b^2 + w)w\}^{1/2}} = \pi p_0 b K(e) \tag{3.62c}$$

The total load acting on the ellipse is given by

$$P = 2\pi a b p_0/3 \tag{3.63}$$

Finding the components of displacement and stress at a general point in the solid from equation (3.59) is not straightforward, firstly because the limit λ_1 is the root of a cubic equation (3.53) and secondly because, for certain stress components, it is necessary to determine the auxiliary function $\psi_1 = \int_z^{\infty} \psi \, dz$.

The difficulties are least when finding the stresses along the z-axis. In this case $\lambda_1 = z^2$, and the integration with respect to z to determine the derivatives of ψ_1 is straightforward. These calculations have been performed by Thomas & Hoersch (1930), Belajev (1917) and Lundberg & Sjövall (1958) with the results that along the z-axis

$$\frac{\sigma_x}{p_0} = \frac{2b}{e^2 a}(\Omega_x + v\Omega'_x) \tag{3.64a}$$

$$\frac{\sigma_y}{p_0} = \frac{2b}{e^2 a}(\Omega_y + v\Omega'_y) \tag{3.64b}$$

$$\frac{\sigma_z}{p_0} = -\frac{b}{e^2 a}\left(\frac{1-T^2}{T}\right) \tag{3.64c}$$

where

$$\Omega_x = -\tfrac{1}{2}(1-T) + \zeta\{F(\phi, e) - E(\phi, e)\}$$
$$\Omega'_x = 1 - (a^2 T/b^2) + \zeta\{(a^2/b^2)E(\phi, e) - F(\phi, e)\}$$
$$\Omega_y = \tfrac{1}{2} + (1/2T) - (Ta^2/b^2) + \zeta\{(a^2/b^2)E(\phi, e) - F(\phi, e)\}$$
$$\Omega'_y = -1 + T + \zeta\{F(\phi, e) - E(\phi, e)\}$$
$$T = \left(\frac{b^2 + z^2}{a^2 + z^2}\right)^{1/2}, \quad \zeta = \frac{z}{a} = \cot\phi$$

The elliptic integrals $F(\phi, e)$ and $E(\phi, e)$ are tabulated.[†] Within the surface of contact, along the x-axis

$$\frac{\sigma_x}{p_0} = -2v\gamma - (1 - 2v)\frac{b}{ae^2}\left\{(1 - b\gamma/a) - \frac{x}{ae}\tanh^{-1}\left(\frac{ex}{a + b\gamma}\right)\right\} \tag{3.65a}$$

$$\frac{\sigma_y}{p_0} = -2v\gamma - (1 - 2v)\frac{b}{ae^2}\left\{\left(\frac{a\gamma}{b} - 1\right) + \frac{x}{ae}\tanh^{-1}\left(\frac{ex}{a + b\gamma}\right)\right\} \tag{3.65b}$$

and along the y-axis

$$\frac{\sigma_x}{p_0} = -2v\gamma - (1 - 2v)\frac{b}{ae^2}\left\{(1 - b\gamma/a) - \frac{y}{ae}\tan^{-1}\left(\frac{aey}{b(a\gamma + b)}\right)\right\} \tag{3.66a}$$

$$\frac{\sigma_y}{p_0} = -2v\gamma - (1 - 2v)\frac{b}{ae^2}\left\{\left(\frac{a\gamma}{b} - 1\right) + \frac{y}{ae}\tan^{-1}\left(\frac{aey}{b(a\gamma + b)}\right)\right\} \tag{3.66b}$$

where $\gamma = \{1 - (x/a)^2 - (y/b)^2\}^{1/2}$.

At the centre ($x = y = 0$)

$$\frac{\sigma_x}{p_0} = -2v - (1 - 2v)\frac{b}{a + b} \tag{3.67a}$$

[†] Abramowitz, M. & Stegun, I. A., *Handbook of Math. Functions*, Dover, 1965.

$$\frac{\sigma_y}{p_0} = -2\nu - (1 - 2\nu)\frac{a}{a + b} \tag{3.67b}$$

Outside the loaded ellipse the surface stresses are equal and opposite, i.e. there is a state of pure shear:

$$\frac{\sigma_x}{p_0} = -\frac{\sigma_y}{p_0} = -(1 - 2\nu)\frac{b}{ae^2}\left\{1 - \frac{x}{ae}\tanh^{-1}\left(\frac{ex}{a}\right) - \frac{y}{ae}\tan^{-1}\left(\frac{aey}{b^2}\right)\right\} \tag{3.68a}$$

and

$$\tau_{xy} = -(1 - 2\nu)\frac{b}{ae^2}\left\{\frac{y}{ae}\tanh^{-1}\left(\frac{ex}{a}\right) - \frac{x}{ae}\tan^{-1}\left(\frac{aey}{b^2}\right)\right\} \tag{3.68b}$$

Fessler & Ollerton (1957) determined the shear stresses τ_{zx} and τ_{yz} in the planes of symmetry $y = 0$, and $x = 0$ respectively, with the result

$$\frac{\tau_{zx}}{p_0} = -\frac{b}{a}\frac{x}{a}\left(\frac{z}{a}\right)^2 \frac{\left\{\left(1 + \frac{\lambda_1}{a^2}\right)\frac{\lambda_1}{a^2}\right\}^{-3/2}\left(\frac{b^2}{a^2} + \frac{\lambda_1}{a^2}\right)^{-1/2}}{\left(\frac{ax}{a^2 + \lambda_1}\right)^2 + \left(\frac{az}{\lambda_1}\right)^2} \tag{3.69a}$$

and

$$\frac{\tau_{yz}}{p_0} = -\frac{a}{b}\frac{y}{b}\left(\frac{z}{b}\right)^2 \frac{\left\{\left(1 - \frac{\lambda_1}{b^2}\right)\frac{\lambda_1}{b^2}\right\}^{-3/2}\left(\frac{a^2}{b^2} + \frac{\lambda_1}{b^2}\right)^{-1/2}}{\left(\frac{by}{b^2 + \lambda_1}\right)^2 + \left(\frac{bz}{\lambda_1}\right)^2} \tag{3.69b}$$

where λ_1 is the positive root of equation (3.53). Expressions from which the stress components at a general point can be computed have been obtained by Sackfield & Hills (1983a).

(c) General pressure $(n = m - \frac{1}{2})$

In this case it follows from equation (3.52) that the surface displacements within the ellipse are given by

$$\bar{u}_z(x, y) = \frac{1 - \nu^2}{\pi E}[\psi]_{z=0} = \frac{1 - \nu^2}{E}\frac{\Gamma(n + 1)\Gamma(\frac{1}{2})}{\Gamma(n + \frac{3}{2})}p_0ab$$

$$\times \int_0^\infty \left(1 - \frac{x^2}{a^2 + w} - \frac{y^2}{b^2 + w}\right)^m \frac{dw}{\{(a^2 + w)(b^2 + w)w\}^{1/2}} \tag{3.70}$$

By expanding the bracket under the integral sign it is apparent that the expressions for the displacement will take the form

$$\bar{u}_z = C_0 + \sum_{l=1}^{l=m} C_l x^{2l} y^{2(m-l)} \tag{3.71}$$

Alternatively if the displacements within the ellipse are specified by a polynomial of the form (3.71), which would be the case if the half-space were being indented by a rigid frictionless punch, then the pressure distribution on the elliptical face of the punch would take the form

$$p(x, y) = p_0 (ab)^{-m} \sum_{l=1}^{l=m} C_l' x^{2l} y^{2(m-l)} \{1 - (x/a)^2 - (y/b)^2\}^{-1/2}$$

$$(3.72)$$

General expressions for finding the relation between the coefficient C_l and C_l' are given by Shail (1978) and Gladwell (1980).

3.6 Concentrated tangential force

In this and the next section we shall investigate the displacements and stresses due to a tangential traction $q_x(\xi, \eta)$ acting over the loaded area S. The tangential traction parallel to the y-axis q_y and the normal pressure p are both taken to be zero. Thus in equations (3.2) to (3.7) we put

$$G_1 = H_1 = G = H = 0$$

thus

$$\psi_1 = \frac{\partial F_1}{\partial x}, \quad \psi = \frac{\partial^2 F_1}{\partial x \partial z}$$

whence

$$u_x = \frac{1}{4\pi G} \left\{ 2 \frac{\partial^2 F_1}{\partial z^2} + 2v \frac{\partial^2 F_1}{\partial x^2} - z \frac{\partial^3 F_1}{\partial x^2 \partial z} \right\} \tag{3.73a}$$

$$u_y = \frac{1}{4\pi G} \left\{ 2v \frac{\partial^2 F_1}{\partial x \partial y} - z \frac{\partial^3 F_1}{\partial x \partial y \partial z} \right\} \tag{3.73b}$$

$$u_z = \frac{1}{4\pi G} \left\{ (1 - 2v) \frac{\partial^2 F_1}{\partial x \partial z} - z \frac{\partial^3 F_1}{\partial x \partial z^2} \right\} \tag{3.73c}$$

where

$$F_1 = \int_S \int q_x(\xi, \eta) \{z \ln (\rho + z) - \rho\} \, d\xi \, d\eta$$

and

$$\rho^2 = (\xi - x)^2 + (\eta - y)^2 - z^2$$

When the appropriate derivatives are substituted in equations (3.73), we get

$$u_x = \frac{1}{4\pi G} \int_S \int q_x(\xi, \eta)$$

$$\times \left\{ \frac{1}{\rho} + \frac{1-2\nu}{\rho+z} + \frac{(\xi-x)^2}{\rho^3} - \frac{(1-2\nu)(\xi-x)^2}{\rho(\rho+z)^2} \right\} d\xi \, d\eta \qquad (3.74a)$$

$$u_y = \frac{1}{4\pi G} \int_S \int q_x(\xi, \eta)$$

$$\times \left\{ \frac{(\xi-x)(\eta-y)}{\rho^3} - (1-2\nu)\frac{(\xi-x)(\eta-y)}{\rho(\rho+z)^2} \right\} d\xi \, d\eta \qquad (3.74b)$$

$$u_z = -\frac{1}{4\pi G} \int_S \int q_x(\xi, \eta) \left\{ \frac{(\xi-x)z}{\rho^3} + (1-2\nu)\frac{(\xi-x)}{\rho(\rho+z)} \right\} d\xi \, d\eta$$

$$(3.74c)$$

The tangential traction is now taken to be concentrated on a vanishingly small area at the origin, so that $\int_S \int q_x(\xi, \eta) \, d\xi \, d\eta$ reduces to a concentrated force Q_x acting at the origin ($\xi = \eta = 0$) in a direction parallel to the x-axis. Equations (3.74) for the displacements throughout the solid reduce to

$$u_x = \frac{Q_x}{4\pi G} \left[\frac{1}{\rho} + \frac{x^2}{\rho^3} + (1-2\nu) \left\{ \frac{1}{\rho+z} - \frac{x^2}{\rho(\rho+z)^2} \right\} \right] \qquad (3.75a)$$

$$u_y = \frac{Q_x}{4\pi G} \left[\frac{xy}{\rho^3} - (1-2\nu)\frac{xy}{\rho(\rho+z)^2} \right] \qquad (3.75b)$$

$$u_z = \frac{Q_x}{4\pi G} \left[\frac{xz}{\rho^3} + (1-2\nu)\frac{x}{\rho(\rho+z)} \right] \qquad (3.75c)$$

where now $\rho^2 = x^2 + y^2 + z^2$.

By differentiating equations (3.75) the strain components and hence the stress components are found, with the results

$$\frac{2\pi\sigma_x}{Q_x} = -\frac{3x^3}{\rho^5} + (1-2\nu)$$

$$\times \left\{ \frac{x}{\rho^3} - \frac{3x}{\rho(\rho+z)^2} + \frac{x^3}{\rho^3(\rho+z)^2} + \frac{2x^3}{\rho^2(\rho+z)^3} \right\} \qquad (3.76a)$$

$$\frac{2\pi\sigma_y}{Q_x} = -\frac{3xy^2}{\rho^5} + (1-2\nu)$$

$$\times \left\{ \frac{x}{\rho^3} - \frac{x}{\rho(\rho+z)^2} + \frac{xy^2}{\rho^3(\rho+z)^2} + \frac{2xy^2}{\rho^2(\rho+z)^3} \right\} \qquad (3.76b)$$

$$\frac{2\pi\sigma_z}{Q_x} = -\frac{3xz^2}{\rho^5} \tag{3.76c}$$

$$\frac{2\pi\tau_{xy}}{Q_x} = -\frac{3x^2 y}{\rho^5} + (1 - 2\nu)$$

$$\times \left\{ -\frac{y}{\rho(\rho+z)^2} + \frac{x^2 y}{\rho^3(\rho+z)^2} + \frac{2x^2 y}{\rho^2(\rho+z)^3} \right\} \tag{3.76d}$$

$$\frac{2\pi\tau_{yz}}{Q_x} = -\frac{3xyz}{\rho^5} \tag{3.76e}$$

$$\frac{2\pi\tau_{zx}}{Q_x} = -\frac{3x^2 z}{\rho^5} \tag{3.76f}$$

$$\frac{2\pi}{Q_x}(\sigma_x + \sigma_y + \sigma_z) = -2(1+\nu)x/\rho^3 \tag{3.76g}$$

The stresses and displacement on the surface, excluding the origin, are obtained by putting $z = 0$, and $\rho = r$. These expressions can be used to build up the stress components within the solid due to any known distribution of tangential traction by superposition.

3.7 Uni-directional tangential tractions on elliptical and circular regions

The influence of tangential traction has not been studied so extensively as that of normal pressure, but tractions of the form

$$q_x(x, y) = q_0 \{1 - (x/a)^2 - (y/b)^2\}^n \tag{3.77}$$

which act parallel to the x-axis on an elliptical area bounded by the curve

$$(x/a)^2 + (y/b)^2 - 1 = 0 \tag{3.78}$$

are important in the theory of contact stresses. These tractions are comparable with the pressure distributions considered in §5 (see equation (3.50)). To examine the extent to which the problems are analogous we should compare equations (3.73) for the displacements due to tangential traction with equations (3.11) for the displacements due to normal pressure. Remembering that $\psi_1 = \partial H_1/\partial z$ and $\psi = \partial^2 H_1/\partial z^2$, it is immediately apparent that there is no complete analogy between the two sets of equations: the displacements due to a tangential traction cannot be written down directly from the known displacements due to a similar distribution of normal pressure. However a very restricted analogy does exist. In the case where Poisson's ratio is zero, the surface displacements due to a tangential traction may be written:

$$\bar{u}_x = \frac{1}{2\pi G} \frac{\partial^2 F_1}{\partial z^2}, \quad \bar{u}_y = 0$$

(3.79)

$$\bar{u}_z = \frac{1}{4\pi G} \frac{\partial^2 F_1}{\partial x \partial z}$$

whilst the corresponding displacements due to normal pressure are

$$\bar{u}_x = \bar{u}_y = -\frac{1}{4\pi G} \frac{\partial^2 H_1}{\partial x \partial z}$$

(3.80)

$$\bar{u}_z = \frac{1}{2\pi G} \frac{\partial^2 H_1}{\partial z^2}$$

We recall from the definitions given in (3.2) that F_1 and H_1 are analogous so that, under the action of identical distributions of tangential and normal traction q_x and p,

$$(\bar{u}_x)_q = (\bar{u}_z)_p$$

and

(3.81)

$$(\bar{u}_z)_q = -(\bar{u}_x)_p$$

provided that $\nu = 0$. It will be remembered that this analogy exists for two-dimensional loading of a half-space whatever the value of Poisson's ratio (cf. equation (2.30)).

For non-zero values of Poisson's ratio it is not possible to use equation (3.52) from potential theory to find the function $\partial^2 F_1/\partial z^2$ corresponding to ψ, and we must proceed instead from equations (3.74).

(a) Circular region, $n = -\frac{1}{2}$
We will consider first the distribution of traction:

$$q_x(x, y) = q_0(1 - r^2/a^2)^{-1/2}$$

(3.82)

acting parallel to Ox on the circular region of radius a shown in Fig. 3.6. A pressure distribution of this form (equation (3.34)) produces a constant normal displacement of the surface within the circle. By the analogy we have just been discussing, for $\nu = 0$, the tangential traction given by (3.82) would produce a uniform tangential displacement of the surface $\bar{u}_x$ in the direction of the traction. We shall now proceed to show that the given traction still results in a uniform tangential displacement for non-zero values of ν.

Restricting the discussion to surface displacements within the loaded circle $(r \leqslant a)$ equations (3.74) reduce to

$$\bar{u}_x = \frac{1}{2\pi G} \int\!\!\int_S q_x(\xi, \eta) \left\{ \frac{1-\nu}{s} + \nu \frac{(\xi - x)^2}{s^3} \right\} d\xi \, d\eta$$

(3.83a)

$$\bar{u}_y = \frac{\nu}{2\pi G} \int_S \int q_x(\xi, \eta) \, \frac{(\xi - x)(\eta - y)}{s^3} \, d\xi \, d\eta \qquad (3.83b)$$

$$\bar{u}_z = -\frac{1 - 2\nu}{4\pi G} \int_S \int q_x(\xi, \eta) \, \frac{\xi - x}{s^2} \, d\xi \, d\eta \qquad (3.83c)$$

where $s^2 = (\xi - x)^2 + (\eta - y)^2$.

These expressions for the surface displacements could also have been derived by superposition, using equations (3.75) for the displacements at a general point $B(x, y)$ due to a concentrated tangential force $Q_x = q_x \, d\xi \, d\eta$ acting at $C(\xi, \eta)$.

In order to perform the surface integration we change the coordinates from (ξ, η) to (s, ϕ) as shown, where

$$\xi^2 + \eta^2 = (x + s \cos \phi)^2 + (y + s \sin \phi)^2$$

Putting $\alpha^2 = a^2 - x^2 - y^2$ and $\beta = x \cos \phi + y \sin \phi$

$$q_x(s, \phi) = q_0 a(\alpha^2 - 2\beta s - s^2)^{-1/2} \qquad (3.84)$$

Equations (3.83) then become

$$\bar{u}_x = \frac{1}{2\pi G} \int_0^{2\pi} \int_0^{s_1} q_x(s, \phi) \{(1 - \nu) + \nu \cos^2 \phi\} \, d\phi \, ds \qquad (3.85a)$$

Fig. 3.6

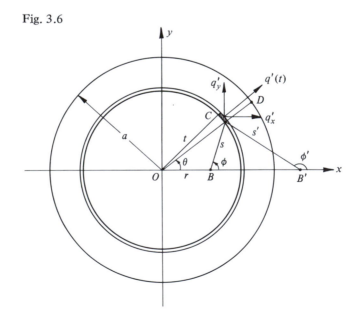

$$\bar{u}_y = \frac{\nu}{2\pi G} \int_0^{2\pi} \int_0^a q_x(s, \phi) \sin\phi \cos\phi \, d\phi \, ds \qquad (3.85b)$$

$$\bar{u}_z = -\frac{1-2\nu}{4\pi G} \int_0^{2\pi} \int_0^{s_1} q_x(s, \phi) \cos\phi \, d\phi \, ds \qquad (3.85c)$$

The limit s_1 is given by point D lying on the boundary of the circle, for which

$$s_1 = -\beta + (\alpha^2 + \beta^2)^{1/2}$$

Integrating first with respect to s, we have

$$\int_0^{s_1} (\alpha^2 - 2\beta s - s^2)^{-1/2} \, ds = \pi/2 - \tan^{-1}(\beta/\alpha)$$

When performing the integration with respect to ϕ between the limits 0 and 2π, we note that $\beta(\phi) = -\beta(\phi + \pi)$, so that for $(r \leqslant a)$

$$\bar{u}_x = \frac{q_0 a}{4G} \int_0^{2\pi} \{(1-\nu) + \nu \cos^2\phi\} \, d\phi$$

$$= \frac{\pi(2-\nu)}{4G} q_0 a = \text{constant} \qquad (3.86a)$$

$$\bar{u}_y = 0 \qquad (3.86b)$$

$$\bar{u}_z = \frac{(1-2\nu)q_0 a}{4\pi G} \int_0^{2\pi} \cos\phi \, \tan^{-1}(\beta/\alpha) \, d\phi$$

$$= -\frac{(1-2\nu)q_0 a}{2G} \left\{ \frac{a}{r} - \frac{(a^2 - r^2)^{1/2}}{r} \right\} \qquad (3.86c)$$

The normal traction given by (3.34) which produces a constant normal displacement of the surface within the circle $(r \leqslant a)$ was interpreted physically as the pressure which would be exerted on the flat face of a rigid frictionless cylindrical punch pressed into contact with the surface of an elastic half-space. By analogy, therefore, we are tempted to ask whether the tangential traction we have just been considering (3.82) represents the shear stress in the adhesive when a rigid cylindrical punch, whose flat face adheres to the surface of an elastic half-space, is given a tangential displacement parallel to the x-axis. However difficulty arises due to the non-zero normal displacements given by (3.86c), so that the punch face would not fit flush with the surface of the half-space without introducing additional tractions, both normal and tangential, at the interface.

The tractions acting on the surface of a rigid cylindrical punch which adheres to the surface of a half-space and is given a displacement in the *normal* direction have been found by Mossakovski (1954) and Spence (1968). The pressure distribution on the face of the punch with adhesion is not very different from that when the punch is frictionless. This was also shown to be the case for a two-dimensional punch (§2.8). The problem of finding the tractions for an adhesive cylindrical punch which is given a tangential displacement has been solved by Ufliand (1967). By analogy with the case of normal displacements, the shear traction on the punch face was found to be close to that given by (3.82). This approximation amounts to neglecting the mismatch of normal displacements with the flat face of the punch.

Obtaining expressions in closed form for the stress components within the solid is extremely involved. The variation of τ_{xz} along the z-axis due to the traction (3.82) is given by

$$\tau_{xz} = -p_0(1 - z^2/a^2)^{-2} \tag{3.87}$$

(b) Elliptical region, $n = -\frac{1}{2}$
When a tangential traction

$$q_x = q_0 \left(1 - \frac{x^2}{a^2} - \frac{y^2}{b^2}\right)^{-1/2} \tag{3.88}$$

acts parallel to Ox on an elliptical region of semi-axes a and b Mindlin (1949) has shown that the tangential displacement of the surface is again constant and in the Ox direction. Within the elliptical region

$$\bar{u}_x = \begin{cases} \dfrac{q_0 b}{G} \left[K(e) - \dfrac{\nu}{e^2} \{(1 - e^2)K(e) - E(e)\} \right], & a > b \\[3mm] \dfrac{q_0 a}{G} \left[K(e) - \dfrac{\nu}{e^2} \{K(e) - E(e)\} \right], & a < b \end{cases} \tag{3.89a}$$

$$\bar{u}_y = 0 \tag{3.89b}$$

(c) Circular region, $n = \frac{1}{2}$
The distribution of traction

$$q_x = q_0(1 - r^2/a^2)^{1/2} \tag{3.90}$$

acting on a circular region of radius a may be treated in the same way, by substituting (3.90) into equations (3.85) to find the surface displacements within the circle ($r \leqslant a$). The integration with respect to s is the same as in (3.40). Using that result and omitting the terms which do not contribute to the integration with respect to ϕ, we find

$$\bar{u}_x = \frac{\pi q_0}{32Ga} \{4(2 - \nu)a^2 - (4 - 3\nu)x^2 - (4 - \nu)y^2\} \tag{3.91a}$$

Similarly

$$\bar{u}_y = \frac{\pi q_0}{32Ga} 2\nu xy \tag{3.91b}$$

In this case $\bar{u}_x$ is not constant throughout the loaded circle and $\bar{u}_y$ does not vanish. Again we note that the normal displacements are not zero although we shall not evaluate them explicitly.

The tangential surface displacements outside the loaded circle $(r > a)$ due to the traction given by (3.90) have been found by Illingworth (see Johnson, 1955) with the results:

$$\bar{u}_x = \frac{q_0}{8Ga} [(2 - \nu)\{(2a^2 - r^2) \sin^{-1}(a/r) + ar(1 - a^2/r^2)^{1/2}\}$$
$$+ \tfrac{1}{2}\nu\{r^2 \sin^{-1}(a/r) + (2a^2 - r^2)(1 - a^2/r^2)^{1/2}(a/r)\}(x^2 - y^2)] \tag{3.92a}$$

$$\bar{u}_y = \frac{q_0\nu}{8Ga} \{r^2 \sin^{-1}(a/r) + (2a^2 - r^2)(1 - a^2/r^2)^{1/2}(a/r)\} xy \tag{3.92b}$$

A comprehensive investigation of the stresses within the solid due to the traction (3.90) has been carried out by Hamilton & Goodman (1966) and Hamilton (1983). Their results are discussed in §7.1(*b*) in relation to sliding contact.

(d) Elliptical region, $n = \tfrac{1}{2}$
The corresponding traction

$$q_x = q_0 \left(1 - \frac{x^2}{a^2} - \frac{y^2}{b^2}\right)^{1/2} \tag{3.93}$$

acting on an elliptical region of semi-axes a and b gives rise to tangential diplacements within the ellipse given by

$$\bar{u}_x = \frac{q_0 a}{2G} (C - Ax^2 - By^2) \tag{3.94a}$$

$$\bar{u}_y = \frac{q_0 a}{2G} Dxy \tag{3.94b}$$

where A, B, C and D are functions of shape and size of the ellipse. They have been expressed in terms of tabulated elliptic integrals by Vermeulen & Johnson (1964). Stresses within the half-space have been found by Bryant & Keer (1982) and Sackfield & Hills (1983*b*).

(e) Elliptical region, $n = m - \frac{1}{2}$

Finally we consider a general distribution of traction of the form

$$q_x(x,y) = q_0 \left(1 - \frac{x^2}{a^2} - \frac{y^2}{b^2}\right)^{m-1/2} \tag{3.95}$$

acting on an elliptical region of the surface. We saw in §5 that a pressure distribution of this form gave rise to normal displacements which varied throughout
the elliptical region as a polynomial in x and y of order $2m$. The two examples
of tangential traction which we have investigated, i.e. $m = 0$ and $m = 1$, have
resulted in surface displacements which, in the first case are constant – a polynomial of zero order – and in the second case (equation (3.94)) vary through
the elliptical region as a polynomial of second order in x and y. Kalker (1967*a*)
has proved that the general tangential traction (3.95) does give rise to tangential
surface displacements of order $2m$ in x and y, and has shown how the coefficients
of the polynomial can be computed. Thus if the displacements within the ellipse
are specified and can be approximated by a finite number of terms in a polynomial series, then the resultant tangential traction which would give rise to
those specified displacements can be found as the sum of an equal number of
terms having the form of (3.95).

3.8 Axi-symmetrical tractions

An important special case of 'point' loading arises when the half-space
is loaded over a circular region by surface tractions, both normal and tangential,
which are rotationally symmetrical about the z-axis. The magnitudes of the
tractions are thereby independent of θ and the direction of the tangential
traction is radial at all points. The system of stresses induced in solids of revolution by an axially symmetrical distribution of surface tractions is discussed by
Timoshenko & Goodier (1951, Chapter 13). It follows from the symmetry that
the components of shearing stress $\tau_{r\theta}$ and $\tau_{\theta z}$ vanish, whilst the remaining
stress components are independent of θ.

A number of examples of axi-symmetric distributions of normal traction
were considered in §3.4. In this section we shall approach the problem somewhat
differently and include radial tangential tractions.

Referring to Fig. 3.6, we start by considering a normal line load of intensity
p' per unit length acting on a ring of radius t. The normal and tangential displacements at a surface point $B(r, 0)$ are found from equations (3.18*a* and *c*)
for a concentrated normal force, with the result:

$$\bar{u}_z'(r) = \frac{1 - \nu^2}{\pi E} \, 2 \int_0^\pi \frac{p't \, d\theta}{s}$$

$$= \frac{1-v^2}{\pi E} \, 4p't \int_0^\pi \{(t+r)^2 - 4tr \sin^2(\theta/2)\}^{-1/2} \, d(\theta/2)$$

$$= \frac{1-v^2}{\pi E} \frac{4p't}{t+r} \mathbf{K}(k) \tag{3.96a}$$

where $k^2 = 4tr/(t+r)^2$.

$$\bar{u}'_r(r) = \frac{(1-2v)(1+v)}{2\pi E} \, 2p' \int_0^\pi \frac{t \cos\phi \, d\theta}{s}$$

$$= \begin{cases} -\dfrac{(1-2v)(1+v)}{E} \, p't/r, & r > t \tag{3.96b} \\[2mm] 0, & r < t \tag{3.96c} \end{cases}$$

Now consider a tangential line load of intensity q' per unit length acting radially at radius t. At any point such as C, q' is resolved into $q'_x = q' \cos\theta$ and $q'_y = q' \sin\theta$. The displacements at a surface point B due to concentrated forces $q'_x t \, d\theta$ and $q'_y t \, d\theta$ acting at C are found to be from equations (3.75). Integrating for a complete ring gives

$$\bar{u}'_r(r) = \frac{2(1-v^2)}{\pi E} \, q'(t)t \int_0^\pi \frac{\cos\theta}{s} \, d\theta$$

$$= \frac{4(1-v^2)}{\pi E} \, q'(t) \frac{t}{t+r} \left\{ \left(\frac{2}{k^2} - 1 \right) \mathbf{K}(k) - \frac{2}{k^2} \mathbf{E}(k) \right\} \tag{3.97a}$$

and

$$\bar{u}'_z(r) = \begin{cases} -\dfrac{(1-2v)(1+v)}{E} \, q'(t), & r \leqslant t \tag{3.97b} \\[2mm] 0, & r > t \tag{3.97c} \end{cases}$$

The surface displacement due to a distributed pressure $p(t)$ and traction $q(t)$ can be built up from equations (3.96) and (3.97) with the result

$$\bar{u}_z = \frac{4(1-v^2)}{\pi E} \int_0^a \frac{t}{t+r} \, p(t) \mathbf{K}(k) \, dt - \frac{(1-2v)(1+v)}{\pi E} \int_r^a q(t) \, dt \tag{3.98a}$$

$$\bar{u}_r = \frac{4(1-v^2)}{\pi E} \int_0^a \frac{t}{t+r} \, q(t) \left\{ \left(\frac{2}{k^2} - 1 \right) \mathbf{K}(k) - \frac{2}{k^2} \mathbf{E}(k) \right\} \, dt$$

$$- \frac{(1-2v)(1+v)}{Er} \int_0^r tp(t) \, dt \tag{3.98b}$$

When $r > a$ the second term in equation (3.98a) should be ignored and the upper limit in the second term in (3.98b) becomes a. These expressions enable the surface displacements to be calculated, numerically at least, for any axisymmetric distributions of traction. They are not convenient however when the surface displacements are specified and the surface tractions are unknown. Integral transform methods have been developed for this purpose. This mathematical technique is beyond the scope of this book and the interested reader is referred to the books by Sneddon (1951) and Gladwell (1980) and the work of Spence (1968). However we shall quote the following useful results.

Noble & Spence (1971) introduce the functions

$$L(\lambda) = \frac{1}{2G} \int_\lambda^1 \frac{\rho p(\rho)\, d\rho}{(\rho^2 - \lambda^2)^{1/2}}, \quad M(\lambda) = \frac{\lambda}{2G} \int_\lambda^1 \frac{q(\rho)\, d\rho}{(\rho^2 - \lambda^2)^{1/2}} \tag{3.99}$$

which can be evaluated if the pressure $p(\rho)$ and traction $q(\rho)$ are known within the loaded circle $\rho(=r/a) \leqslant 1$. Alternatively, if the normal surface displacement $\bar{u}_z(\rho)$ is known within the circle due to $p(\rho)$ *acting alone*, then

$$L(\lambda) = \frac{1}{2(1-\nu)a} \frac{d}{d\lambda} \int_0^\lambda \frac{\rho \bar{u}_z(\rho)\, d\rho}{(\lambda^2 - \rho^2)^{1/2}} \tag{3.100}$$

or if the tangential displacement $\bar{u}_r(\rho)$ due to $q(\rho)$ *acting alone* is known, then

$$M(\lambda) = \frac{1}{2(1-\nu)a} \frac{d}{d\lambda} \int_0^\lambda \frac{\bar{u}_r(\rho)\, d\rho}{(\lambda^2 - \rho^2)^{1/2}} \tag{3.101}$$

The displacements and stresses throughout the surface of the half-space may now be expressed in terms of $L(\lambda)$ and $M(\lambda)$, thus

$$\frac{\bar{u}_r(\rho)}{a} = \begin{cases} \dfrac{2(1-2\nu)}{\pi\rho} \left\{ \displaystyle\int_\rho^1 \dfrac{\lambda L(\lambda)\, d\lambda}{(\lambda^2 - \rho^2)^{1/2}} - \displaystyle\int_\rho^1 L(\lambda)\, d\lambda \right\} \\[4mm] \quad + \dfrac{4(1-\nu)}{\pi\rho} \displaystyle\int_0^\rho \dfrac{\lambda M(\lambda)\, d\lambda}{(\rho^2 - \lambda^2)^{1/2}}, \quad \rho \leqslant 1 \\[4mm] \dfrac{2(1-2\nu)}{\pi\rho} \displaystyle\int_0^1 L(\lambda)\, d\lambda + \dfrac{4(1-\nu)}{\pi\rho} \displaystyle\int_0^1 \dfrac{\lambda M(\lambda)\, d\lambda}{(\rho^2 - \lambda^2)^{1/2}}, \\[4mm] \quad\quad\quad \rho > 1 \end{cases} \tag{3.102a}$$

$$\frac{\bar{u}_z(\rho)}{a} = \begin{cases} \dfrac{4(1-\nu)}{\pi} \displaystyle\int_0^\rho \dfrac{L(\lambda)\, d\lambda}{(\rho^2 - \lambda^2)^{1/2}} - \dfrac{2(1-2\nu)}{\pi} \displaystyle\int_\rho^1 \dfrac{M(\lambda)\, d\lambda}{(\lambda^2 - \rho^2)^{1/2}}, \\[4mm] \quad\quad\quad \rho \leqslant 1 \\[4mm] \dfrac{4(1-\nu)}{\pi} \displaystyle\int_0^\rho \dfrac{L(\lambda)\, d\lambda}{(\rho^2 - \lambda^2)^{1/2}}, \quad \rho > 1 \end{cases}$$

$$\tag{3.102b}$$

$$\frac{\tau_{rz}(\rho)}{2G} = \begin{cases} -\dfrac{q(\rho)}{2G} = \dfrac{2}{\pi\rho}\dfrac{d}{d\rho}\displaystyle\int_\rho^1 \dfrac{M(\lambda)\,d\lambda}{(\lambda^2-\rho^2)^{1/2}}, & \rho \le 1 \\[3mm] 0, & \rho > 1 \end{cases} \tag{3.103a}$$

$$\frac{\sigma_z(\rho)}{2G} = \begin{cases} -\dfrac{p(\rho)}{2G} = \dfrac{2}{\pi\rho}\dfrac{d}{d\rho}\displaystyle\int_\rho^1 \dfrac{\lambda L(\lambda)\,d\lambda}{(\lambda^2-\rho^2)^{1/2}}, & \rho \le 1 \\[3mm] 0, & \rho > 1 \end{cases} \tag{3.103b}$$

$$\frac{\sigma_r(\rho)}{2G} = \begin{cases} -\dfrac{p(\rho)}{2G} - \dfrac{2(1-2\nu)}{\pi\rho^2}\left\{\displaystyle\int_\rho^1 \dfrac{\lambda L(\lambda)\,d\lambda}{(\lambda^2-\rho^2)^{1/2}} - \int_0^1 L(\lambda)\,d\lambda\right\} \\[3mm] +\dfrac{4}{\pi\rho}\left(\dfrac{d}{d\rho}-\dfrac{1-\nu}{\rho}\right)\displaystyle\int_0^\rho \dfrac{\lambda M(\lambda)\,d\lambda}{(\rho^2-\lambda^2)^{1/2}}, \qquad \rho \le 1 \\[3mm] \dfrac{2(1-2\nu)}{\pi\rho^2}\displaystyle\int_0^1 L(\lambda)\,d\lambda + \dfrac{4}{\pi\rho}\left(\dfrac{d}{d\rho}-\dfrac{1-\nu}{\rho}\right)\int_0^1 \dfrac{\lambda M(\lambda)\,d\lambda}{(\rho^2-\lambda^2)^{1/2}}, \\[3mm] \hspace{8cm} \rho > 1 \end{cases}$$

$$\tag{3.103c}$$

$$\frac{\sigma_\theta(\rho)}{2G} = \begin{cases} -\dfrac{2\nu}{2G}p(\rho) + \dfrac{2(1-2\nu)}{\pi\rho^2}\left\{\displaystyle\int_\rho^1 \dfrac{\lambda L(\lambda)\,d\lambda}{(\lambda^2-\rho^2)^{1/2}} - \int_0^1 L(\lambda)\,d\lambda\right\} \\[3mm] +\dfrac{4}{\pi\rho}\left(\nu\dfrac{d}{d\rho}+\dfrac{1-\nu}{\rho}\right)\displaystyle\int_0^\rho \dfrac{\lambda L(\lambda)\,d\lambda}{(\rho^2-\lambda^2)^{1/2}}, \qquad \rho \le 1 \\[3mm] -\dfrac{2(1-2\nu)}{\pi\rho^2}\displaystyle\int_0^1 L(\lambda)\,d\lambda + \dfrac{4}{\pi\rho}\left(\nu\dfrac{d}{d\rho}+\dfrac{1-\nu}{\rho}\right) \\[3mm] \times\displaystyle\int_0^1 \dfrac{\lambda M(\lambda)\,d\lambda}{(\rho^2-\lambda^2)^{1/2}}, \qquad \rho > 1 \end{cases}$$

$$\tag{3.103d}$$

In the case where both $\bar{u}_r$ and $\bar{u}_z$ are specified within the loaded circle, equations (3.102a and b) are coupled integral equations for $L(\lambda)$ and $M(\lambda)$. They have been reduced to a single integral equation by Abramian *et al.* (1966) and Spence (1968), from which $L(\lambda)$, $M(\lambda)$ and hence $p(\rho)$ and $q(\rho)$ can be found.

A problem of this type arises when a rigid flat-ended cylindrical punch of radius a is pressed normally in contact with an elastic half-space under conditions in which the flat face of the punch adheres to the surface. This is the axially symmetrical analogue of the two-dimensional rigid punch discussed in §2.8(*b*). The punch indents the surface with a uniform displacement δ. Thus the boundary conditions within the circle of contact, $r \leqslant a$, are

$$\bar{u}_z = \delta, \quad \bar{u}_r = 0 \tag{3.104}$$

Mossakovski (1954) and Spence (1968) solve this problem and show that the load on the punch P is related to the indentation δ by

$$P = 4Ga\delta \ \ln \{(3 - 4\nu)/(1 - 2\nu)\} \tag{3.105}$$

The load on the face of a *frictionless* punch is given by equations (3.36) and (3.37) which, for comparison with (3.105), may be written

$$P = 4Ga\delta/(1 - \nu)$$

The 'adhesive' load is greater than the 'frictionless' load by an amount which varies from 10% when $\nu = 0$, to zero when $\nu = 0.5$.

Spence (1975) has also examined the case of partial slip. During monotonic loading the contact circle is divided into a central region of radius c which does not slip and an annulus $c \leqslant r \leqslant a$ where the surface of the half-space slips radially inwards under the face of the punch. Turner (1979) has examined the behaviour on unloading. As the load is reduced the inner boundary of the slip zone $r = c$ shrinks in size with the slip there maintaining its *inward* direction. At the same time a thin annulus at the periphery adheres to the punch without slip until, when the load has decreased to about half its maximum value, *outward* slip begins at $r = a$ and rapidly spreads across the contact surface.

The surface displacements produced by an axi-symmetric distribution of pressure, calculated in §3 by the classical method, could equally well have been found by substituting $p(\rho)$ into equation (3.99) and then (3.102). The surface stress could also be found directly from equations (3.103).

3.9   Torsional loading

In this section we examine tangential tractions which act in a circumferential direction, that is perpendicular to the radius drawn from the origin. Such tractions induce a state of torsion in the half-space.

(a) Circular region

For the circular region shown in Fig. 3.7 we shall assume that the magnitude of the traction $q(r)$ is a function of r only. Thus

$$q_x = -q(r) \sin \theta = -q(t)\eta/t \tag{3.106a}$$
$$q_y = q(r) \cos \theta = q(t)\xi/t \tag{3.106b}$$

The expressions for the displacements u_x, u_y and u_z can be written in the form of equations (3.7), where $H = 0$ and F and G are given by

$$F = -\int_S\int \frac{q(t)}{t} \eta \ln (\rho + z) \, d\xi \, d\eta \tag{3.107a}$$

and

$$G = \int_S \int \frac{q(t)}{t} \xi \ln (\rho + z) \, d\xi \, d\eta \qquad (3.107b)$$

In this case, from the reciprocal nature of F and G with respect to the coordinates, it follows that $\partial G/\partial y = -\partial F/\partial x$, so that the expressions for the displacements on the surface reduce to

$$\bar{u}_x = \frac{1}{2\pi G} \frac{\partial F}{\partial z} = -\frac{1}{2\pi G} \int_0^{2\pi} \int_0^{s_1} \frac{q(t)}{t} \eta \, ds \, d\phi \qquad (3.108a)$$

$$\bar{u}_y = \frac{1}{2\pi G} \frac{\partial G}{\partial z} = \frac{1}{2\pi G} \int_0^{2\pi} \int_0^{s_1} \frac{q(t)}{t} \xi \, ds \, d\phi \qquad (3.108b)$$

$$\bar{u}_z = 0 \qquad (3.108c)$$

If we consider the displacement of the point $B(x, 0)$, as shown in Fig. 3.7, then $\eta/t = \sin \theta$ and it is apparent that the surface integral in (3.108a) vanishes. So we are left with the circumferential component $\bar{u}_y$ as the only non-zero displacement, which was to be expected in a purely torsional deformation.

Now consider the traction

$$q(r) = q_0 r (a^2 - r^2)^{-1/2}, \quad r \le a \qquad (3.109)$$

Substituting in (3.108b) for the surface displacement

$$\bar{u}_y = \frac{q_0}{2\pi G} \int_0^{2\pi} \int_0^{s_1} (a^2 - x^2 - 2xs \cos \phi - s^2)^{-1/2}(x + s \cos \phi) \, ds \, d\phi$$

The integral is of the form met previously and gives

$$\bar{u}_y = \pi q_0 x / 4G$$

Fig. 3.7

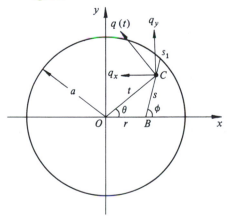

In view of the circular symmetry we can write

$$\bar{u}_\theta = \pi q_0 r / 4G$$
$$\bar{u}_r = \bar{u}_z = 0 \tag{3.110}$$

Thus the traction (3.109) produces a rigid rotation of the loaded circle through an angle $\beta = \pi q_0 / 4G$. The traction gives rise to a resultant twisting moment

$$M_z = \int_0^a q(r) 2\pi r \, dr$$

$$= 4\pi a^3 q_0 / 3 \tag{3.111}$$

Hence equation (3.109) gives the traction acting on the surface of a flat-ended cylindrical punch which adheres to the surface of a half-space when given a twist about its axis. Since the normal displacements $\bar{u}_z$ due to the twist are zero, the pressure distribution on the face of the punch is not influenced by the twist. This is in contrast to the behaviour of a punch which is given a uni-axial tangential displacement, where we saw (in §7) that the normal pressure and tangential tractions are not independent.

Hetenyi & McDonald (1958) have considered the distribution of traction

$$q(r) = \bar{\tau}_{z\theta} = q_0 (1 - r^2/a^2)^{1/2}, \quad r \leqslant a \tag{3.112}$$

Expressions have been found for u_θ, $\tau_{r\theta}$ and $\tau_{z\theta}$ and values of the stress components $\tau_{r\theta}$ (r, z) have been tabulated. The maximum value is $0.73q_0$ on the surface at $r = a$.

(b) Elliptical region

We turn now to a loaded region of elliptical shape in order to find the distribution of traction which will again result in a rigid rotation of the loaded ellipse. In this case there is no rotational symmetry and we tentatively put

$$q_x = -q_0' y \{1 - (x/a)^2 - (y/b)^2\}^{-1/2} \tag{3.113a}$$

and

$$q_y = q_0'' x \{1 - (x/a)^2 - (y/b)^2\}^{-1/2} \tag{3.113b}$$

These expressions are substituted in equations (3.2) to obtain the potential functions F_1 and G_1, which in turn are substituted in (3.7) to obtain the tangential displacements of a general surface point (x, y). Performing the integrations in the usual way Mindlin (1949) showed that the displacements correspond to a rigid rotation of the elliptical region through a small angle β i.e. $\bar{u}_x = -\beta y$ and $\bar{u}_y = \beta x$, provided that

$$q_0' = \frac{G\beta}{2a} \frac{B - 2\nu(1 - e^2)C}{BD - \nu CE} \tag{3.114a}$$

and

$$q_0'' = \frac{G\beta}{2a} \frac{D - 2\nu C}{BD - \nu CE} \tag{3.114b}$$

where $D(e)$, $B(e)$ and $C(e)$ can be expressed in terms of the standard elliptic integrals $E(e)$ and $K(e)$ and $e = (1 - a^2/b^2)^{1/2}$ is the eccentricity of the ellipse, viz.:

$$D = (K - E)/e^2$$
$$B = \{E - (1 - e^2)K\}/e^2$$
$$C = \{(2 - e^2)K - 2E\}/e^4$$

The twisting moment M_z is given by

$$M_z = \tfrac{2}{3} \pi b^3 G\beta \frac{E - 4\nu(1 - e^2)C}{BD - \nu CE} \tag{3.115}$$

4

Normal contact of elastic solids: Hertz theory†

4.1 Geometry of smooth, non-conforming surfaces in contact

When two non-conforming solids are brought into contact they touch initially at a single point or along a line. Under the action of the slightest load they deform in the vicinity of their point of first contact so that they touch over an area which is finite though small compared with the dimensions of the two bodies. A theory of contact is required to predict the shape of this area of contact and how it grows in size with increasing load; the magnitude and distribution of surface tractions, normal and possibly tangential, transmitted across the interface. Finally it should enable the components of deformation and stress in both bodies to be calculated in the vicinity of the contact region.

Before the problem in elasticity can be formulated, a description of the geometry of the contacting surfaces is necessary. In Chapter 1 we agreed to take the point of first contact as the origin of a rectangular coordinate system in which the x–y plane is the common tangent plane to the two surfaces and the z-axis lies along the common normal directed positively into the lower solid (see Fig. 1.1). Each surface is considered to be topographically smooth on both micro and macro scale. On the micro scale this implies the absence or disregard of small surface irregularities which would lead to discontinuous contact or highly local variations in contact pressure. On the macro scale the profiles of the surfaces are continuous up to their second derivative in the contact region. Thus we may express the profile of each surface in the region close to the origin approximately by an expression of the form

$$z_1 = A_1 x^2 + B_1 y^2 + C_1 xy + \ldots \tag{4.1}$$

where higher order terms in x and y are neglected. By choosing the orientation of the x and y axes, x_1 and y_1, so that the term in xy vanishes, (4.1) may be written:

† A summary of Hertz elastic contact stress formulae is given in Appendix 3, p. 427.

$$z_1 = \frac{1}{2R_1'} x_1{}^2 + \frac{1}{2R_1''} y_1{}^2 \qquad (4.2a)$$

where R_1' and R_1'' are the principal radii of curvature of the surface at the origin. They are the maximum and minimum values of the radius of curvature of all possible cross-sections of the profile. If a cross-sectional plane of symmetry exists one principal radius lies in that plane. A similar expression may be written for the second surface:

$$z_2 = -\left(\frac{1}{2R_2'} x_2{}^2 + \frac{1}{2R_2''} y_2{}^2\right) \qquad (4.2b)$$

The separation between the two surfaces is then given by $h = z_1 - z_2$. We now transpose equation (4.1) and its counterpart to a common set of axes x and y, whereby

$$h = Ax^2 + By^2 + Cxy$$

By a suitable choice of axes we can make C zero, hence

$$h = Ax^2 + By^2 = \frac{1}{2R'} x^2 + \frac{1}{2R''} y^2 \qquad (4.3)$$

where A and B are positive constants and R' and R'' are defined as the principal *relative* radii of curvature. If the axes of principal curvature of each surface, i.e. the x_1 axis and the x_2 axis, are inclined to each other by an angle α, then it is shown in Appendix 2 that

$$(A+B) = \tfrac{1}{2}\left(\frac{1}{R'} + \frac{1}{R''}\right) = \tfrac{1}{2}\left(\frac{1}{R_1'} + \frac{1}{R_1''} + \frac{1}{R_2'} + \frac{1}{R_2''}\right) \qquad (4.4)$$

and

$$|B - A| = \tfrac{1}{2}\left\{\left(\frac{1}{R_1'} - \frac{1}{R_1''}\right)^2 + \left(\frac{1}{R_2'} - \frac{1}{R_2''}\right)^2 \right.$$
$$\left. + 2\left(\frac{1}{R_1'} - \frac{1}{R_1''}\right)\left(\frac{1}{R_2'} - \frac{1}{R_2''}\right)\cos 2\alpha\right\}^{1/2} \qquad (4.5)$$

We introduce an equivalent radius R_e defined by

$$R_e = (R'R'')^{1/2} = \tfrac{1}{2}(AB)^{-1/2}$$

In this description of the initial separation between the two surfaces in terms of their principal radii of curvature we have taken a *convex* surface to have a *positive* radius. Equations (4.4) and (4.5) apply equally to concave or saddle-shaped surfaces by ascribing a negative sign to the concave curvatures.

It is evident from equation (4.3) that contours of constant gap h between the undeformed surfaces are ellipses the length of whose axes are in the ratio

$(B/A)^{1/2} = (R'/R'')^{1/2}$. Such elliptical contours are displayed by the interference fringes between two cylindrical lenses, each of radius R, with their axes inclined at $45°$, shown in Fig. 4.1(a). In this example $R'_1 = R'_2 = R$; $R''_1 = R''_2 = \infty$; $\alpha = 45°$, for which equations (4.4) and (4.5) give $A + B = 1/R$ and $B - A = 1/\sqrt{(2)}R$, i.e. $A = (1 - 1/\sqrt{2})/2R$ and $B = (1 + 1/\sqrt{2})/2R$. The relative radii of curvature are thus: $R' = 1/2A = 3.42R$ and $R'' = 1/2B = 0.585R$. The equivalent radius $R_e = (R'R'')^{1/2} = \sqrt{(2)}R$ and $(R'/R'')^{1/2} =$

Fig. 4.1. Interference fringes at the contact of two equal cylindrical lenses with their axes inclined at $45°$: (a) unloaded, (b) loaded.

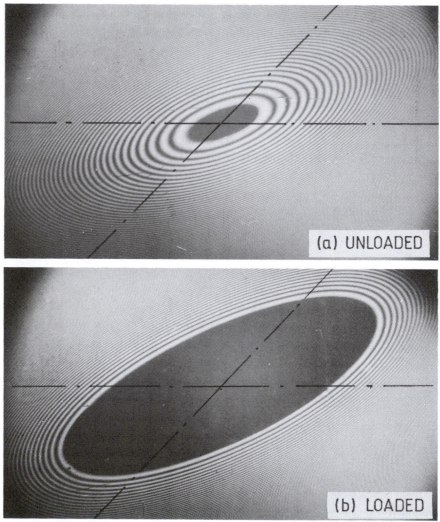

(a) UNLOADED

(b) LOADED

$(B/A)^{1/2} = 2.41$. This is the ratio of the major to minor axes of the contours of constant separation shown in Fig. 4.1(a).

We can now say more precisely what we mean by non-conforming surfaces: the relative curvatures $1/R'$ and $1/R''$ must be sufficiently large for the terms Ax^2 and By^2 on the right-hand side of equation (4.3) to be large compared with the higher order terms which have been neglected. The question of conforming surfaces is considered in §5.3.

A normal compressive load is now applied to the two solids and the point of contact spreads into an area. If the two bodies are solids of revolution, then $R'_1 = R''_1 = R_1$ and $R'_2 = R''_2 = R_2$, whereupon $A = B = \frac{1}{2}(1/R_1 + 1/R_2)$. Thus contours of constant separation between the surfaces before loading are circles centred at O. After loading, it is evident from the circular symmetry that the contact area will also be circular. Two cylindrical bodies of radii R_1 and R_2 in contact with their axes parallel to the y-axis have $R'_1 = R_1, R''_1 = \infty$, $R'_2 = R_2, R''_2 = \infty$ and $\alpha = 0$, so that $A = \frac{1}{2}(1/R_1 + 1/R_2), B = 0$. Contours of constant separation are straight lines parallel to the y-axis and, when loaded, the surfaces will make contact over a narrow strip parallel to the y-axis. In the case of general profiles it follows from equation (4.3) that contours of constant separation are ellipses in plan. We might expect, therefore, that under load the contact surface would be elliptical in shape. It will be shown in due course that this is in fact so. A special case arises when two equal cylinders both of radius R are in contact with their axes perpendicular. Here $R'_1 = R, R''_1 = \infty, R'_2 = R$, $R''_2 = \infty, \alpha = \pi/2$, from which $A = B = \frac{1}{2}R$. In this case the contours of constant separation are circles and identical to those due to a sphere of the same radius R in contact with a plane surface ($R'_2 = R''_2 = \infty$).

We shall now consider the deformation as a normal load P is applied. Two solids of general shape (but chosen convex for convenience) are shown in cross-section after deformation in Fig. 4.2. Before deformation the separation between two corresponding surface points $S_1(x, y, z_1)$ and $S_2(x, y, z_2)$ is given by equation (4.3). From the symmetry of this expression about O the contact region must extend an equal distance on either side of O. During the compression distant points in the two bodies T_1 and T_2 move towards O, parallel to the z-axis, by displacements δ_1 and δ_2 respectively. If the solids did not deform their profiles would overlap as shown by the dotted lines in Fig. 4.2. Due to the contact pressure the surface of each body is displaced parallel to Oz by an amount $\bar{u}_{z1}$ and $\bar{u}_{z2}$ (measured positive into each body) relative to the distant points T_1 and T_2. If, after deformation, the points S_1 and S_2 are coincident within the contact surface then

$$\bar{u}_{z1} + \bar{u}_{z2} + h = \delta_1 + \delta_2 \tag{4.6}$$

Writing $\delta = \delta_1 + \delta_2$ and making use of (4.3) we obtain an expression for the elastic displacements:

$$\bar{u}_{z1} + \bar{u}_{z2} = \delta - Ax^2 - By^2 \tag{4.7}$$

where x and y are the common coordinates of S_1 and S_2 projected onto the x-y plane. If S_1 and S_2 lie outside the contact area so that they do not touch it follows that

$$\bar{u}_{z1} + \bar{u}_{z2} > \delta - Ax^2 - By^2 \tag{4.8}$$

To solve the problem, it is necessary to find the distribution of pressure transmitted between the two bodies at their surface of contact, such that the resulting elastic displacements normal to that surface satisfy equation (4.7) within the contact area and equation (4.8) outside it.

Fig. 4.2.

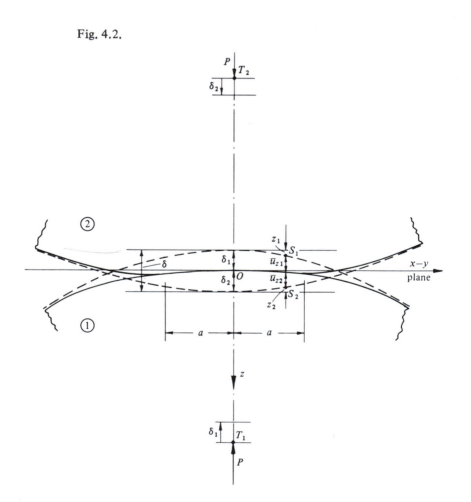

Before proceeding to examine the problem in elasticity, however, it is instructive to see how the deformations and stresses grow as the load is applied on the basis of elementary dimensional reasoning. For simplicity we shall restrict the discussion to (a) solids of revolution in which the contact area is a circle of radius a and (b) two-dimensional bodies in which the contact area is an infinite strip of width $2a$.

We note that in Fig. 4.2, $\delta_1 = \bar{u}_{z1}(0)$ and $\delta_2 = \bar{u}_{z2}(0)$, so that equation (4.6) can be written in non-dimensional form

$$\left\{ \frac{\bar{u}_{z1}(0)}{a} - \frac{\bar{u}_{z1}(x)}{a} \right\} + \left\{ \frac{\bar{u}_{z2}(0)}{a} - \frac{\bar{u}_{z2}(x)}{a} \right\} = \tfrac{1}{2}(1/R_1 + 1/R_2)x^2/a \quad (4.9)$$

Putting $x = a$ and writing $\bar{u}_z(0) - u_z(a) = d$, the 'deformation' within the contact zone, (4.9) becomes

$$\frac{d_1}{a} + \frac{d_2}{a} = \frac{a}{2}\left(\frac{1}{R_1} + \frac{1}{R_2} \right) \quad (4.10)$$

Provided that the deformation is small, i.e. $d \ll a$, the state of *strain* in each solid is characterised by the ratio d/a. Now the magnitude of the strain will be proportional to the contact pressure divided by the elastic modulus; therefore, if p_m is the average contact pressure acting mutually on each solid, (4.10) becomes†

$$p_m/E_1 + p_m/E_2 \propto a(1/R_1 + 1/R_2)$$

i.e.

$$p_m \propto \frac{a(1/R_1 + 1/R_2)}{1/E_1 + 1/E_2} \quad (4.11)$$

Thus, for a given geometry and materials, the contact pressure and the associated stresses increase in direct proportion to the linear dimension of the contact area. To relate the growth of the contact to the load, two and three-dimensional contacts must be examined separately.

(a) In the contact of cylinders, the load per unit axial length $P = 2ap_m$, whence from (4.11)

$$a \propto \{P(1/E_1 + 1/E_2)/(1/R_1 + 1/R_2)\}^{1/2} \quad (4.12)$$

and

$$p_m \propto \{P(1/R_1 + 1/R_2)/(1/E_1 + 1/E_2)\}^{1/2} \quad (4.13)$$

† It transpires that the 'plane-strain modulus' $E/(1 - \nu^2)$ is the correct elastic modulus to use in contact problems, but Young's modulus E is used here to retain the simplicity of the argument.

from which we see that the contact width and contact pressure increase as the square root of the applied load.

(*b*) In the contact of spheres, or other solids of revolution, the compressive load $P = \pi a^2 p_m$. Hence from (4.11)

$$a \propto \{P(1/E_1 + 1/E_2)/(1/R_1 + 1/R_2)\}^{1/3} \tag{4.14}$$

and

$$p_m \propto \{P(1/R_1 + 1/R_2)^2/(1/E_1 + 1/E_2)^2\}^{1/3} \tag{4.15}$$

In this case the radius of the contact circle and the contact pressure increase as the cube root of the load.

In the case of three-dimensional contact the compressions of each solid δ_1 and δ_2 are proportional to the local indentations d_1 and d_2, hence the approach of distant points

$$\delta = \delta_1 + \delta_2 \propto d_1 + d_2$$
$$\propto \{P^2(1/E_1 + 1/E_2)^2(1/R_1 + 1/R_2)\}^{1/3} \tag{4.16}$$

The approach of two bodies due to elastic compression in the contact region is thus proportional to (load)$^{2/3}$.

In the case of two-dimensional contact the displacements δ_1 and δ_2 are not proportional to d_1 and d_2 but depend upon the arbitrarily chosen datum for elastic displacements. An expression similar to (4.16) cannot be found in this case.

We have shown how the contact area, stress and deformation might be expected to grow with increasing load and have also found the influence of curvature and elastic moduli by simple dimensional reasoning. To obtain absolute values for these quantities we must turn to the theory of elasticity.

4.2   Hertz theory of elastic contact

The first satisfactory analysis of the stresses at the contact of two elastic solids is due to Hertz (1882*a*). He was studying Newton's optical inter-ference fringes in the gap between two glass lenses and was concerned at the possible influence of elastic deformation of the surfaces of the lenses due to the contact pressure between them. His theory, worked out during the Christmas vacation 1880 at the age of twenty-three, aroused considerable interest when it was first published and has stood the test of time. In addition to static loading he also investigated the quasi-static impacts of spheres (see §11.4). Hertz (1882*b*) also attempted to use his theory to give a precise definition of hardness of a solid in terms of the contact pressure to *initiate* plastic yield in the solid by pressing a harder body into contact with it. This definition has proved unsatisfactory because of the difficulty of detecting the point of first yield under the action of contact stress. A satisfactory theory of hardness had to wait for the develop-ment of the theory of plasticity. This question is considered in Chapter 6.

Hertz formulated the conditions expressed by equations (4.7) and (4.8) which must be satisfied by the normal displacements on the surface of the solids. He first made the hypothesis that the contact area is, in general, elliptical, guided no doubt by his observations of interference fringes such as those shown in Fig. 4.1(*b*). He then introduced the simplification that, for the purpose of calculating the local deformations, each body can be regarded as an elastic half-space loaded over a small elliptical region of its plane surface. By this simplification, generally followed in contact stress theory, the highly concentrated contact stresses are treated separately from the general distribution of stress in the two bodies which arises from their shape and the way in which they are supported. In addition, the well developed methods for solving boundary-value problems for the elastic half-space are available for the solution of contact problems. In order for this simplification to be justifiable two conditions must be satisfied: the significant dimensions of the contact area must be small compared (*a*) with the dimensions of each body and (*b*) with the relative radii of curvature of the surfaces. The first condition is obviously necessary to ensure that the stress field calculated on the basis of a solid which is infinite in extent is not seriously influenced by the proximity of its boundaries to the highly stressed region. The second condition is necessary to ensure firstly that the surfaces just outside the contact region approximate roughly to the plane surface of a half-space, and secondly that the strains in the contact region are sufficiently small to lie within the scope of the linear theory of elasticity. Metallic solids loaded within their elastic limit inevitably comply with this latter restriction. However, caution must be used in applying the results of the theory to low modulus materials like rubber where it is easy to produce deformations which exceed the restriction to small strains.

Finally, the surfaces are assumed to be frictionless so that only a normal pressure is transmitted between them. Although physically the contact pressure must act perpendicular to the interface which will not necessarily be planar, the linear theory of elasticity does not account for changes in the boundary forces arising from the deformation they produce (with certain special exceptions). Hence, in view of the idealisation of each body as a half-space with a plane surface, normal tractions at the interface are taken to act parallel to the z-axis and tangential tractions to act in the x–y plane.

Denoting the significant dimension of the contact area by a, the relative radius of curvature by R, the significant radii of each body by R_1 and R_2 and the significant dimensions of the bodies both laterally and in depth by l, we may summarise the assumptions made in the Hertz theory as follows:

(i) The surfaces are continuous and non-conforming: $a \ll R$;
(ii) The strains are small: $a \ll R$;

(iii) Each solid can be considered as an elastic half-space: $a \ll R_{1,2}, a \ll l$;

(iv) The surfaces are frictionless: $q_x = q_y = 0$.

The problem in elasticity can now be stated: the distribution of mutual pressure $p(x, y)$ acting over an area S on the surface of two elastic half-spaces is required which will produce normal displacements of the surfaces $\bar{u}_{z1}$ and $\bar{u}_{z2}$ satisfying equation (4.7) within S and (4.8) outside it.

(a) Solids of revolution

We will consider first the simpler case of solids of revolution $(R'_1 = R''_1 = R_1; R'_2 = R''_2 = R_2)$. The contact area will be circular, having a radius a, say. From equations (4.4) and (4.5) it is clear that $A = B = \frac{1}{2}(1/R_1 + 1/R_2)$, so that the boundary condition for displacements within the contact expressed in (4.7) can be written

$$\bar{u}_{z1} + \bar{u}_{z2} = \delta - (1/2R)r^2 \tag{4.17}$$

where $(1/R) = (1/R_1 + 1/R_2)$ is the relative curvature.

A distribution of pressure which gives rise to displacements which satisfy (4.17) has been found in §3.4, where the pressure distribution proposed by Hertz (equation (3.39))

$$p = p_0\{1 - (r/a)^2\}^{1/2}$$

was shown to give normal displacements (equation (3.41a))

$$\bar{u}_z = \frac{1-\nu^2}{E} \frac{\pi p_0}{4a}(2a^2 - r^2), \quad r \leqslant a$$

The pressure acting on the second body is equal to that on the first, so that by writing

$$\frac{1}{E^*} = \frac{1-\nu_1^2}{E_1} + \frac{1-\nu_2^2}{E_2}$$

and substituting the expressions for $\bar{u}_{z1}$ and $\bar{u}_{z2}$ into equation (4.17) we get

$$\frac{\pi p_0}{4aE^*}(2a^2 - r^2) = \delta - (1/2R)r^2 \tag{4.18}$$

from which the radius of the contact circle is given by

$$a = \pi p_0 R/2E^* \tag{4.19}$$

and the mutual approach of distant points in the two solids is given by

$$\delta = \pi a p_0/2E^* \tag{4.20}$$

The total load compressing the solids is related to the pressure by

$$P = \int_0^a p(r)2\pi r \, dr = \tfrac{2}{3}p_0\pi a^2 \tag{4.21}$$

Hence the maximum pressure p_0 is 3/2 times the mean pressure p_m. In a practical problem, it is usually the total load which is specified, so that it is convenient to use (4.21) in combination with (4.19) and (4.20) to write

$$a = \left(\frac{3PR}{4E^*} \right)^{1/3} \tag{4.22}$$

$$\delta = \frac{a^2}{R} = \left(\frac{9P^2}{16RE^{*2}} \right)^{1/3} \tag{4.23}$$

$$p_0 = \frac{3P}{2\pi a^2} = \left(\frac{6PE^{*2}}{\pi^3 R^2} \right)^{1/3} \tag{4.24}$$

These expressions have the same form as (4.14), (4.15) and (4.16) which were obtained by dimensional reasoning. However they also provide absolute values for the contact size, compression and maximum pressure.

Before this solution to the problem can be accepted, we must ask whether (4.17) is satisfied uniquely by the assumed pressure distribution and also check whether condition (4.8) is satisfied to ensure that the two surfaces do not touch or interfere *outside* the loaded circle. By substituting equation (3.42a) for the normal displacement $(r > a)$ into equation (4.8) and making use of (4.19), it may be verified that the Hertz distribution of pressure does not lead to contact outside the circle $r = a$.

On the question of uniqueness, we note from §3.4 that a pressure distribution of the form (equation (3.34))

$$p = p_0' \{1 - (r/a)^2\}^{-1/2}$$

produces a uniform normal displacement within the loaded circle. Thus such a pressure could be added to or subtracted from the Hertz pressure while still satisfying the condition for normal displacements given by (4.17). However, this pressure distribution also gives rise to an infinite gradient of the surface immediately outside the loaded circle in the manner of a rigid cylindrical punch. Clearly two elastic bodies having smooth continuous profiles could not develop a pressure distribution of this form without interference outside the circle $r = a$. On the other hand, if such a pressure distribution were subtracted from the Hertz pressure, the normal traction just inside the loaded circle would be *tensile* and of infinite magnitude. In the absence of adhesion between the two surfaces, they cannot sustain tension, so that both positive and negative tractions of the form given above are excluded. No other distribution of normal traction produces displacements which satisfy (4.17) so that we conclude that the Hertz pressure distribution is the unique solution to the problem.

The stresses within the two solids due to this pressure distribution have been found in §3.4, and are shown in Fig. 4.3. At the surface, within the contact area, the stress components are given by equation (3.43); they are all compressive except at the very edge of contact where the radial stress is tensile having a maximum value $(1 - 2\nu)p_0/3$. This is the greatest tensile stress anywhere and it is held responsible for the ring cracks which are observed to form when brittle materials like glass are pressed into contact. At the centre the radial stress is compressive and of value $(1 + 2\nu)p_0/2$. Thus for an incompressible material ($\nu = 0.5$) the stress at the origin is hydrostatic. Outside the contact area the radial and circumferential stresses are of equal magnitude and are tensile and compressive respectively (equation (3.44)).

Fig. 4.3. Stress distributions at the surface and along the axis of symmetry caused by (left) uniform pressure and (right) Hertz pressure acting on a circular area radius a.

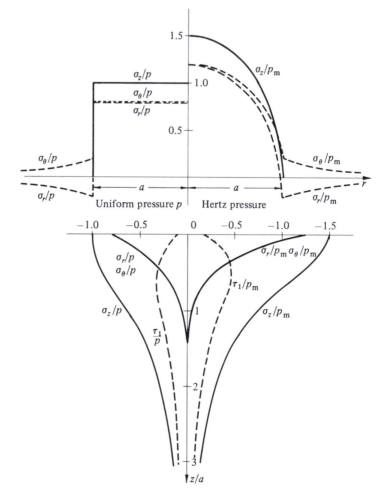

Expressions for the stresses beneath the surface along the z-axis are given in equations (3.45). They are principal stresses and the principal shear stress ($\tau_1 = \frac{1}{2}$ (principal stress difference)) has a value of approximately $0.31p_0$ at a depth of $0.48a$ (for $\nu = 0.3$). This is the maximum shear stress in the field, exceeding the shear stress at the origin $= \frac{1}{2}|\sigma_z - \sigma_r| = 0.10p_0$, and also the shear stress in the surface at the edge of the contact $= \frac{1}{2}|\sigma_r - \sigma_\theta| = 0.13p_0$. Hence plastic yielding would be expected to initiate beneath the surface. This question is considered in detail in the next chapter.

(b) General profiles

In the general case, where the separation is given by equation (4.3), the shape of the contact area is not known with certainty in advance. However we assume tentatively that S is elliptical in shape, having semi-axes a and b. Hertz recognised that the problem in elasticity is analogous to one of electrostatic potential. He noted that a charge, whose intensity over an elliptical region on the surface of a conductor varies as the ordinate of a semi-ellipsoid, gives rise to a variation in potential throughout that surface which is parabolic. By analogy, the pressure distribution given by equation (3.58)

$$p = p_0\{1 - (x/a)^2 - (y/b)^2\}^{1/2}$$

produces displacements within the ellipse given by equation (3.61):

$$\bar{u}_z = \frac{1 - \nu^2}{\pi E}(L - Mx^2 - Ny^2)$$

Thus for both bodies,

$$\bar{u}_{z1} + \bar{u}_{z2} = (L - Mx^2 - Ny^2)/\pi E^* \qquad (4.25)$$

which satisfies the condition (4.7): $\bar{u}_{z1} + \bar{u}_{z2} = \delta - Ax^2 - By^2$ provided that (from equations (3.62))

$$A = M/\pi E^* = (p_0/E^*)(b/e^2 a^2)\{K(e) - E(e)\} \qquad (4.26a)$$

$$B = N/\pi E^* = (p_0/E^*)(b/a^2 e^2)\{(a^2/b^2)E(e) - K(e)\} \qquad (4.26b)$$

$$\delta = L/\pi E^* = (p_0/E^*)bK(e) \qquad (4.26c)$$

where $E(e)$ and $K(e)$ are complete elliptic integrals of argument $e = (1 - b^2/a^2)^{1/2}$, $b < a$. The pressure distribution is semi-ellipsoidal and, from the known volume of an ellipsoid, we conclude that the total load P is given by

$$P = (2/3)p_0 \pi ab \qquad (4.27)$$

from which the average pressure $p_m = (2/3)p_0$.

To find the shape and size of the ellipse of contact, we write

$$\frac{B}{A} = \left(\frac{R'}{R''}\right) = \frac{(a/b)^2 E(e) - K(e)}{K(e) - E(e)} \qquad (4.28)$$

and

$$(AB)^{1/2} = \tfrac{1}{2}(1/R'R'')^{1/2} = 1/2R_e$$

$$= \frac{p_0}{E^*} \frac{b}{a^2 e^2} [\{(a/b)^2 E(e) - K(e)\}\{K(e) - E(e)\}]^{1/2} \qquad (4.29)$$

We now write $c = (ab)^{1/2}$ and substitute for p_0 from (4.27) into (4.29) to obtain

$$c^3 \equiv (ab)^{3/2} = \left(\frac{3PR_e}{4E^*}\right) \frac{4}{\pi e^2} (b/a)^{3/2}$$

$$\times [\{(a/b)^2 E(e) - K(e)\}\{K(e) - E(e)\}]^{1/2}$$

i.e.

$$c = (ab)^{1/2} = \left(\frac{3PR_e}{4E^*}\right)^{1/3} F_1(e) \qquad (4.30)$$

The compression is found from equations (4.26c) and (4.27):

$$\delta = \frac{3P}{2\pi abE^*} bK(e)$$

$$= \left(\frac{9P^2}{16E^{*2}R_e}\right)^{1/3} \frac{2}{\pi} \left(\frac{b}{a}\right)^{1/2} \{F_1(e)\}^{-1}K(e)$$

$$= \left(\frac{9P^2}{16E^{*2}R_e}\right)^{1/3} F_2(e) \qquad (4.31)$$

and the maximum pressure is given by

$$p_0 = \frac{3P}{2\pi ab} = \left(\frac{6PE^{*2}}{\pi^3 R_e^2}\right)^{1/3} \{F_1(e)\}^{-2} \qquad (4.32)$$

The eccentricity of the contact ellipse, which is independent of the load and depends only on the ratio of the relative curvatures (R'/R''), is given by equation (4.28). It is apparent from equation (4.3) that, before deformation, contours of constant separation h are ellipses in which $(b/a) = (A/B)^{1/2} = (R''/R')^{1/2}$. Equation (4.28) has been used to plot the variation of $(b/a)(B/A)^{1/2}$ as a function of $(B/A)^{1/2}$ in Fig. 4.4. If the contact ellipse had the same shape as contours of equal separation when the surfaces just touch, $(b/a)(B/A)^{1/2}$ would always have the value 1.0. It may be seen from the figure that $(b/a)(B/A)^{1/2}$ decreases from unity as the ratio of relative curvatures (R'/R'') increases. Thus the contact ellipse is somewhat more slender than the ellipse of constant separation. The broken line in Fig. 4.4 shows that $(b/a)(B/A)^{1/2} \approx (B/A)^{-1/6}$, i.e.

$$b/a \approx (B/A)^{-2/3} = (R'/R'')^{-2/3} \qquad (4.33)$$

We have introduced an equivalent contact radius $c \ (= (ab)^{1/2})$ and an equivalent relative curvature R_e $(= (R'R'')^{1/2})$ and obtained expressions for c, the maximum contact pressure p_0 and the compression δ in equations (4.30), (4.31) and (4.32). Comparison with the corresponding equations (4.22), (4.23) and (4.24) for solids of revolution shows that the first term is the same in each case; the second term may be regarded as a 'correction factor' to allow for the eccentricity of the ellipse. These correction factors – $F_1(e)$, $\{F_1(e)\}^{-2}$ and $F_2(e)$ – are also plotted against $(R'/R'')^{1/2}$ in Fig. 4.4; they depart rather slowly from unity with increasing ellipticity.

As an example, consider the contact of the cylinders, each of radius R, with their axes inclined at $45°$, illustrated by the interference fringes shown in Fig. 4.1(*b*). As shown in §4.1 the ratio of relative curvatures $(R'/R'')^{1/2} = (B/A)^{1/2} = 2.41$ and the equivalent radius $R_e = (R'R'')^{1/2} = \sqrt{(2)}R$. Under load, the ratio of major to minor axis $a/b = 3.18$ from the curve in Fig. 4.4 or ≈ 3.25 from equation (4.33). Also from Fig. 4.4, $F_1 \approx F_2 = 0.95$ and $F_1^{-2} \approx 1.08$ from which the contact size c $(= (ab)^{1/2})$, the compression δ and the contact pressure p_0 can be found using equations (4.30), (4.31) and (4.32) respectively. Even

Fig. 4.4. Contact of bodies with general profiles. The shape of the ellipse b/a and the functions F_1, F_2 and F_3 $(= F_1^{-2})$ in terms of the ratio (R'/R'') of relative curvatures, for use in equations (4.30), (4.31) and (4.32).

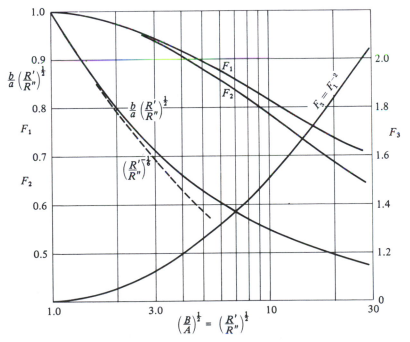

though the ellipse of contact has a $3:1$ ratio of major to minor axes, taking $F_1 = F_2 = F_3 = 1.0$, i.e. using the formulae for circular contact with an equivalent radius R_e, leads to overestimating the contact size c and the compression δ by only 5% and to underestimating the contact pressure p_0 by 8%.

For ease of numerical computation various authors, e.g. Dyson (1965) and Brewe & Hamrock (1977) have produced approximate algebraic expressions in terms of the ratio (A/B) to replace the elliptic integrals in equations (4.30), (4.31) and (4.32). Tabular data have been published by Cooper (1969).

It has been shown that a semi-ellipsoidal pressure distribution acting over an elliptical region having the dimensions defined above satisfies the boundary conditions (4.7) within the ellipse. To confirm the hypothesis that the contact area is in fact elliptical it is necessary that condition (4.8) also be satisfied: that there should be no contact outside the prescribed ellipse. From equation (3.59), the displacements on the surface outside the loaded ellipse are given by

$$\bar{u}_z = \frac{1 - \nu^2}{E} \frac{\pi ab}{2} p_0 \int_{\lambda_1}^{\infty} \left(1 - \frac{x^2}{a^2 + w} - \frac{y^2}{b^2 + w}\right) \frac{dw}{\{(a^2 + w)(b^2 + w)w\}^{1/2}}$$

where λ_1 is the positive root of equation (3.53). We write $\int_{\lambda_1}^{\infty} \equiv [I]_{\lambda_1}^{\infty} = [I]_0^{\infty} - [I]_0^{\lambda_1}$. In the region in question: $z = 0, x^2/a^2 + y^2/b^2 > 1$, from which it appears that $[I]_0^{\lambda_1}$ is negative. But the pressure distribution and contact dimensions have been chosen such that

$$\frac{ab}{2E^*} p_0 [I]_0^{\infty} = \delta - Ax^2 - By^2.$$

Hence in the region outside the contact

$$\bar{u}_{z1} + \bar{u}_{z2} > \delta - Ax^2 - By^2$$

i.e. condition (4.8) is satisfied and the assumption of an elliptical contact area is justified.

Expressions for the stresses within the solids are given by equations (3.64)–(3.69). The general form of the stress field is similar to that in which the contact region is circular. If a and b are taken in the x and y directions respectively with $a > b$, at the centre of the contact surface

$$\sigma_x = -p_0\{2\nu + (1 - 2\nu)b/(a + b)\} \qquad (4.34a)$$

$$\sigma_y = -p_0\{2\nu + (1 - 2\nu)a/(a + b)\} \qquad (4.34b)$$

At the ends of the major and minor axes, which coincide with the edge of the contact region, there is equal tension and compression in the radial and circumferential directions respectively, thus at $x = \pm a, y = 0$,

$$\sigma_x = -\sigma_y = p_0(1 - 2\nu) \frac{b}{ae^2} \left\{\frac{1}{e} \tanh^{-1} e - 1\right\} \qquad (4.35a)$$

Table 4.1

b/a	0	0.2	0.4	0.6	0.8	1.0
z/b	0.785	0.745	0.665	0.590	0.530	0.480
$(\tau_1)_{max}/p_0$	0.300	0.322	0.325	0.323	0.317	0.310

and at $x = 0, y = \pm b$,

$$\sigma_y = -\sigma_x = p_0(1 - 2\nu) \frac{b}{ae^2} \left\{ 1 - \frac{b}{ae} \tan^{-1} \left(\frac{ea}{b} \right) \right\} \qquad (4.35b)$$

The maximum shear stress occurs on the z-axis at a point beneath the surface whose depth depends upon the eccentricity of the ellipse as given in Table 4.1. Numerical values of the stresses along the z-axis have been evaluated by Thomas & Hoersch (1930) for $\nu = 0.25$ and by Lundberg & Sjövall (1958).

The simplest experimental check on the validity of the Hertz theory is to measure the growth in size of the contact ellipse with load which, by (4.30), is a cube-root relationship. Hertz performed this experiment using glass lenses coated with lampblack. A thorough experimental investigation has been carried out by Fessler & Ollerton (1957) in which the principal shear stresses on the plane of symmetry, given by equations (3.64) and (3.69) have been measured using the frozen stress method of photo-elasticity. The ratio of the major axis of the contact ellipse a to the minimum radius of curvature R was varied from 0.05 to 0.3 with araldite models having different combinations of positive and negative curvature. At the smallest values of a/R the measured contact size was somewhat greater than the theory predicts. This discrepancy is commonly observed at light loads and is most likely due to the topographical roughness of the experimental surfaces (see Chapter 13). At high loads there was good agreement with the theoretical predictions of both contact area and internal stress up to the maximum value of a/R used ($= 0.3$). This reassuring conclusion is rather surprising since this value of (a/R) corresponds to strains in the contact region rising to about 10%.

(c) Two-dimensional contact of cylindrical bodies

When two cylindrical bodies with their axes both lying parallel to the y-axis in our coordinate system are pressed in contact by a force P per unit length, the problem becomes a two-dimensional one. They make contact over a long strip of width $2a$ lying parallel to the y-axis. Hertz considered this case as the limit of an elliptical contact when b was allowed to become large compared with a. An alternative approach is to recognise the two-dimensional nature of the problem from the outset and to make use of the results developed in Chapter 2 for line loading of a half-space.

Equation (4.3) for the separation between corresponding points on the unloaded surfaces of the cylinders becomes

$$h = z_1 + z_2 = Ax^2 = \tfrac{1}{2}(1/R_1 + 1/R_2)x^2 = \tfrac{1}{2}(1/R)x^2 \tag{4.36}$$

where the relative curvature $1/R = 1/R_1 + 1/R_2$. For points lying within the contact area after loading, equation (4.7) becomes

$$\bar{u}_{z1} + \bar{u}_{z2} = \delta - Ax^2 = \delta - \tfrac{1}{2}(1/R)x^2 \tag{4.37}$$

whilst for points outside the contact region

$$\bar{u}_{z1} + \bar{u}_{z2} > \delta - \tfrac{1}{2}(1/R)x^2 \tag{4.38}$$

We are going to use Hertz' approximation that the displacements $\bar{u}_{z1}$ and $\bar{u}_{z2}$ can be obtained by regarding each body as an elastic half-space, but a difficulty arises here which is absent in the three-dimensional cases discussed previously. We saw in Chapter 2 that the value of the displacement of a point in an elastic half-space loaded two-dimensionally could not be expressed relative to a datum located at infinity, in view of the fact that the displacements decrease with distance r from the loaded zone as $\ln r$. Thus $\bar{u}_{z1}$ and $\bar{u}_{z2}$ can only be defined relative to an arbitrarily chosen datum. The approach of distant points in the two cylinders, denoted by δ in equation (4.37), can take any value depending upon the choice of datum. In physical terms this means that the approach δ cannot be found by consideration of the local contact stresses alone; it is also necessary to consider the stress distribution within the bulk of each body. This is done for circular cylinders in §5.6.

For the present purpose of finding the local contact stresses the difficulty is avoided by differentiating (4.37) to obtain a relation for the surface *gradients*. Thus

$$\frac{\partial \bar{u}_{z1}}{\partial x} + \frac{\partial \bar{u}_{z2}}{\partial x} = -(1/R)x \tag{4.39}$$

Referring to Chapter 2, we see that the surface gradient due to a pressure $p(x)$ acting on the strip $-a \leqslant x \leqslant a$ is given by equation (2.25b). The pressure on each surface is the same, so that

$$\frac{\partial \bar{u}_{z1}}{\partial x} + \frac{\partial \bar{u}_{z2}}{\partial x} = -\frac{2}{\pi E^*} \int_{-a}^{a} \frac{p(s)}{x - s}\, ds$$

Substituting in equation (4.39)

$$\int_{-a}^{a} \frac{p(s)}{x - s}\, ds = \frac{\pi E^*}{2R}\, x \tag{4.40}$$

This is an integral equation for the unknown pressure $p(x)$ of the type (2.39) in which the right-hand side $g(x)$ is a polynomial in x of first order. The solution of this type of equation is discussed in §2.7. If, in equation (2.45), we put $n = 1$ and write $\pi E(n + 1)B/2(1 - \nu^2) = \pi E^*/2R$, the required distribution of pressure is given by equation (2.48) in which, by (2.47),

$$I_n = I_1 = \pi(x^2/a^2 - \tfrac{1}{2})$$

Thus

$$p(x) = -\frac{\pi E^*}{2R} \frac{x^2 - a^2/2}{\pi(a^2 - x^2)^{1/2}} + \frac{P}{\pi(a^2 - x^2)^{1/2}} \tag{4.41}$$

This expression for the pressure is not uniquely defined until the semi-contact-width a is related to the load P. First we note that the pressure must be positive throughout the contact for which

$$P \geqslant \pi a^2 E^*/4R \tag{4.42}$$

If P exceeds the value given by the right-hand side of (4.42) then the pressure rises to an infinite value at $x = \pm a$. The profile of an elastic half-space which is loaded by a pressure distribution of the form $p_0(1 - x^2/a^2)^{-1/2}$ is discussed in §2.7(a). The surface gradient just outside the loaded region is infinite (see Fig. 2.12). Such a deformed profile is clearly inconsistent with the condition of our present problem, expressed by equation (4.38), that contact should not occur outside the loaded area. We must conclude therefore that

$$P = \pi a^2 E^*/4R$$

i.e.

$$a^2 = \frac{4PR}{\pi E^*} \tag{4.43}$$

whereupon

$$p(x) = \frac{2P}{\pi a^2}(a^2 - x^2)^{1/2} \tag{4.44}$$

which falls to zero at the edge of the contact.

The maximum pressure

$$p_0 = \frac{2P}{\pi a} = \frac{4}{\pi} p_m = \left(\frac{PE^*}{\pi R}\right)^{1/2} \tag{4.45}$$

where p_m is the mean pressure.

The stresses within the two solids can now be found by substituting the pressure distribution (4.44) into equation (2.23). At the contact interface $\sigma_x = \sigma_y = -p(x)$; outside the contact region all the stress components at the surface are zero. Along the z-axis the integration is straightforward giving

$$\sigma_x = -\frac{p_0}{a} \{(a^2 + 2z^2)(a^2 + z^2)^{-1/2} - 2z\} \qquad (4.46a)$$

$$\sigma_z = -\frac{p_0}{a}(a^2 + z^2)^{-1/2} \qquad (4.46b)$$

These are principal stresses so that the principal shear stress is given by

$$\tau_1 = -\frac{p_0}{a} \{z - z^2(a^2 + z^2)^{-1/2}\}$$

from which

$$(\tau_1)_{\max} = 0.30p_0, \quad \text{at } z = 0.78a \qquad (4.47)$$

These stresses are all independent of Poisson's ratio although, for plane strain, the third principal stress $\sigma_y = \nu(\sigma_x + \sigma_z)$. The variations of σ_x, σ_z and τ_1 with depth below the surface given by equations (4.46) are plotted in Fig. 4.5(a), which may be seen to be similar to the variations of stress beneath the surface in a circular contact (Fig. 4.3). Contours of principal shear stress τ_1 are plotted in Fig. 4.5(b), which may be compared with the photo-elastic fringes shown in Fig. 4.6(d). McEwen (1949) expresses the stresses at a general point (x, z) in terms of m and n, defined by

$$m^2 = \tfrac{1}{2}[\{(a^2 - x^2 + z^2)^2 + 4x^2z^2\}^{1/2} + (a^2 - x^2 + z^2)] \qquad (4.48a)$$

Fig. 4.5. Contact of cylinders: (a) subsurface stresses along the axis of symmetry, (b) contours of principal shear stress τ_1.

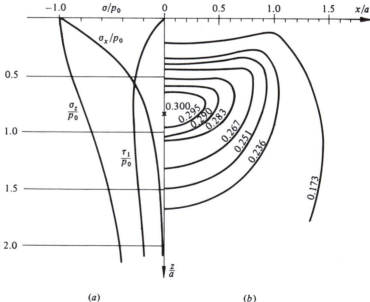

(a)   (b)

and

$$n^2 = \tfrac{1}{2}\left[\{(a^2 - x^2 + z^2)^2 + 4x^2 z^2\}^{1/2} - (a^2 - x^2 + z^2)\right] \qquad (4.48b)$$

where the signs of m and n are the same as the signs of z and x respectively. Whereupon

$$\sigma_x = -\frac{p_0}{a}\left\{ m\left(1 + \frac{z^2 + n^2}{m^2 + n^2}\right) - 2z\right\} \qquad (4.49a)$$

$$\sigma_z = -\frac{p_0}{a}\, m\left(1 - \frac{z^2 + n^2}{m^2 + n^2}\right) \qquad (4.49b)$$

Fig. 4.6. Two-dimensional photo-elastic fringe patterns (contours of principal shear stress): (a) point load (§2.2); (b) uniform pressure (§2.5(a)); (c) rigid flat punch (§2.8); (d) contact of cylinders (§4.2(c)).

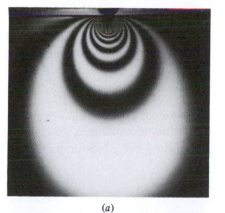

(a)

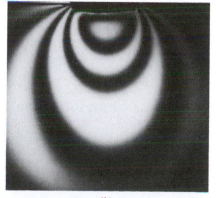

(b)

(c)

(d)

and

$$\tau_{xz} = -\frac{p_0}{a} \, n \left(\frac{m^2 - z^2}{m^2 + n^2} \right) \tag{4.49c}$$

Alternative expressions have been derived by Beeching & Nicholls (1948), Poritsky (1950), Sackfield & Hills (1983a). A short table of values is given in Appendix 4. The variation of stress with x at a constant depth $z = 0.5a$ is shown in Fig. 9.3.

4.3 Elastic foundation model

The difficulties of elastic contact stress theory arise because the displacement at any point in the contact surface depends upon the distribution of pressure throughout the whole contact. To find the pressure at any point in the contact of solids of given profile, therefore, requires the solution of an integral equation for the pressure. This difficulty is avoided if the solids can be modelled by a simple Winkler elastic foundation or 'mattress' rather than an elastic half-space. The model is illustrated in Fig. 4.7. The elastic foundation, of depth h, rests on a rigid base and is compressed by a rigid indenter. The profile of the indenter, $z(x,y)$, is taken as the sum of the profiles of the two bodies being modelled, i.e.

$$z(x,y) = z_1(x,y) + z_2(x,y) \tag{4.50}$$

There is no interaction between the springs of the model, i.e. shear between adjacent elements of the foundation is ignored. If the penetration at the origin is denoted by δ, then the normal elastic displacements of the foundation are given by

$$\bar{u}_z(x,y) = \begin{cases} \delta - z(x,y), & \delta > z \\ 0, & \delta \leqslant z \end{cases} \tag{4.51}$$

The contact pressure at any point depends only on the displacement at that

Fig. 4.7

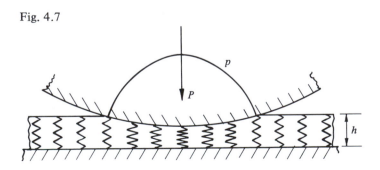

point, thus

$$p(x, y) = (K/h)\bar{u}_z(x, y) \tag{4.52}$$

where K is the elastic modulus of the foundation.

For two bodies of curved profile having relative radii of curvature R' and R'', $z(x, y)$ is given by equation (4.3) so that we can write

$$\bar{u}_z = \delta - (x^2/2R') - (y^2/2R'') \tag{4.53}$$

inside the contact area. Since $\bar{u}_z = 0$ outside the contact, the boundary is an ellipse of semi-axes $a = (2\delta R')^{1/2}$ and $b = (2\delta R'')^{1/2}$. The contact pressure, by (4.52), is

$$p(x, y) = (K\delta/h)\{1 - (x^2/a^2) - (y^2/b^2)\} \tag{4.54}$$

which is paraboloidal rather than ellipsoidal as given by the Hertz theory. By integration the total load is

$$P = K\pi ab\delta/2h \tag{4.55}$$

In the axi-symmetric case $a = b = (2\delta R)^{1/2}$ and

$$P = \frac{\pi}{4}\left(\frac{Ka}{h}\right)\frac{a^3}{R} \tag{4.56}$$

For the two-dimensional contact of long cylinders, by equation (4.37)

$$\bar{u}_z = \delta - x^2/2R = (a^2 - x^2)/2R \tag{4.57}$$

so that

$$p(x) = (K/2Rh)(a^2 - x^2) \tag{4.58}$$

and the load

$$P = \tfrac{2}{3}\left(\frac{Ka}{h}\right)\frac{a^2}{R} \tag{4.59}$$

Equations (4.56) and (4.59) express the relationship between the load and the contact width. Comparing them with the corresponding Hertz equations (4.22) and (4.43), agreement can be obtained, if in the axi-symmetric case we choose $K/h = 1.70E^*/a$ and in the two-dimensional case we choose $K/h = 1.18E^*/a$. For K to be a material constant it is necessary to maintain geometrical similarity by increasing the depth of the foundation h in proportion to the contact width a. Alternatively, thinking of h as fixed requires K to be reduced in inverse proportion to a. It is a consequence of the approximate nature of the model that the values of K required to match the Hertz equations are different for the two configurations. However, if we take $K/h = 1.35E^*/a$, the value of a under a given load will not be in error by more than 7% for either line or point contact.

The compliance of a point contact is not so well modelled. Due to the neglect of surface displacements outside the contact, the foundation model gives

$\delta = a^2/2R$ which is half of that given by Hertz (equation (4.23)). If it were more important in a particular application to model the compliance accurately we should take $K/h = 0.60E^*/a$; the contact size a would then be too large by a factor of $\sqrt{2}$.

The purpose of the foundation model, of course, is to provide simple approximate solutions in complex situations where half-space theory would be very cumbersome. For example, the normal frictionless contact of bodies whose arbitrary profiles cannot be represented adequately by their radii of curvature at the point of first contact can be handled easily in this way (see §5.3). The contact area is determined directly in shape and size by the profiles $z(x, y)$ and the penetration δ. The pressure distribution is given by (4.52) and the corresponding load by straight summation of pressure. For a contact area of arbitrary shape a representative value of a must be chosen to determine (K/h).

The foundation model is easily adapted for tangential loading (see §8.7); also to viscoelastic solids (see §9.4).

5

Non-Hertzian normal contact of elastic bodies

The assumptions and restrictions made in the Hertz theory of elastic contact were outlined in the previous chapter: parabolic profiles, frictionless surfaces, elastic half-space theory. In this chapter some problems of normal elastic contact are considered in which we relax one or more of these restrictions. Before looking at particular situations, however, it is instructive to examine the stress conditions which may arise close to the edge of contact.

5.1 Stress conditions at the edge of contact

We have seen in Chapter 4 that, when two non-conforming elastic bodies having continuous profiles are pressed into contact, the pressure distribution between them is not determined uniquely by the profiles of the bodies *within* the contact area. Two further conditions have to be satisfied: (i) that the interface should not carry any tension and (ii) that the surfaces should not interfere *outside* the contact area. These conditions eliminate terms in the pressure distribution of the form $C(1 - x^2/a^2)^{-1/2}$ which give rise to an infinite tension or compression at the edge of the contact area ($x = \pm a$) (see equation (4.41)). The resulting pressure distribution was found to be semi-ellipsoidal, i.e. of the form $p_0(1 - x^2/a^2)^{1/2}$, which falls to zero at $x = \pm a$.

If we now recall the stresses produced in line loading by a *uniform* distribution of pressure (§2.5), they are everywhere finite, but the gradient of the surface is infinite at the edge of the contact (eq. (2.30b) and Fig. 2.8). This infinite gradient of the surface is associated with the jump in pressure from zero outside to p inside the contact. It is clear that two surfaces, initially smooth and continuous, could not deform in this way without interference outside the loaded area. These observations lead to an important principle: *the pressure distribution between two elastic bodies, whose profiles are continuous through the boundary of the contact area, falls continuously to zero at the boundary.*†

The examples cited in support of this statement were for frictionless surfaces, but it may be shown that the principle is still true if there is slipping friction at the edge of the contact such that $q = \mu p$ and also if friction is sufficient to prevent slip entirely.

If one or both of the bodies has a discontinuous profile at the edge of the contact the situation is quite different and, in general, a high stress concentration would be expected at the edge. The case of a rigid flat punch with square corners was examined in §2.8. For a frictionless punch the pressure distribution was of the form $p_0(1 - x^2/a^2)^{-1/2}$ which, at a small distance ρ from one of the corners, may be written $p_0(2\rho/a)^{-1/2}$. It is recognised, of course, that this infinite stress cannot exist in reality. Firstly, the linear theory of elasticity which gave rise to that result is only valid for small strains and, secondly, real materials will yield plastically at a finite stress. Nevertheless, as developments in linear elastic fracture mechanics have shown, the strength of stress singularities calculated by linear elastic theory is capable of providing useful information about the intensity of stress concentrations and the probable extent of plastic flow.

The conditions at the edge of the contact of a rigid punch with an elastic half-space are influenced by friction on the face of the punch and also by the value of Poisson's ratio for the half-space. If friction prevents slip entirely the pressure and traction on the face of the punch are given by equation (2.69). Close to a corner ($\rho = a - x \ll a$) the pressure distribution may be written

$$p(\rho) = \frac{2(1 - \nu)}{\pi(3 - 4\nu)} (2a\rho)^{-1/2} \cos \{\eta \ln (2a/\rho)\} \tag{5.1}$$

where $\eta = (1/2\pi) \ln (3 - 4\nu)$. This remarkable singularity exhibits an oscillation in pressure at the corner of the punch ($\rho \to 0$).

For an incompressible half-space, however, $\nu = 0.5$, $\eta = 0$ and the pressure distribution reverts to that without friction. It was shown in §2.8 that, in the absence of an adhesive, the surfaces must slip. The form of the pressure distribution close to the edge of the punch may then be obtained from equation (2.75) to give

$$p(\rho) = \frac{P \cos (\pi\gamma)}{\pi} (2a\rho)^{-1/2} (2a/\rho)^{\pm\gamma} \tag{5.2}$$

where $\tan (\pi\gamma) = -\mu(1 - 2\nu)/2(1 - \nu)$. When either the coefficient of friction is zero or Poisson's ratio is 0.5, $\gamma = 0$ and the pressure distribution reverts to the frictionless form.

---

† This principle was appreciated by Boussinesq (1885).

In the above discussion we have examined the stress concentration produced by a rigid punch with a square corner. The question now arises: how would that stress concentration be influenced if the punch were also elastic and the angle at the corner were other than 90°? This question has been investigated by Dundurs & Lee (1972) for frictionless contact, by Gdoutos & Theocaris (1975) and Comninou (1976) for frictional contacts, and by Bogy (1971) for no slip. They analyse the state of stress in a two-dimensional elastic wedge of angle ϕ, which has one face pressed into contact with an elastic half-space as shown in Fig. 5.1. The half-space itself may be thought of as a second wedge of angle π. The variation of the stress components with ρ close to the apex of the wedge may take one of the following forms:

(*a*) ρ^{s-1}, if s is real and $0 < s < 1$;
(*b*) $\rho^{\xi-1} \cos(\eta \ln \rho)$ or $\rho^{\xi-1} \sin(\eta \ln \rho)$, if $s = \xi + i\eta$ is complex and $0 < \xi < 1$;
(*c*) $\ln \rho$;
(*d*) constant (including zero);

depending upon the elastic constants of the wedge and half-space, the angle of the wedge ϕ and the frictional conditions at the interface. Dundurs shows that the influence of the elastic constants is governed by only two independent variables:

$$\alpha \equiv \frac{\{(1-\nu_1)/G_1\} - \{(1-\nu_2)/G_2\}}{\{(1-\nu_1)/G_1\} + \{(1-\nu_2)/G_2\}} \tag{5.3a}$$

Fig. 5.1

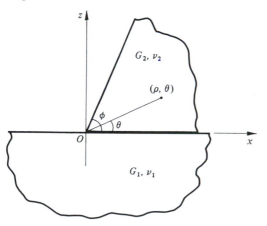

and

$$\beta \equiv \tfrac{1}{2} \left[\frac{\{(1-2\nu_1)/G_1\} - \{(1-2\nu_2)/G_2\}}{\{(1-\nu_1)/G_1\} + \{(1-\nu_2)/G_2\}} \right] \tag{5.3b}$$

α is a measure of the difference in 'plane strain modulus' $\{(1-\nu^2)/E\}$; it varies from -1.0 when the half-space is rigid to $+1.0$ when the wedge is rigid. β has extreme values $\pm\tfrac{1}{2}$ when one body is rigid and the other has zero Poisson's ratio. If both bodies are incompressible $\beta = 0$. Some typical values of α and β are given in Table 5.1 from which it may be seen that $|\beta|$ seldom exceeds 0.25.

When there is slip between the wedge and the half-space the stresses at the apex may be of the form (a), (c) or (d) as defined above, but complex values of s which lead to oscillating stresses do not arise. For the pressure to be finite at O (case (d)) it is found that

$$\alpha \leqslant \frac{(\pi+\phi)\cos\phi + (\mu\pi - 1)\sin\phi}{(\pi-\phi)\cos\phi + (\mu\pi + 1)\sin\phi} \tag{5.4}$$

Unless the pressure actually falls to zero at O, however, the tangential stress σ in the half-space has a logarithmic singularity at O (case (c)). This was found to be the case with a uniform pressure as shown in equation (2.31a) and Fig. 2.9. If α exceeds the right-hand side of equation (5.4) there is a power law singularity in pressure at O with the value of s depending upon the values of α, β, ϕ and μ. In this expression the wedge is taken to be slipping relative to the half-space in the positive direction of x, i.e. from left to right in Fig. 5.1. For slip in the opposite direction *negative* values of μ should be used in equation (5.4). The stress concentration at O is reduced by positive sliding and increased by negative sliding. As might be expected, the stress concentration increases with increasing wedge angle ϕ. When the wedge is effectively bonded to the half-space then the stress is always infinite at O. For larger values of $|\alpha|$ and $|\beta|$, s may be complex (case (b)) and the pressure and shear traction both

Table 5.1

Body 1	Body 2	G_1 (GPa)	ν_1	G_2 (GPa)	ν_2	α	β
Rubber	metal	$\ll G_2$	0.50	$\gg G_1$	–	1.00	0
Perspex	steel	0.97	0.38	80	0.30	0.97	0.19
Glass	steel	22	0.25	80	0.30	0.57	0.21
Duralumin	steel	28	0.32	80	0.30	0.61	0.12
Cast iron	steel	45	0.25	80	0.30	0.31	0.12
Tungsten carbide	steel	300	0.22	80	0.30	−0.54	−0.24

oscillate close to O. For smaller values of $|\alpha|$ and $|\beta|$, s is real and a power singularity arises (case (a)). To find the value of s in any particular case the reader is referred to the papers cited.

By way of example we shall consider a rectangular elastic block, or an elastic cylinder with flat ends, compressed between two half-spaces. The distributions of pressure and frictional traction on the faces of the block or cylinder have been found by Khadem & O'Connor (1969a, b) for (a) no slip (bonded) and (b) no friction at the interface. Close to the edges of contact the stress conditions for both the rectangular block and the cylinder can be determined by reference to the two-dimensional wedge discussed above, with a wedge angle $\phi = 90°$. If the block is rigid and the half-spaces are elastic with $\nu = 0.3$ ($\alpha = 1.0; \beta = 0.286$) the situation is that of a rigid punch discussed in §2.8. In the absence of friction the pressure near the corner varies as $\rho^{-0.5}$ as given by equation (2.64). Points on the interface move tangentially inwards towards the centre of the punch, corresponding to *negative* slip as defined above so that, if the motion is resisted by finite friction ($\mu = -0.5$ say), the stress near the corner varies as $\rho^{-0.45}$, given by equation (5.2). With an infinite friction coefficient, so that all slip is prevented, s is complex and the pressure oscillates as given by equations (2.69) and (5.1).

If now we consider the reverse situation, in which the block is elastic ($\nu = 0.3$) and the half-spaces are rigid ($\alpha = -1, \beta = -0.286$), in the absence of friction the pressure on a face of the block will be uniform. Through Poisson's ratio it will expand laterally so that the slip at a corner is again negative. When this slip is resisted by friction ($\mu = -0.5$) the stress at a corner varies as $\rho^{-0.43}$; if slip is completely prevented it varies as $\rho^{-0.29}$.

Finally we consider block and half-spaces of identical materials so that $\alpha = \beta = 0$. For all frictional conditions the pressure is infinite at the edges: without friction it varies as $\rho^{-0.23}$; with slipping friction, taking $\mu = -0.5$, it varies as $\rho^{-0.44}$. With no slip s is again real and the pressure varies as $\rho^{-0.45}$.

5.2 Blunt wedges and cones

The Hertz theory of contact is restricted to surfaces whose profiles are smooth and continuous; in consequence the stresses are finite everywhere. A rigid punch having sharp square corners, on the other hand, was shown in §2.8 to introduce an infinite pressure at the edges of the contact. In this section we examine the influence of a sharp discontinuity in the *slope* of the profile within the contact area by reference to the contact of a wedge or cone with plane surface. In order for the deformations to be sufficiently small to lie within the scope of the linear theory of elasticity, the semi-angle α of the wedge or cone must be close to $90°$.

If we take a two-dimensional wedge indenting a flat surface such that the width of the contact strip is small compared with the size of the two solids then we can use the elastic solutions for a half-space for both wedge and plane surface. The deformation is shown in Fig. 5.2(a). The normal displacements are related to the wedge profile by

$$\bar{u}_{z1} + \bar{u}_{z2} = \delta - \cot \alpha \, |x|, \quad -a < x < a \tag{5.5}$$

Fig. 5.2. Indentation by a blunt wedge: (a) pressure distribution; (b) photo-elastic fringes.

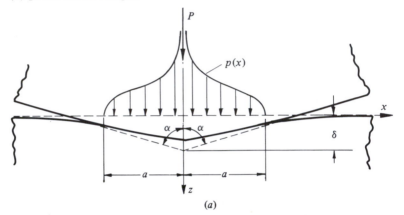

(a)

(b)

Thus

$$\bar{u}'_{z1} + \bar{u}'_{z2} = -(\text{sign } x) \cot \alpha \qquad (5.6)$$

where $(\text{sign } x) = +1$ or -1 when x is $+$ve or $-$ve respectively. Neglecting friction, the normal pressure acting at the interface is found by substituting in equation (2.25b) to give

$$\frac{2}{\pi E^*} \int_{-a}^{a} \frac{p(s)}{x-s} \, ds = (\text{sign } x) \cot \alpha \qquad (5.7)$$

This is an integral equation for $p(x)$ of the type (2.39) with the general solution (2.41).

Now

$$\int_{-a}^{a} \frac{(a^2-s^2)^{1/2}(\text{sign } s) \, ds}{x-s} = \int_{0}^{a} \frac{(a^2-s^2)^{1/2} \, ds}{x-s} - \int_{-a}^{0} \frac{(a^2-s^2)^{1/2} \, ds}{x-s}$$

$$= \int_{0}^{a} (a^2-s^2)^{1/2} \left\{ \frac{1}{x-s} - \frac{1}{x+s} \right\} ds = 2 \int_{0}^{a} \frac{(a^2-s^2)^{1/2} s \, ds}{x^2-s^2}$$

$$= 2a - (a^2-s^2)^{1/2} \ln \left\{ \frac{a+(a^2-x^2)^{1/2}}{a-(a^2-x^2)^{1/2}} \right\} \qquad (5.8)$$

Substituting in (2.41) and using (2.42), we get

$$p(x) = \frac{E^* \cot \alpha}{2\pi} \left[\frac{2a}{(a^2-x^2)^{1/2}} - \ln \left\{ \frac{a+(a^2-x^2)^{1/2}}{a-(a^2-x^2)^{1/2}} \right\} \right]$$

$$+ \frac{P}{\pi(a^2-x^2)^{1/2}} \qquad (5.9)$$

If the smooth faces of the wedge extend beyond the edges of the contact the pressure must fall to zero at the edges to avoid tension or interference outside the contact, whence

$$P = aE^* \cot \alpha \qquad (5.10)$$

The pressure distribution is then

$$p(x) = \frac{E^* \cot \alpha}{2\pi} \ln \left\{ \frac{a+(a^2-x^2)^{1/2}}{a-(a^2-x^2)^{1/2}} \right\} = \frac{E^* \cot \alpha}{\pi} \cosh^{-1}(a/x) \quad (5.11)$$

This pressure distribution is plotted in Fig. 5.2(a); it rises to an infinite value at the apex to the wedge. It would appear that the discontinuity in slope of the profiles within the contact region leads to a logarithmic singularity in pressure. Although the pressure is infinite at the apex of the wedge the principal shear stress in the x-z plane is not so. The stress components within the solids due to the pressure distribution (5.11) may be calculated using equation (2.23). Along the z-axis, where σ_x and σ_z are principal stresses it turns out that

$$\tau_1 = \tfrac{1}{2}|\sigma_x - \sigma_z| = (E^*a/\pi) \cot \alpha \, (a^2 + z^2)^{-1/2} \tag{5.12}$$

which has a finite maximum value beneath the apex:

$$(\tau_1)_{max} = (E^*/\pi) \cot \alpha$$

The indentation of a flat surface by a blunt cone gives similar results. Love (1939) used the classical approach outlined in §3.1 to find the appropriate potential function. He showed that the pressure on the face of the cone is given by

$$p(r) = \tfrac{1}{2}E^* \cot \alpha \cosh^{-1}(a/r) \tag{5.13}$$

and the total force

$$P = \tfrac{1}{2}\pi a^2 E^* \cot \alpha \tag{5.14}$$

Sneddon (1948) used the integral transform technique to obtain the same result and to evaluate all the components of stress within the solid. The tractions at the surface of a cone which adheres to the contacting solid have been found by Spence (1968).

The distribution of pressure on the cone is similar to that on the wedge: it rises to a theoretically infinite value at the apex. At that point the tangential stress in the surface is given by

$$\sigma_r = \sigma_\theta = -\tfrac{1}{2}(1 + 2\nu)p_0$$

which is also infinite. The special case of an incompressible material ($\nu = 0.5$) is noteworthy. By considering the surface stresses we find an infinite hydro-static pressure exists at the apex. By considering the variations in stress along the z-axis, we find that the principal shear stress is given by

$$\tau_1 = \tfrac{1}{2}|\sigma_r - \sigma_z| = \tfrac{1}{2}E^*a^2 \cot \alpha \, (a^2 + z^2)^{-1}$$

which has a maximum yet finite value $\tfrac{1}{2}E^* \cot \alpha$ at the apex. The state of stress at the apex therefore comprises a finite shear stress in a radial plane superposed on an infinite hydrostatic pressure. Contours of τ_1 beneath the wedge are shown by the photo-elastic fringes in Fig. 5.2(b).

5.3 Conforming surfaces

In the previous chapter smooth non-conforming surfaces in contact were defined: the initial separation between such surfaces in the contact region can be represented to an adequate approximation by a second-order polynomial. Non-conforming surfaces can therefore be characterised completely by their radii of curvature at the point of first contact. However when the undeformed profiles conform rather closely to each other a different description of their initial separation may be necessary. Conforming surfaces in contact frequently depart in another way from the conditions in which the Hertz theory applies. Under the application of load the size of the contact area grows rapidly and may become comparable with the significant dimensions of the contacting bodies themselves. A pin in a hole with a small clearance is an obvious case in point. When the arc of contact occupies an appreciable fraction of the circumference of the hole neither the pin nor the hole can be regarded as an elastic half-space so that the Hertz treatment is invalid.

We will consider first the contact of bodies whose profiles in the contact region cannot be adequately represented by a second-order polynomial but, nevertheless, can still be regarded as half-spaces for the purpose of calculating elastic deformations and stresses.

The profiles are represented by a polynomial to the required degree of approximation. Thus for a two-dimensional contact (assuming symmetry about the point of first contact) we can express the initial separation by

$$h = z_1 + z_2 = A_1 x^2 + A_2 x^4 + \cdots + A_n x^{2n} + \cdots \tag{5.15}$$

and for a contact with axial symmetry

$$h = A_1 r^2 + A_2 r^4 + \cdots + A_n r^{2n} \tag{5.16}$$

Substituting (5.15) or (5.16) in equation (4.6) gives the condition which has to be satisfied by the normal displacements of each surface within the contact region. Steuermann (1939) has found the distributions of pressure $p_n(x)$ and $p_n(r)$ for profiles having the form $A_n x^{2n}$ and $A_n r^{2n}$ respectively. In the two-dimensional case we can use equations (2.47) and (2.48). Taking the index n as defined in (5.15) and allowing for the elasticity of both surfaces, equation (2.48) for the contact pressure becomes

$$p_n(x) = \frac{P_n}{\pi(a^2 - x^2)^{1/2}} - \frac{E^* n A_n a^{2n}}{(a^2 - x^2)^{1/2}} \left\{ \left(\frac{x}{a}\right)^{2n} - \frac{1}{2}\left(\frac{x}{a}\right)^{2n-2} \right.$$
$$\left. - \frac{1 \cdot 3 \ldots (2n-3)}{2 \cdot 4 \ldots 2n} \right\} \tag{5.17}$$

If the profiles are smooth and continuous there cannot be an infinite pressure at $x = \pm a$, from which it follows that

$$P_n = n\pi E^* A_n a^{2n} \frac{1 \cdot 3 \ldots (2n-1)}{2 \cdot 4 \ldots 2n} \tag{5.18}$$

and

$$p_n(x) = n E^* A_n a^{2n-2} \left\{ \left(\frac{x}{a}\right)^{2n-2} + \frac{1}{2}\left(\frac{x}{a}\right)^{2n-4} + \cdots \right.$$
$$\left. + \frac{1 \cdot 3 \ldots (2n-3)}{2 \cdot 4 \ldots (2n-2)} \right\} (a^2 - x^2)^{1/2} \tag{5.19}$$

The second-order profiles assumed in the Hertz theory correspond to $n = 1$, in which case equations (5.17) and (5.19) reduce to (4.43) and (4.44) respectively. For higher values of n the pressure has its maximum values away from the centre of the contact. As $n \to \infty$ the configuration approaches that of two flat surfaces which make contact over a strip $|x| \leqslant a$, and are separated by thin gaps or cracks on either side. The pressure distribution in this case approaches that for a flat rigid punch in which the pressure is infinite at the edges. When the profiles can be represented by a single term in equation (5.15) the size of the contact strip is related to the load by equation (5.18). With a more general profile, the pressure distribution and total load are given by the superposition of expressions such as (5.18) and (5.19).

In the axi-symmetric case Steuermann finds the equivalent expressions:

$$P_n = \frac{4A_n E^* n a^{2n+1}}{(2n+1)} \frac{2 \cdot 4 \ldots 2n}{1 \cdot 3 \ldots (2n-1)} \tag{5.20}$$

$$p_n(r) = \frac{n A_n E^* a^{2n-2}}{\pi} \left\{ \frac{2 \cdot 4 \ldots 2n}{1 \cdot 4 \ldots (2n-1)} \right\}^2 \left\{ \left(\frac{r}{a}\right)^{2n-2} + \frac{1}{2} \left(\frac{r}{a}\right)^{2n-4} \right.$$
$$\left. + \cdots + \frac{1 \cdot 3 \ldots (2n-3)}{2 \cdot 4 \ldots (2n-2)} \right\} (a^2 - r^2)^{1/2} \tag{5.21}$$

The compression can also be found in this case:

$$\delta = \frac{2 \cdot 4 \ldots 2n}{1 \cdot 3 \ldots (2n-1)} A_n a^{2n} \tag{5.22}$$

The two examples given above were capable of analytical solution because the shape of the area of contact was known in advance. For more general profiles this shape is not known; only for second-order surfaces will it be an ellipse. Some idea of the shape may be obtained from contours of equal separation before deformation, but it cannot be assumed that this shape will be maintained after deformation. Numerical methods of finding the contact stresses between conforming bodies of arbitrary profile are described in §9.

We turn now to two problems of contact between conforming solids which cannot be represented by half-spaces: (a) the two-dimensional contact of a pin in a hole in an infinite plate and (b) the contact of a sphere with a conforming spherical cavity. The geometry for a pin and hole is shown in Fig. 5.3. The difference in radii ΔR ($= R_2 - R_1$) is small compared with either R_1 or R_2. The external load P is applied to the pin effectively at its centre C and causes C to displace by δ; the reaction is taken by a uniform stress in the plate at a large distance from the hole. The deformation in the contact region is shown in Fig. 5.3(b). Points on the two surfaces S_1 and S_2 which come into contact on the interface at S experience both radial and tangential elastic displacements $\bar{u}_r$ and

$\bar{u}_\theta$. Since ΔR and δ are both small compared with R_1 and R_2,

$$(R_2 + \bar{u}_{r2}) - (R_1 + \bar{u}_{r1}) = (\Delta R + \delta) \cos \phi \qquad (5.23)$$

i.e.

$$\bar{u}_{r2} - \bar{u}_{r1} = \delta \cos \phi - \Delta R (1 - \cos \phi) \qquad (5.24)$$

When the contact arc subtends an angle $\pm \alpha$ which is not small, expression (5.24) differs significantly from the Hertz approximation given by equation (4.37). It is now required to find the distribution of normal pressure (neglecting friction) which, when acting over the arc $\pm \alpha$, produces displacements in the surface of the pin and hole which satisfy (5.24) in the interval $-\alpha < \phi < \alpha$. This problem has been studied in detail by Persson (1964) who has used stress functions appropriate to a circular disc and to a circular hole in an infinite plate to obtain the complete stress field for both pin and hole.

Pressure distributions for different values of α are shown in Fig. 5.4(a). The variation of the contact arc α with load P is shown in Fig. 5.4(b), both for a pin with clearance (ΔR positive) and a pin with interference (ΔR negative). With small loads or large clearance the relationship approaches that of Hertz (eq. (4.42)). The relationship given by Steuermann's theory is also shown. Here the gap between the pin and the hole has been represented by a power series and the contact pressure and load calculated from equations (5.17)–(5.19). The result, though better than that of Hertz by the inclusion of higher terms in the description of the profile, is still in error through Steuermann's assumption that both solids can be regarded as elastic half-spaces.

The analogous problem of a frictionless sphere in a conforming cavity has been analysed by Goodman & Keer (1965) using methods appropriate to spherical bodies. They find that the contact is up to 25% stiffer in compression than would be predicted by the Hertz theory.

Fig. 5.3. Pin in a conforming hole.

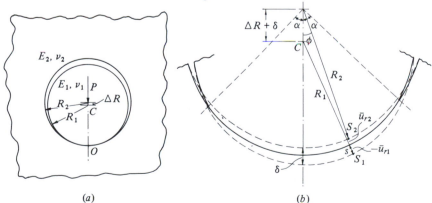

(a) (b)

Fig. 5.4. Pin in a conforming hole: $E_1 = E_2 = E$, $\nu_1 = \nu_2 = 0.30$. (a) contact pressure; (b) contact arc. (From Persson, 1964.)

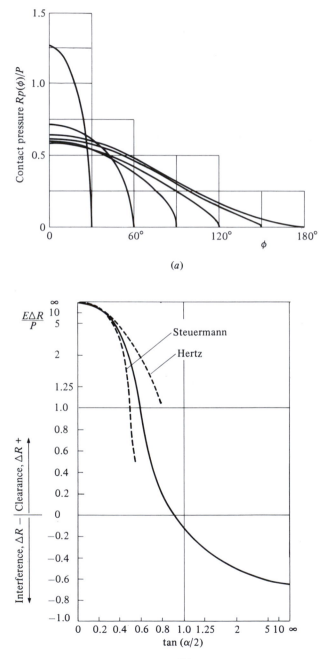

(a)

(b)

5.4 Influence of interfacial friction

Friction at the interface of two non-conforming bodies brought into normal contact plays a part only if the elastic constants of the two materials are different. The mutual contact pressure produces tangential displacements at the interface as well as normal compression (see equation (3.41*b*) for spheres in contact). If the materials of the two solids are dissimilar, the tangential displacements will, in general, be different so that slip will take place. Such slip will be opposed by friction and may, to some extent, be prevented. We might expect therefore a central region where the surfaces stick together and regions of slip towards the edge of the contact. If the coefficient of limiting friction were sufficiently high slip might be prevented entirely.

In the first studies of this problem (Mossakovski, 1954, 1963; Goodman, 1962) the build-up of tangential traction at the interface was developed incrementally for a growth in contact size from a to $a + da$. However, as Spence (1968) pointed out, under appropriate conditions the stress field is self-similar at all stages of loading so that the solution may be obtained directly without recourse to an incremental technique.

In setting up the boundary conditions of the problem we start by assuming that where there is slip the tangential traction q is related to the normal pressure p by

$$|q| = \mu p \tag{5.25}$$

where μ is a constant coefficient of friction. The direction of q opposes the direction of slip. In a two-dimensional contact q acts in a direction parallel to the x-axis, inwards on one surface and outwards on the other. In an axisymmetrical contact the slip, and hence q, must be radial and axi-symmetrical. For non-conforming surfaces having quadratic profiles we deduced in §4.1 (eq. (4.11)) that the magnitude of the stress and strain at any point increases in proportion to the contact size a. In consequence of (5.25) the stresses and strains due to the shear traction also increase in proportion to a, and the boundary between the slipped and adhered regions will be located at a constant fraction of a. In this way self-similarity of the stress field is maintained at all stages of loading.

As the load is increased and the contact size grows, mating points on the two surfaces, which initially lie outside the adhesion zone, undergo different tangential displacements. After they are enveloped by the adhesion zone, they cease to experience any further relative displacement. Such points will then maintain the relative tangential displacement $(\bar{u}_{x1} - \bar{u}_{x2})$ and relative strain $(\partial \bar{u}_{x1}/\partial x - \partial \bar{u}_{x2}/\partial x)$ which they had acquired up to that instant. Now the magnitude of the strain grows in direct proportion to a so that, for two contacting points lying in the adhesion zone at a distance x from the centre, we can write

$$\frac{\partial \bar{u}_{x1}}{\partial x} - \frac{\partial \bar{u}_{x2}}{\partial x} = C|x| \tag{5.26}$$

where C is a constant to be determined.

Consider first the contact of two parallel cylinders. At a particular stage in loading the contact width is $2a$ and we will assume that friction prevents slip over a central region of width $2c$. A symmetrical normal pressure $p(x)$ and an anti-symmetrical tangential traction $q(x)$ act at the interface. The normal displacement gradients within the whole contact region are given by equation (4.39). Substituting into equation (2.25b) and remembering that the tractions on each surface are equal and opposite, we find

$$\int_{-a}^{a} \frac{p(s)}{x-s} \, ds - \pi \beta q(x) = \pi E^* x / 2R, \quad -a \leqslant x \leqslant a \tag{5.27}$$

where $1/R = 1/R_1 + 1/R_2$ and the constant β is a measure of the difference in elastic constants of the two materials defined by equation (5.3). In the adhesion region, substituting (2.25a) into the condition of no-slip given by (5.26) gives

$$\pi \beta p(x) + \int_{-a}^{a} \frac{q(s) \, ds}{x-s} = -\tfrac{1}{2}\pi E^* C|x|, \quad |x| \leqslant c \tag{5.28}$$

Also for no-slip

$$|q| \leqslant \mu p \tag{5.29}$$

In the slip regions

$$q = \pm \mu p, \quad c < |x| < a \tag{5.30}$$

where the sign q is determined by the direction of slip. If equations (5.27) to (5.30) are divided by the contact size a they are transformed into equations for (p/a), (q/a) which are independent of the actual value of a, thereby confirming the previous argument that similarity of the stress field is maintained during loading.

As a first step to solving equations (5.27) and (5.28) for the tractions $p(x)$ and $q(x)$ we simplify the problem by assuming that there is no slip throughout the contact area. Equation (5.28) then applies over the interval $(|x| \leqslant a)$ and, together with (5.27), provides dual integral equations for $p(x)$ and $q(x)$ of the type discussed in §2.7, having boundary conditions of class III. A further simplification results from neglecting the influence of the shear traction upon the normal pressure, i.e. by neglecting the second term on the left-hand side of (5.27). The equations are now uncoupled. The pressure distribution is given by the Hertz theory (equation (4.44)) and equation (5.28) for the tangential traction may be written

$$\int_{-a}^{a} \frac{q(s) \, ds}{s-x} = -\pi \beta p_0 (1 - x^2/a^2)^{1/2} - \tfrac{1}{2}\pi E^* C|x| \tag{5.31}$$

This is an integral equation of the type (2.39) having a general solution given by (2.41).

It is convenient to imagine the traction $q(x)$ as being made up of two components: $q'(x)$ and $q''(x)$, which satisfy (5.31) with each of the terms on the right-hand side taken in turn. Thus $q'(x)$ is the tangential traction necessary to cancel the difference in tangential displacements arising from the normal pressure, and $q''(x)$ is the traction necessary to produce the additional displacements, proportion to $|x|$, which are necessary to satisfy the no-slip condition. Substituting into (2.41) and integrating we find

$$q'(x) = 2\beta \frac{p_0}{\pi} \left[\frac{x}{(a^2 - x^2)^{1/2}} + \frac{1}{2a} (a^2 - x^2)^{1/2} \ln \left| \frac{a+x}{a-x} \right| \right] \qquad (5.32)$$

and

$$q''(x) = \frac{CE^*}{2\pi} \left[-\frac{2x}{(a^2 - x^2)^{1/2}} + x \ln \left\{ \frac{a + (a^2 - x^2)^{1/2}}{a - (a^2 - x^2)^{1/2}} \right\} \right] \qquad (5.33)$$

The constant C is determined by the fact that the traction should fall to zero at the edges of the contact. Therefore the term $(a^2 - x^2)^{-1/2}$ must vanish when $q'(x)$ and $q''(x)$ are added, whereupon

$$C = 2\beta p_0 / E^* a$$

and

$$q(x) = \frac{\beta p_0}{\pi a} \left[(a^2 - x^2)^{1/2} \ln \left| \frac{a+x}{a-x} \right| + x \ln \left\{ \frac{a + (a^2 - x^2)^{1/2}}{a - (a^2 - x^2)^{1/2}} \right\} \right] \qquad (5.34)$$

If the ratio of $q(x)$ to $p(x)$ is examined we find that it rises to infinity at the edges of the contact, which shows that some slip is inevitable. The realistic circumstances in which slip takes place on each side of a central no-slip zone of width $2c$ have been studied by Spence (1975).

On similarity grounds Spence has shown that, for the same elastic constants and coefficient of friction, the extent of the slip region is the same for any indenter having the profile $z = Ax^n$ and is equal to that for a flat-ended punch. The value of c is therefore given by equation (2.73) in which $(1 - 2v)/(1 - v)$ is replaced by 2β to take into account the elasticity of both bodies. This relationship is plotted in Fig. 5.5 (curve A). The traction has been evaluated by Spence and is shown in Fig. 5.6, for $\mu/\beta = 0.99$, which gives $c = 0.7a$.

The contact of dissimilar spheres without slip has been analysed by Goodman (1962) on the basis of neglecting the influence of tangential traction on normal pressure. If the traction is again separated into the two components: to cancel the tangential displacements due to normal pressure requires

Fig. 5.5. Normal contact of dissimilar solids: the extent of the slip region: curve A – line contact, eq. (2.73); curve B – axi-symmetric point contact, eq. (5.38).

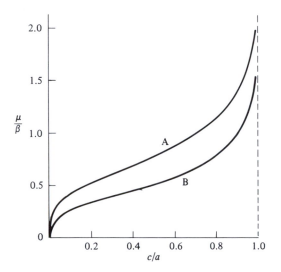

Fig. 5.6. Tangential tractions at contact of dissimilar solids. (*a*) Line contact: A – no slip, eq. (5.34); B – partial slip, $\mu/\beta = 0.99$. (*b*) Point contact: C – no slip, eq. (5.37); D – partial slip, $\mu/\beta = 0.66$. Broken lines – $\mu p/\beta p_0$.

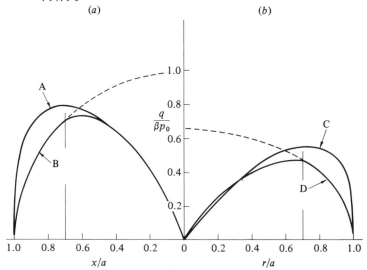

$$q'(r) = \frac{\beta p_0}{\pi} \frac{1}{r} \left[a^2 (a^2 - r^2)^{-1/2} - 2(a^2 - r^2)^{1/2} \right.$$

$$\left. + \int_r^a \frac{t^2}{(t^2 - r^2)^{1/2}} \ln \left| \frac{t+r}{t-r} \right| dt \right] \tag{5.35}$$

and to satisfy the no-slip condition (5.26) requires

$$q''(r) = \frac{3CE^*}{8} \left[-r(a^2 - r^2)^{-1/2} + \frac{r}{a} \ln \left\{ \frac{a + (a^2 - r^2)^{1/2}}{r} \right\} \right] \tag{5.36}$$

The unknown constant C is again determined by the condition that the resultant traction should be zero at $r = a$, whereupon

$$q(r) = \frac{\beta p_0}{\pi} \left[-\frac{1}{r} (a^2 - r^2)^{1/2} + \frac{r}{a} \ln \left\{ \frac{a + (a^2 - r^2)^{1/2}}{r} \right\} \right.$$

$$\left. + \frac{2}{ra} \int_r^a \frac{t^2}{(t^2 - r^2)^{1/2}} \ln \left| \frac{t+r}{t-r} \right| dt \right] \tag{5.37}$$

This traction is plotted in Fig. 5.6. The ratio $q(r)/p(r)$ is again infinite at $r = a$, so that some slip must occur. The extent of the slip region in monotonic loading is the same as for a rigid flat punch, and is given by

$$\frac{a}{2c} \ln \left(\frac{a+c}{a-c} \right) = \frac{\beta}{\mu} \mathbf{K}'(c/a) \tag{5.38}$$

where $\mathbf{K}'(c/a)$ is the complete elliptical integral of argument $(1 - c^2/a^2)^{1/2}$. This relationship is also shown in Fig. 5.5 (curve B), and the traction when $c = 0.7a$ is plotted in Fig. 5.6. As the coefficient of friction is decreased the no-slip circle shrinks towards the central point and the traction approaches $\mu p(r)$.

Complete solutions to the problem which include the influence of tangential traction on the pressure have been obtained by Mossakovski (1963) and Spence (1968, 1975). They show that, depending upon the value of β, friction can increase the total load required to produce a contact of given size by at most 5% compared with Hertz.

Distributions of frictional traction for line contact and axi-symmetrical point contact are plotted in Fig. 5.6. They act outwards on the more deformable surface and inwards on the more rigid one. Without slip the magnitude of the traction is proportional to the parameter β which characterises the difference in elastic properties of the materials. Clearly β vanishes not only when the materials are identical but also when they are both incompressible ($\nu = 0.5$).

In these cases frictional tractions are absent and the Hertz solution applies. Some values of β for typical pairs of materials are given in Table 5.1. The maximum possible value of β is 0.5 and practical values rarely exceed 0.2. Thus the frictional traction is much smaller than the normal pressure and its influence on the internal stresses in not great. It does have an important effect, however, on the tangential stress $\bar{\sigma}_x$ or $\bar{\sigma}_r$ at the surface just outside the contact area. In the case of line contact, without slip, the traction of equation (5.34) gives rise to stresses at the edge of contact given by

$$\bar{\sigma}_x(-a) = \bar{\sigma}_x(a) = -2\beta p_0 \tag{5.39}$$

compressive on the compliant surface and tensile on the rigid one. Slip will have the effect of reducing this stress. If slip were complete, so that $|q| = \mu p$ everywhere

$$\bar{\sigma}_x(-a) = \bar{\sigma}_x(a) = -(4/\pi)\mu p_0 \tag{5.40}$$

In reality slip will only be partial and there will be a no-slip zone in the centre of width $2c$. However equation (5.40) is a good approximation provided $c/a < 0.7$, i.e. when $\mu/\beta < 1.0$.

For the axi-symmetric case the radial stress in the surface may be calculated using equation (3.103c). With no slip the traction of equation (5.37) gives rise to a radial stress in the surface at $r = a$ given by

$$\bar{\sigma}_r(a) = -1.515(1 - 0.16\nu)\beta p_0 \tag{5.41}$$

If, on the other hand, complete slip takes place

$$\bar{\sigma}_r(a) = -1.185(1 - 0.23\nu)\mu p_0 \tag{5.42}$$

Again equation (5.42) is a good approximation for partial slip provided $c/a < 0.7$, when $\mu/\beta < 0.66$. In the axi-symmetric case the normal pressure itself gives rise to a radial tension outside the contact circle which has a maximum value $\frac{1}{3}(1 - 2\nu)p_0$ at $r = a$ and decreases as r^{-2} (eq. (3.44)). On the more compliant surface ($\beta + \mathrm{ve}$) the compressive stress produced by the frictional traction attenuates the tension coming from the normal pressure and has the effect of pushing the location of the maximum tension to a radius somewhat greater than a. On the more rigid surface ($\beta - \mathrm{ve}$) the radial stress due to friction is tensile and adds to that due to pressure to give a maximum tension at $r = a$. Johnson, O'Connor & Woodward (1973) have investigated this effect and have shown that it influences the resistance of brittle materials to Hertzian fracture when the material of the indenter is different from that of the specimen.

The frictional traction which develops when the load is reduced is also of interest in view of the observation that ring cracks frequently occur during unloading. Some appreciation of this behaviour may be obtained from Turner's analysis (1979) of unloading a flat-ended punch.

5.5   Adhesion between elastic bodies

So far in this book we have taken it for granted that, while a mutual pressure is exerted at the interface between two bodies in contact, no tensile traction can be sustained. This assumption accords with common experience that, in the absence of a specific adhesive, the contact area between non-conforming elastic solids falls to zero when the load is removed and that no tensile force is required to separate them. On the other hand, a physicist describing the interaction between the ideal surfaces of two solids would tell a different story (see, for example, Tabor, 1975). As a result of the competing forces of attraction and repulsion between individual atoms or molecules in both bodies, two ideally flat solid surfaces will have an equilibrium separation z_0; at a separation less than z_0 they will repel each other and at a separation greater than z_0 they will attract. The variation of force per unit area as a function of separation z is usually represented by a law of the form

$$p(z) = -Az^{-n} + Bz^{-m}, \quad m > n \tag{5.43}$$

as shown in Fig. 5.7 where the pressure (repulsion) is taken to be positive. In these circumstances it is clear that a tensile force – the 'force of adhesion' – has to be exerted to separate the surfaces. The magnitude of the maximum tensile traction is large, but the effective range of action is very small. In view of the difficulty in measuring surface forces directly, it is usual to measure the work 2γ required to separate the surfaces from $z = z_0$ to $z = \infty$ and to ascribe a *surface energy* γ to each newly created free surface. If the solids are dissimilar, the work to separate the surfaces becomes $\gamma_1 + \gamma_2 - 2\gamma_{12}$ where γ_1 and γ_2 are the intrinsic surface energies of the two solids and γ_{12} is the energy of the interface.

The reason why this expected adhesion between solids is not usually observed, even when great care is taken to remove contaminant films, lies in the inevitable

Fig. 5.7. Force–separation curve and surface energy for ideal surfaces, eq. (5.43).

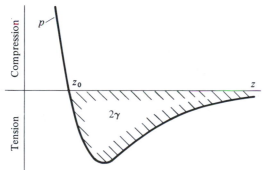

roughness of real surfaces, whose asperity heights are large compared with the range of action of the adhesive forces. The real area of contact, which occurs at the crests of the high spots, is much smaller than the apparent area (see Chapter 13). Adhesive junctions formed between the lower asperities on loading are elbowed apart on unloading by the compression between the higher asperities. In this way adhesion developed at the points of real contact is progressively broken down. Exceptions to this state of affairs arise with (a) cleaved mica which can be prepared with an atomically smooth surface and (b) low modulus solids such as gelatine or rubber which can mould themselves to accommodate a modest amount of surface roughness. In these circumstances the real area of contact is identical with the apparent area.

To study the effect of adhesive forces in the absence of surface roughness we shall consider two non-conforming axi-symmetric solids which make contact over a circular area, of radius a. Frictional tractions of the sort discussed in the previous section will be ignored (see Johnson *et al.*, 1971; and Johnson, 1976). The normal elastic displacement in the contact circle produced by the normal traction must satisfy equation (4.17), i.e.

$$\bar{u}_{z1} + \bar{u}_{z2} = \delta - r^2/2R$$

We found in §4.2(a) that this condition was satisfied by a pressure distribution of the form

$$p(r) = p_0(1 - r^2/a^2)^{1/2} + p_0'(1 - r^2/a^2)^{-1/2} \tag{5.44}$$

where $p_0 = 2aE^*/\pi R$. A positive value of p_0' was rejected since an infinite pressure at $r = a$ implied interference between the two surfaces outside the contact area: a negative value of p_0' was rejected on the grounds that tension could not be sustained. In the presence of adhesive (attractive) forces, however, we cannot exclude the possibility of a negative p_0'. By considering the work done in compression by the pressure of (5.44), the elastic strain energy stored in the two bodies is easily shown to be

$$U_E = \frac{\pi^2 a^3}{E^*}\left(\tfrac{2}{15}p_0^2 + \tfrac{2}{3}p_0 p_0' + p_0'^2\right) \tag{5.45}$$

The total compression is found from equations (3.36) and (4.20) to be

$$\delta = (\pi a/2E^*)(p_0 + 2p_0') \tag{5.46}$$

We now consider the variation in strain energy U_E with contact radius a, keeping the overall relative displacement of the two bodies δ constant. With p_0 as specified above we find

$$\left[\frac{\partial U_E}{\partial a}\right]_\delta = \frac{\pi^2 a^2}{E^*}p_0'^2 \tag{5.47}$$

Since δ is kept constant no external work is done, so that for equilibrium we would expect $\partial U_E/\partial a$ to vanish, giving $p_0' = 0$, as indeed it is in the Hertz theory.

In the present problem, adhesive forces introduce a surface energy U_S which is decreased when the surfaces come into intimate contact and increased when they separate. Therefore we can write

$$U_S = -2\gamma\pi a^2$$

where γ is the surface energy per unit area of each surface. The total free energy of the system is now

$$U_T = U_E + U_S$$

For equilibrium $[\partial U_T/\partial a]_\delta$ vanishes giving

$$\frac{\pi^2 a^2}{E^*} p_0'^2 = -\frac{\partial U_S}{\partial a} = 4\pi\gamma a$$

i.e.

$$p_0' = -(4\gamma E^*/\pi a)^{1/2} \tag{5.48}$$

where the minus sign is chosen since compressive stresses at $r = a$ have been excluded. The net contact force is given by

$$P = \int_0^a 2\pi r p(r)\,\mathrm{d}r = (\tfrac{2}{3}p_0 + 2p_0')\pi a^2$$

Substituting for p_0 and p_0' and rearranging give a relationship between a and P:

$$\left(P - \frac{4E^*a^3}{3R}\right)^2 = 16\pi\gamma E^*a^3 \tag{5.49}$$

This relationship is plotted in Fig. 5.8 where it is compared with experimental measurements using gelatine spheres in contact with perspex. When the bodies are loaded by a compressive (positive) force the adhesive forces pull the surfaces into contact over an area which exceeds that given by the Hertz theory. Reducing the load to zero leaves the surfaces adhering together with a radius given by point C in Fig. 5.8. The application of a tensile (negative) load causes the contact radius to shrink further. At point B, when

$$P = -P_c = -3\pi\gamma R \tag{5.50}$$

and

$$a = a_c = (9\gamma R^2/4E^*)^{1/3} \tag{5.51}$$

the situation becomes unstable and the surfaces separate. Thus P_c given by equation (5.50) is the 'force of adhesion'. If, instead of controlling the load, we control the relative displacement δ between the solids, the adhesive contact is stable down to point A in Fig. 5.8 ($P = -5P_c/9, a = a_c/3^{2/3}$). Beyond this point the adhesive junction breaks.

The traction distribution given by (5.44) and the shape of the deformed surface outside the contact (from equations (3.38) and (3.42a)) are shown in Fig. 5.9 for an elastic sphere in contact with a rigid flat surface. There is an infinite tensile traction and the deformed profile meets the flat surface in a sharp corner at $r = a$. In reality the stress will not be infinite nor the corner perfectly sharp, but there will be some rounding of the corner until the surface traction is consistent with the force-separation law illustrated in Fig. 5.7. Provided that the elastic displacements are large compared with the effective range of action of the surface forces, the analysis outlined above will give a good measure of the influence of adhesion on the deformation of elastic bodies in contact. The idealisation is the same as that of a Griffith crack in linear elastic fracture mechanics. Indeed the gap just outside the contact of two separating bodies may be thought of as an opening crack. Maugis *et al.* (1976, 1978) and also Greenwood & Johnson (1981) have made use of the 'stress intensity factor'

Fig. 5.8. Variation of contact radius with load, eq. (5.49), compared with measurements on gelatine spheres in contact with perspex. Radius R: circle – 24.5 mm, cross – 79 mm, square – 255 mm.

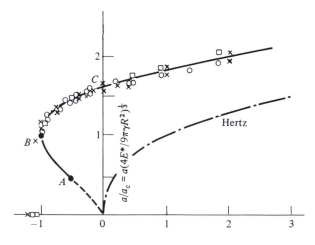

Fig. 5.9. Surface traction and deformation of an elastic sphere in contact with a rigid plane, solid line – with adhesion, eq. (5.44), broken line – without adhesion (Hertz).

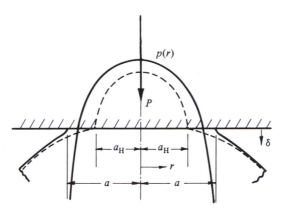

concept of fracture mechanics to analyse the adhesive contact of elastic and viscoelastic solids.

5.6 Contact of cylindrical bodies

The elastic compression of two-dimensional bodies in contact cannot be calculated solely from the contact stresses given by the Hertz theory. Some account must be taken of the shape and size of the bodies themselves and the way in which they are supported. In most practical circumstances such calculations are difficult to perform, which has resulted in a variety of approximate formulae for calculating the elastic compression of bodies in line contact such as gear teeth and roller bearings in line contact (Roark, 1965; Harris, 1966). However the compression of a long circular cylinder which is in non-conformal contact with two other surfaces along two generators located at opposite ends of a diameter can be analysed satisfactorily.

Such a cylinder is shown in cross-section in Fig. 5.10. The compressive load per unit axial length P gives rise to a Hertzian distribution of pressure at O_1

$$p = \frac{2P}{\pi a_1}(1 - x^2/a_1^2)^{1/2} \tag{5.52}$$

where the semi-contact-width is given by

$$a_1^2 = 4PR/\pi E_1^* \tag{5.53}$$

where E_1^* is the composite modulus of the roller and the contacting body.

The stress distribution in a cylinder due to diametrically opposed concentrated loads is given by Timoshenko & Goodier (1951) p. 107. It comprises the superposition of the stress fields due to two concentrated forces P acting on the plane boundaries of two half-spaces tangential to the cylinder at O_1 and O_2 (see eq. (2.14)), together with a uniform bi-axial tension:

$$\sigma_x = \sigma_z = P/\pi R \tag{5.54}$$

which frees the circular boundary of the cylinder from stress.

Since $a \ll R$ we can consider the cylinder in Fig. 5.10 as being subjected to a combination of diametrically opposed forces distributed according to (5.52). We now require the radial component of strain ϵ_z at a point A lying between O_1 and C on the axis of symmetry. The state of stress at A is made up of three contributions: (i) the stress due to the Hertzian distribution of pressure on the contact at O_1, given by equation (5.52); (ii) the stress due to the contact pressure at O_2, which, in view of the large distance of A from O_2, can be taken to be that due to a concentrated force P, given by equation (2.16); and (iii) the bi-axial tension given by (5.54). Adding these three contributions we obtain for

Fig. 5.10

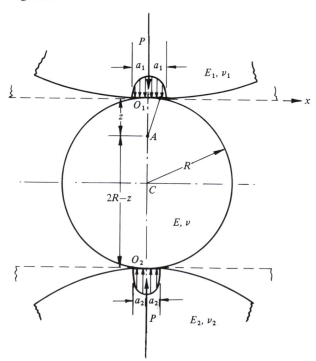

the stresses at A:

$$\sigma_x = \frac{P}{\pi}\left\{\frac{1}{R} - \frac{2(a_1^2 + 2z^2)}{a_1^2(a_1^2 + z^2)^{1/2}} + \frac{4z}{a_1^2}\right\} \tag{5.55a}$$

$$\sigma_z = \frac{P}{\pi}\left\{\frac{1}{R} - \frac{2}{2R - z} - \frac{2}{(a_1^2 + z^2)^{1/2}}\right\} \tag{5.55b}$$

In plane strain

$$\epsilon_z = \{(1 - \nu^2)/E\}\{\sigma_z - \sigma_x\nu/(1 - \nu)\}$$

The compression of the upper half of the cylinder O_1C is then found by integrating ϵ_z from $z = 0$ to $z = R$, where $a \ll R$, to give

$$\delta_1 = P\frac{(1 - \nu^2)}{\pi E}\{2\ln(4R/a_1) - 1\} \tag{5.56}$$

A similar expression is obtained for the compression of the lower half of the cylinder so that the total compression of the diameter through the mid-points of the contact areas O_1O_2 is

$$\delta = 2P\frac{(1 - \nu^2)}{\pi E}\{\ln(4R/a_1) + \ln(4R/a_2) - 1\} \tag{5.57}†$$

For comparison we can calculate the compression of a half-space relative to a point at a depth d below the centre of a Hertzian contact pressure distribution, with the result:

$$\delta = P\frac{(1 - \nu^2)}{\pi E}\{2\ln(2d/a) - \nu/(1 - \nu)\} \tag{5.58}$$

Taking $d = R$ the true compression of the half-cylinder (5.56) exceeds the compression based upon a half-space (5.58) by less than 10% within the practical range of loads.

When one of the contacting bodies roughly takes the form of a rectangular block of thickness t, then the compression of the block through its thickness may be obtained with reasonable approximation by putting $d = t$ in equation (5.58), provided that the thickness of the block is large compared with the contact width ($t \gg a$).

Another important feature of the contact of cylindrical bodies falls outside the scope of the Hertz theory. Real cylinders are of finite length and, although the contact stresses over the majority of the length of the cylinder are predicted accurately by the Hertz theory, significant deviations occur close to the ends.

---

† This expression differs from a much quoted result due to Föeppl (*Drang und Zwang*, vol. I, p. 319, 1924) on account of Föeppl's use of a *parabolic* contact pressure distribution.

In most circumstances there is a concentration of contact stress at the ends which makes the effect of practical importance. In the design of roller bearings, for example, the axial profile of the rollers is modified with a view to eliminating the stress concentration at the ends. The different possible end conditions which may arise when a uniform cylinder is in contact with another surface are shown diagrammatically in Fig. 5.11. In case (a) both surfaces come to an end at the same cross-sectional plane. On cross-sections away from the ends an axial compressive stress $\sigma_y = \nu(\sigma_x + \sigma_z)$ exists to maintain the condition of plane strain. At the free ends this compressive stress is relaxed, permitting the solids to expand slightly in the axial direction and thereby *reducing* the contact pressure at the end.

An estimate of the reduction in pressure at the end may be obtained by assuming that the end of the cylinder is in a state of *plane stress*. Equation (5.56) for the radial compression of the cylinder may be written

$$\delta = (a^2/2R)\{2\ln(4R/a) - 1\} \tag{5.59}$$

which applies for both plane stress and plane strain. If the cylinder does not tilt, the compression δ is uniform along its length, so that the contact width a must also be approximately uniform right up to the end. Now in plane strain $a = 2p_0R(1-\nu^2)/E$ whereas in plane stress $a = 2p_0'R/E$. Hence the pressure at the end $p_0' \approx (1-\nu^2)p_0$.

In case (b) the roller has a square end but the mating surface extends beyond the end. In this case there is a sharp stress concentration at the end of the roller. The nature of the singularity can be assessed from the considerations discussed in §1. For example, with no friction and equal elastic moduli the contact pressure at a small distance y from the end ($y \ll a$) will vary as $y^{-0.23}$.

Case (c) is typical of a cylindrical bearing roller. The track surface extends beyond the end of the roller and the roller itself has a profile of radius r connecting the cylindrical body smoothly with the flat end. Provided r is appreciably larger than the contact width $2a$, then the relief of axial stress σ_y which occurs in case (a) is not possible in either body and both can be regarded as half-spaces for the purpose of estimating the contact stresses. The reason for the stress concentration in this case may be appreciated when it is remembered that the compression of the two surfaces in the centre of the loaded region will not be very different from that with an infinitely long roller. At the ends, however, this same deformation has to be achieved by a load which extends in one direction only; beyond the end of the contact strip the surfaces are unloaded and their compression relaxes as indicated in Fig. 5.11(c). Just inside the end of the contact area the necessary compression is brought about by an increased pressure, which results in an increase in the width of the contact strip. The 'dog bone' shape of contact area has been observed experimentally.

Fig. 5.11. Roller end effects: (*a*) two coincident sharp ends, (*b*) one sharp end, (*c*) rounded end.

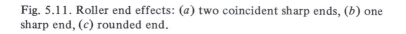

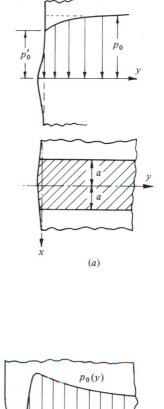

(*a*)

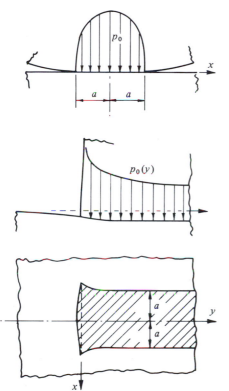

(*b*)

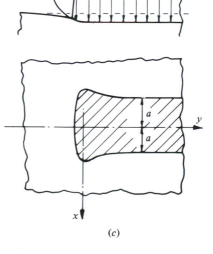

(*c*)

To reduce the stress concentration at the ends, the axial profile of the roller should be slightly barrelled. In theory, optimum conditions would be achieved if the contact pressure were uniform along the length. Lundberg (1939) has investigated this situation. A Hertzian distribution of pressure

$$p(x, y) = (P/\pi al) \{1 - (x/a)^2\}^{1/2} \tag{5.60}$$

is assumed to act on a rectangular contact area of width $2a$ and length $2l$, where $l \gg a$ and a is given by the Hertz theory (eq. (4.43)). The compression at the centre of the rectangle is shown to be

$$\delta(0, 0) = \frac{P}{\pi l E^*} \{1.886 + \ln (l/a)\} \tag{5.61}$$

Along the length of the roller

$$\Delta(y) = \delta(0, 0) - \delta(0, y) \approx \frac{P}{2\pi l E^*} \ln \{1 - (y/l)^2\} \tag{5.62}$$

This expression becomes inaccurate close to the ends of the roller. At the ends themselves $(y = \pm l)$

$$\Delta(l) = \frac{P}{2\pi l E^*} \{1.193 + \ln (l/a)\} \tag{5.63}$$

Equations (5.62) and (5.63) for $\Delta(y)$ express the small correction to the axial profile of the roller required to obtain the uniform axial distribution of pressure of equation (5.60). Internal stresses due to this pressure distribution have been calculated by Kunert (1961). This profile correction, however, is difficult to manufacture and is correct only at the design load. Therefore a more general relationship between axial profile and pressure distribution is of practical interest. Over most of the length of the roller the pressure distribution in the *transverse* direction may be taken to be Hertzian, but the contact width now varies along the length. Nayak & Johnson (1979) have shown that the pressure $p(0, y)$ at any point along most of the length is related to the semi-contact-width $a(y)$ at that point by the Hertz equation (4.43). At the ends, the stress distribution is three-dimensional and must be treated as such for accurate results. Some calculations along these lines have been carried out by Ahmadi, Keer & Mura (1983).

5.7 Anisotropic and inhomogeneous materials
The elastic deformation in the contact region is obtained in the Hertz theory by assuming each solid deforms as an elastic, isotropic, homogeneous half-space. If the material of either solid is anisotropic or inhomogeneous, or if their thicknesses are not large compared with the size of the contact area their compliance under the contact pressure will differ from that assumed in the

classical theory. Practical examples of contact between anisotropic solids are found with single crystals and extruded polymer filaments; between inhomogeneous materials with foundations built on stratified rock or soil.

(a) Anisotropy

Detailed discussion of the contact of anisotropic solids is beyond the scope of this book, but an important result has been demonstrated by Willis (1966) which should be mentioned. Willis considers the contact of two nonconforming bodies of general shape under the conditions for which the Hertz theory applies except that the two solids have general anisotropy. He shows that *the contact area is still elliptical in shape and that the pressure distribution is semi-ellipsoidal* (eq. (3.58)). However, the direction of the axes of the ellipse of contact are not determined solely by the geometry of the surface profiles, but depend also upon the elastic constants. In the special case of transversely anisotropic solids (five independent elastic constants), which are in contact such that their axes of symmetry are both parallel to the common normal at the point of contact, analytical solutions for the contact stresses and deformations can be obtained with hardly more difficulty than for isotropic solids (see Turner, 1980).

Two-dimensional anisotropic contact problems are discussed by Galin (1953) and the indentation of an anisotropic half-space by a rigid punch is solved in Green & Zerna (1954). Equation (5.57) for the compression of a cylinder has been extended by Pinnock *et al.* (1966) to a transversely anisotropic polymer filament and used to determine the values of the appropriate elastic constants by measuring the diametral compression of the filament. A full discussion of anisotropy may be found in the book by Gladwell (1980, Chap. 12).

(b) Inhomogeneity

Inhomogeneous materials are of interest in soil mechanics in the calculation of the settlement of foundations. The elastic modulus of soil usually increases with depth below the surface and a particularly simple analysis is possible for an incompressible elastic half-space ($\nu = 0.5$) whose elastic moduli increase in direct proportion to the depth, i.e.

$$G = \tfrac{1}{3}E = mz \tag{5.64}$$

where m is a material constant. Calladine & Greenwood (1978) show in a simple way that the stress fields produced by a concentrated line load or a concentrated point load are the same as those found in a homogeneous half-space, given by equations (2.14) and (3.19) respectively. The displacements in the inhomogeneous material are different, however, being purely radial, given by $u_r = P/2\pi mr$ for the line load and $u_\rho = P/4\pi m\rho^2$ for the point load. It follows that a half-space of such a material behaves like a simple Winkler elastic foundation in which the normal displacement $\bar{u}_z$ at any point on the surface is directly proportional to the pressure applied at that point, with a stiffness $2m$. Thus a long rigid foundation, of width $2a$ and weight W per unit length, resting on a half-space of this material would depress the surface by $W/4am$. The stress distribution beneath the foundation would be that found in a homogeneous half-space due to a uniform pressure $p = W/2a$ given by equations (2.27).

5.8   Layered solids, plates and shells
(a) The elastic layer

The contact of solids which have surface layers whose elastic properties differ from the substrate frequently occurs in practice; for example, the rubber covered rollers which are widely used in processing machinery. The basic situation is illustrated in Fig. 5.12(a) in which body (2) is in contact with the surface layer (1) on substrate (3). If the thickness b of the layer is large compared with the contact size $2a$, then the substrate has little influence and the contact stresses between (1) and (2) are given by the Hertz theory. In this section we are concerned with the situation in which b is comparable with or less than $2a$. The behaviour then depends on the nature of the attachment of the layer to the substrate. There are various possibilities: (a) the layer may maintain contact with the substrate at all points, but be free to slip without frictional restraint; (b) at the other extreme the layer may be bonded to the substrate; (c) slip may occur when the shear traction at interface exceeds limiting friction; and (d) the layer, initially in complete contact with the substrate, may partially lift from the substrate under load. The non-conforming contact between the layer and body (2) may also be influenced by frictional traction. Even if the elastic constants are the same (i.e. $E_1 = E_2, \nu_1 = \nu_2$) the limited thickness of the layer results in a relative tangential displacement at the interface which will be resisted by friction. Most analyses at the present time, however, assume the contact to be frictionless and are restricted to either the plane-strain conditions of line contact, or the axi-symmetric contact of solids of revolution in which the contact area is circular. We shall discuss the plane-strain situation first.

If the contact width is small compared with the radii of curvature of the bodies, the curvature of the layer can be ignored in analysing its deformation and the solids (2) and (3) can be taken to be elastic half-spaces.

In the case where the layer is everywhere in contact with a rigid frictionless substrate, the boundary conditions at the layer–substrate interface are $\tau_{xz} = 0$ and $u_z = 0$. The stresses in the layer are then the same as in one half of a layer of thickness $2b$ to which identical pressure distributions are applied to the opposing faces (Fig. 5.12(b)). The stresses in the layer are best expressed in terms of Fourier Integral Transforms, for which the reader is referred to the books by Sneddon (1951) and Gladwell (1980, Chap. 11). In this case, with an *even* distribution of pressure applied symmetrically to each surface $z = \pm b$, Sneddon shows that the normal displacement of each surface is given by

$$\bar{u}_z = \frac{4(1 - \nu_1{}^2)}{\pi E_1} \int_0^\infty \left(\frac{2 \sinh^2 \alpha b}{2\alpha b + \sinh 2\alpha b} \right) \bar{p}(\alpha) \frac{\cos \alpha x}{\alpha} \, d\alpha \qquad (5.65)$$

Fig. 5.12

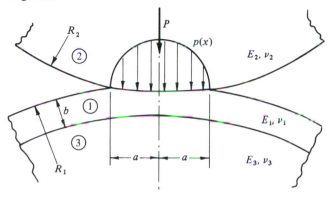

(a)

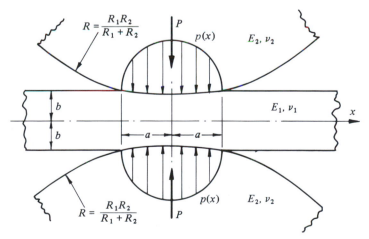

(b)

where $\bar{p}(\alpha)$ is the Fourier Cosine Transform of the pressure $p(x)$, i.e.

$$\bar{p}(\alpha) = \int_0^\infty p(x) \cos \alpha x \, dx \tag{5.66}$$

Comparable expressions to (5.65) in terms of $\bar{p}(\alpha)$ are given by Sneddon for the tangential displacement $\bar{u}_x$ and for the components of stress σ_x, σ_z and τ_{zx} throughout the layer. If the layer is bonded to the substrate the expression for $\bar{u}_z$ corresponding to (5.65) is given by Bentall & Johnson (1968).

For a uniform pressure p distributed over an interval $-c < x < c$, equation (5.66) gives:

$$\bar{p}(\alpha) = (p/\alpha) \sin (\alpha c) \tag{5.67}$$

For a triangular distribution of pressure of peak value p_0,

$$\bar{p}(\alpha) = \frac{2p_0}{c\alpha^2} \sin^2 \left(\frac{c\alpha}{2}\right) \tag{5.68}$$

In the limit, as $c \to 0$, the transform of a concentrated force P is $P/2$. Frictional tractions $q(x)$ on the faces of the layer can be handled in the same way (see Bentall & Johnson, 1968).

The awkward form of the integrand in equation (5.65) and associated expressions has led to serious difficulties in the analysis of contact stresses in strips and layers. Two approaches have been followed. In one the integrand is approximated by an asymptotic form which is appropriate for either thin strips ($b \ll a$), or thick strips ($b \gg a$) (Meijers, 1968; Alblas & Kuipers, 1970). In the other approach the pressure distribution $p(x)$ is built up of discrete elements each of a width $2c$. These may be elements of uniform pressure whose transform is given by equation (5.67) (Conway *et al.*, 1966, 1969) or may be overlapping triangular elements as given by equation (5.68) (Bentall & Johnson, 1968). The use of overlapping triangular elements gives rise to a piecewise linear distribution of pressure; the application of this technique to the contact of solid bodies is described in the next section.

The indentation by a rigid frictionless cylinder of an elastic layer which is supported on a rigid plane surface has been studied by various workers (*a*) for a layer which is bonded to the rigid base and (*b*) for a layer which can slip on the base without friction. The difference between these two cases is significant when the material of the layer is incompressible ($\nu = 0.5$). Solutions for relatively thick layers ($b > a$) have been given by Pao *et al.* (1971) and by Meijers (1968), and for thin layers ($b < a$) by Alblas & Kuipers (1970) and Meijers (1968).

In the limit when $b \ll a$, the state of affairs can be analysed in an elementary way. A thin layer indented by a frictionless rigid cylinder is shown in Fig. 5.13. If $b \ll a$ it is reasonable in the first instance to assume that the deformation through the layer is homogeneous, i.e. plane sections remain plane after compression as shown in Fig. 5.13(*a*), so that the stress σ_x is uniform through the thickness. We will consider first the case of no friction at the interface between the layer and the rigid substrate, whereupon $\sigma_x = 0$ throughout.

In plane strain

$$\epsilon_z = \frac{1-\nu^2}{E} \, \sigma_z = -\frac{1-\nu^2}{E} \, p(x) \qquad (5.69)$$

The compressive strain in the element is given by the geometry of deformation:

$$\epsilon_z = -(\delta - x^2/2R)/b \qquad (5.70)$$

Since the pressure must fall to zero at $x = \pm a$, equations (5.69) and (5.70) give $\delta = a^2/2R$ and

$$p(x) = \frac{E}{1-\nu^2} \, \frac{a^2}{2Rb} \, (1 - x^2/a^2) \qquad (5.71)$$

whence the load

$$P = \tfrac{2}{3} \frac{E}{1-\nu^2} \frac{a^3}{Rb} \qquad (5.72)$$

In the case where the layer is bonded to the substrate and plane sections remain plane, the strain ϵ_x is zero throughout, i.e.

$$\epsilon_x = \frac{1-\nu^2}{E} \left\{ \sigma_x + \frac{\nu}{1-\nu} p(x) \right\} = 0$$

Fig. 5.13. An elastic layer on a rigid substrate indented by a rigid cylinder: (*a*) Poisson's ratio $\nu < 0.45$; (*b*) $\nu = 0.5$.

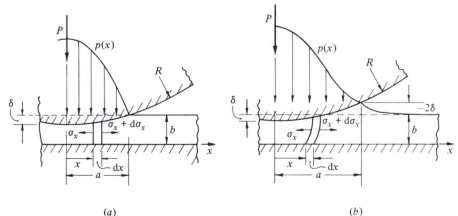

(*a*)　　　　　　　　　　　　　(*b*)

In this case

$$\epsilon_z = \frac{1-\nu^2}{E} \left\{ -p(x) - \frac{\nu}{1-\nu} \sigma_x \right\}$$

Eliminating σ_x and substituting for ϵ_z from (5.70) gives

$$p(x) = \frac{(1-\nu)^2}{1-2\nu} \frac{E}{1-\nu^2} \frac{a^2}{2Rb} (1-x^2/a^2) \tag{5.73}$$

and

$$P = \tfrac{1}{3} \frac{(1-\nu)^2}{1-2\nu} \frac{E}{1-\nu^2} \frac{a^3}{Rb} \tag{5.74}$$

For an incompressible material ($\nu = 0.5$) equation (5.73) implies an infinite contact pressure, showing that the assumption that plane sections remain plane is inappropriate in this case. However, by permitting the cross-sections of the layer to deform into a parabola (i.e. u_x to be second order in z as shown in Fig. 5.13(b)), it may be shown that for a thin layer of an incompressible material:

$$p(x) = \frac{Ea^4}{24Rb^3} (1-x^2/a^2)^2 \tag{5.75}$$

$$P = \frac{2Ea^5}{45Rb^3} \tag{5.76}$$

$$\delta = a^2/6R \tag{5.77}$$

The relationship between contact width and load for different ratios of layer thickness b to semi-contact width a is illustrated in Fig. 5.14, by plotting a/a_∞ against b/a, where $a_\infty^2 = 4PR(1-\nu^2)/\pi E$ = the Hertz semi-contact width ($b/a \to \infty$). The full curves are from Meijers (1968) and show the difference between $\nu = 0.3$ and $\nu = 0.5$. The asymptotic expressions for $b \ll a$, given by equations (5.74) and (5.76) are plotted for comparison.

The pressure distribution given by (5.75) is different from that for a compressible material (5.73). It has a zero gradient at the edges of the contact as shown in Fig. 5.13(b). The change in behaviour takes place quite rapidly in the range of Poisson's ratio between 0.45 and 0.48. The contact pressure varies as $(b/a)^{-3}$ so that it becomes high when the layer is very thin; the deformation of the indenter or the substrate can then no longer be neglected. Stresses in an elastic layer on an *elastic* substrate have been analysed by Gupta *et al.* (1973, 1974) and Barovich *et al.* (1964); see also §10.1.

Axi-symmetrical stresses in a layer or sheet have been expressed in terms of Hankel transforms by Sneddon (1951). The general features are similar to those found in plane deformation; a formal relationship between the two is discussed by Gladwell (1980), Chap. 10. Asymptotic solutions for the stresses in a layer due to a frictionless indenter have been found by Aleksandrov (1968, 1969) for thick and thin layers respectively. The contact of a flat circular punch and sphere with an elastic layer including the effects of friction has been analysed by Conway & Engel (1969) using the numerical method. For layers bonded to a substrate Matthewson (1981) has obtained an asymptotic solution for thin layers including the shear stress in the bond; McCormick (1978) has considered circular and elliptical contact areas for plates of general thickness.

(b) Receding contacts

This book is almost entirely concerned with the contact of non-conforming solids which touch initially at a point or along a line and whose area of contact grows with increasing load. Closely conforming contacts, on the other hand, which touch initially over an appreciable area when loaded may deform such that the contact area decreases. For example a perfectly fitting pin in a hole will initially touch round the whole of its circumference but, when it is loaded perpendicular to its axis, a gap will appear between the pin and the hole on the unloaded side. If the loaded contact area is completely contained within the unloaded contact area, the situation is described as *receding contact*,

Fig. 5.14. Contact width of an elastic layer on a rigid substrate indented by a rigid cylinder. Solid line – Meijers (1968); broken line – asymptotic solutions for $b \ll a$, eq. (5.74) and eq. (5.76); chain line – Hertz ($b \gg a$).

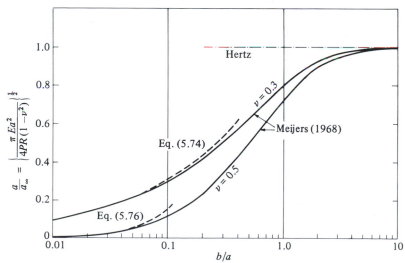

and has been shown by Tsai *et al.* (1974) and Dundurs (1975) to have special properties:

(i) The contact area changes discontinuously from its initial to its loaded shape and size on application of the first increment of load,

(ii) if the load increases in magnitude, but does not change in disposition, the contact area does not change in shape or size, and

(iii) the displacements, strain and stresses increase in direct proportion to the load.

The layer and the substrate shown in Fig. 5.15 will give rise to a receding contact if the layer is free to lift off the substrate under the action of the concentrated load *P*. Gladwell (1976) investigated this problem treating the layer as a simple beam in bending. Neglecting its own weight the beam was shown to make contact with the substrate under load over a distance $2c$, independent of the load, given by

$$(c/b)^3 = 1.845 \left\{ \frac{1-\nu_3{}^2}{E_3} \Big/ \frac{1-\nu_1{}^2}{E_1} \right\} = 1.845 \frac{1-\alpha}{1+\alpha} \tag{5.78}$$

where α is defined by equation (5.3*a*). Keer *et al.* (1972) solved the same problem using the proper elastic equations for a layer. The width of the contact between the layer and the substrate was found to be close to that given by equation (5.78) except when α was close to ± 1.0, i.e. when either the substrate or the layer was comparatively rigid. Ratwani & Erdogan (1973) have examined the situation where a layer which is free to lift is indented by a rigid cylinder ($E_2 \to \infty$) as shown in Fig. 5.12(*a*). At light loads, when $a \ll b$, the situation is much the same as for loading by a concentrated force and the semi-contact width *c* between the layer and substrate is given approximately by equation (5.78). When *a* grows with load to be comparable with *b*, *c* is no longer constant. Keer *et al.* (1972) have also examined axi-symmetric receding contact between a layer and an elastic half-space.

Fig. 5.15

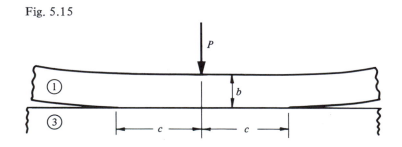

(c) Plates and shells

Thin plates or shell-like bodies in contact react to the contact load by bending. Thus bending stresses are added to the contact stresses. A discussion of the stresses in the vicinity of a concentrated load acting on a beam is given by Timoshenko & Goodier (1951, p. 99). When the beam or plate is relatively thick we can regard the stress field as comprising the superposition of Hertzian contact stresses and simple bending stresses. The bending action introduces a compressive stress in the upper layers of the plate which will add to the longitudinal compo-nent of the contact stress field (eq. (4.46*a*)) which is also compressive. The effect is to reduce the maximum value of the principal shear stress (eq. (4.47)) and to delay the initiation of plastic yield. When the plate is thin compared with the size of the contact area, the stresses are predominantly due to bending. For example, consider a rigid cylinder of radius R which is pressed into contact with a flat plate of length $2l$, width w and thickness $2b$, such that the contact arc is $2a$, where $b \ll a \ll R$ (see Fig. 5.16). The contact loading can be found by the elementary theory of bending. Within the contact arc the plate is bent into a circular arc of radius R, through the action of a uniform bending moment:

$$M = 2Ewb^3/3R(1 - v^2)$$

For the bending moment to be constant within the arc of contact, the contact pressure must comprise two concentrated forces at its edges and be zero else-where, whereupon

$$\tfrac{1}{2}P(l - a) = 2Ewb^3/3R(1 - v^2) \tag{5.79}$$

This equation determines the length of the arc of contact due to a given load P. This simple example shows that, as the load is progressively increased from first contact, so that the arc of contact grows from being small to large compared with the thickness of the plate, the contact pressure distribution changes from having

Fig. 5.16

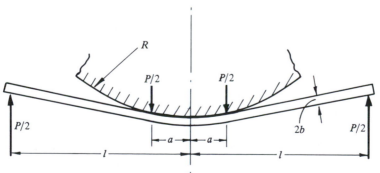

a maximum in the centre to one in which the pressure is concentrated at the edges. When the deformations are large it is necessary to take the changes in geometry into account and to make use of the theory of the 'Elastica' (see Wu & Plunkett, 1965).

Axi-symmetric contact of a paraboloid with a thin plate has been studied by Essenburg (1962) and the compression of a thin spheroidal shell between two rigid flats by Updike & Kalnins (1970, 1972). The use of classical plate and shell theory, in which shear deformation is ignored, leads to the contact pressure being concentrated into a ring of force at the edge of the circle of contact similar to the bent plate shown in Fig. 5.16. To obtain a more realistic distribution of contact pressure it is necessary to include the shear stiffness of the plate or shell. For thin plates, however, the pressure is still a minimum in the centre, rising to a maximum at the edges (see Gladwell & England, 1975).

A spheroidal shell, unlike a cylindrical shell, is not a developable surface. When pressed into contact with a frictionless flat surface, the shell is initially flattened, which introduces a compressive membrane stress. When a critical compression is reached, the shell buckles in the contact zone by the formation of a dimple. Updike & Kalnins (1970, 1972) investigate the onset of instability and discuss the conditions under which buckling will precede plastic yielding and vice versa.

5.9   Numerical methods

Many non-Hertzian contact problems do not permit analytical solutions in closed form. This is particularly true in the case of conforming contacts where the initial separation cannot be described by a simple quadratic expression (eq. (4.3)) and also in problems with friction involving partial slip. It has led to the development of various numerical methods which we shall discuss in this section. The essence of the problem is to determine the distributions of normal and tangential tractions which satisfy the normal and tangential boundary conditions at the interface, both inside and outside the contact area whose shape and size may not be known at the outset. In general the normal and tangential tractions are coupled, but we saw in §4 that considerable simplification can be achieved, with only a small loss of precision, by neglecting the effect upon the normal pressure of the tangential traction which arises when the materials of the two bodies are different. Thus the normal pressure is found on the assumption that the surfaces are *frictionless*. The internal stresses, if required, are found after the surface tractions are known.

The classical method, which has been applied to line contact and axi-symmetric problems in which the shape of the contact area is known, is to represent the pressure distribution by an infinite series of known functions. The series is then truncated to satisfy the boundary conditions approximately. Although a continuous distribution of traction is obtained, this method is basically ill-conditioned and can lead to large errors unless the functions are chosen carefully. The series due to Steuermann described in §3 provide examples of this method.

Modern computing facilities generally favour a different approach in which continuous distributions of traction are replaced by a discrete set of 'traction elements' and the boundary conditions are then satisfied at a discrete number of points – the 'matching points'. The simplest representation of a traction distribution is an array of concentrated normal or tangential forces as shown in Fig. 5.17(*a*). The difficulty with this representation lies in the infinite surface

Fig. 5.17. Discrete pressure elements (*a*) concentrated forces, (*b*) uniform (piecewise constant), (*c*) overlapping triangles (piecewise linear).

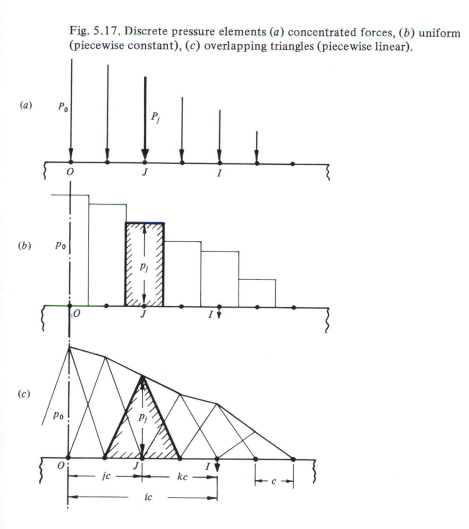

displacement which occurs at the point of application of a concentrated force. This difficulty is avoided if the traction is represented by adjacent columns of *uniform* traction acting on discrete segments of the surface, which give rise to a *stepwise* distribution as shown in Fig. 5.17(*b*). The surface displacements are now finite everywhere, but the displacement gradients are infinite between adjacent elements, where there is a step change in traction. A *piecewise-linear* distribution of traction, on the other hand, produces surface displacements which are everywhere smooth and continuous. Such a distribution of traction in line (two-dimensional) contact may be built up by the superposition of over-lapping triangular traction elements, as shown in Fig. 5.17(*c*). The corresponding traction element in three-dimensional contact is a regular pyramid on an hexagonal base, as shown in Fig. 5.18. An array of such pyramids, erected on an equilateral triangular mesh and overlapping, so that every apex coincides with a mesh point,

Fig. 5.18. Overlapping hexagonal pressure elements on an equilateral triangular (ξ, η) base.

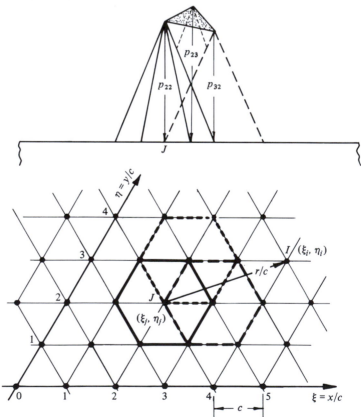

adds up to a resultant distribution of traction comprising plane triangular facets. One such facet based on the points (2, 2), (3, 2) and (2, 3) is shown in Fig. 5.18. The traction distributions are then specified completely by the discrete values p_j of the traction elements.

In order to find the values of the traction elements which best satisfy the boundary conditions two different methods have been developed:

(a) the direct, or Matrix Inversion, method in which the boundary conditions are satisfied exactly at specified 'matching points', usually the mid-points of the boundary elements, and

(b) the Variational method in which the values of the traction elements are chosen to minimise an appropriate energy function.

In describing the two methods we shall consider the normal contact of frictionless solids whose profiles are arbitrary and of such a form that they cannot be adequately characterised by their radii of curvature at their point of first contact. The gap between the two surfaces before deformation, however, is known and is denoted by the function $h(x, y)$ which, in the first instance, we shall assume to be smooth and continuous. It then follows from the principle discussed in §1 that the contact pressure falls continuously to zero at the edge of the contact. The elastic displacements of corresponding points on the two surfaces then satisfy the relationship:

$$\bar{u}_{z1} + \bar{u}_{z2} + h(x, y) - \delta \begin{cases} = 0 \text{ within contact} & (5.80a) \\ > 0 \text{ outside contact} & (5.80b) \end{cases}$$

where δ is the approach of distant reference points in the two bodies.

Whichever method is used, it is first necessary to choose the form of pressure element and to divide the contact surface into segments of appropriate size. Referring to Fig. 5.17, the matrix of influence coefficients C_{ij} is required, which expresses the displacement at a general point I due to a unit pressure element centred at point J. The total displacement at I is then expressed by

$$\{\bar{u}_z\}_i = -\frac{(1 - v^2)c}{E} \sum C_{ij} p_j \tag{5.81}$$

Difficulty arises in line contact (plane strain) where the displacements are undefined to the extent of an arbitrary constant. The difficulty may be overcome by taking displacements relative to a datum point, which is conveniently chosen to be the point of first contact, i.e. the origin. Since $h(0) = 0$, equation (5.80a) may be rewritten for line contact as

$$\{\bar{u}_{z1}(0) - u_{z1}(x)\} + \{\bar{u}_{z2}(0) - \bar{u}_{z2}(x)\}$$

$$-h(x) \begin{cases} = 0 \text{ within contact} & (5.82a) \\ > 0 \text{ outside contact} & (5.82b) \end{cases}$$

and we rewrite equation (5.81):

$$\{\bar{u}_z(0) - \bar{u}_z(x)\}_i = \frac{1 - \nu^2}{E} c \sum B_{ij} p_j \tag{5.83}$$

where $B_{ij} \equiv C_{0j} - C_{ij}$.

For a uniform pressure element in plane strain, the influence coefficients are obtained from equation (2.30d) by replacing a by c and x by kc, with the result:

$$C_{ij}(k) = \frac{1}{\pi} \{(k + 1) \ln (k + 1)^2 - (k - 1) \ln (k - 1)^2\}$$
$$+ \text{const.} \tag{5.84}$$

where $k = i - j$. For a triangular pressure element the influence coefficients are obtained from equations (2.37c), whereby

$$C_{ij}(k) = \frac{1}{2\pi} \{(k + 1)^2 \ln (k + 1)^2 + (k - 1)^2 \ln (k - 1)^2$$
$$- 2k^2 \ln k^2\} + \text{const.} \tag{5.85}$$

For point contacts, the influence coefficients for uniform pressure elements acting on rectangular segments of the surface $(2a \times 2b)$ can be obtained from equations (3.25) by replacing x by $(x_i - x_j)$ and y by $(y_i - y_j)$. Pyramidal pressure elements are based on a grid with axes $x (= \xi c)$ and $y (= \eta c)$ inclined at $60°$ as shown in Fig. 5.18. The distance JI is given by

$$JI \equiv r = ck = c\{(\xi_i - \xi_j)^2 + (\xi_i + \xi_j)(\eta_i - \eta_j)$$
$$+ (\eta_i - \eta_j)^2\}^{1/2} \tag{5.86}$$

The influence coefficients are found by the method described in §3.3. At the centre of the pyramid $(i = j)$: $C_{ij}(0) = (3\sqrt{3}/2\pi) \ln 3 = 0.9085$; at a corner $C_{ij}(1) = \frac{1}{3}C_{ij}(0)$. For values of $k^2 > 1$ the coefficient can be found by replacing the pyramid by a circular cone of the same volume, i.e. which exerts the same load, with the results shown in Table 5.2. For values of $k^2 > 9$ it is sufficiently accurate $(<0.5\%$ error) to regard the pyramid as a concentrated force, so that $C_{ij}(k) = \sqrt{3}/2\pi$.

Table 5.2

$\xi_i - \xi_j, \eta_i - \eta_j$	1,1	2,0	2,1	3,0
k^2	3	4	7	9
$C_{ij}(k)$	0.1627	0.1401	0.1051	0.0925

The total load P carried by the contact is related to the values of the pressure elements by

$$P = A \sum p_j \tag{5.87}$$

where A is a constant depending upon the form and size of the pressure element. For a uniform pressure element A is the surface area of the element; for a pyramidal element, $A = \sqrt{3}c^2/2$. We are now in a position to discuss the methods for finding the values of p_j.

(a) Matrix inversion method

The displacements $\{\bar{u}_z\}_i$ at a general mesh point I are expressed in terms of the unknown pressures p_j by equation (5.81) for point contact and equation (5.83) for line contact. If n is the number of pressure elements, i and j take integral values from 0 to $(n-1)$. Substituting these displacements into equations (5.80a) and (5.82a) respectively gives

$$\sum_{j=0}^{j=n-1} C_{ij} p_j = (E^*/c)(h_i - \delta) \tag{5.88}$$

for point contacts and

$$\sum_{j=0}^{j=n-1} B_{ij} p_j = (E^*/c) h_i \tag{5.89}$$

for line contacts. If the compression δ is specified then equation (5.88) can be solved directly by matrix inversion for the n unknown values of p_j. It is more likely, however, that the total load P is specified. The compression δ then constitutes an additional unknown, but an additional equation is provided by (5.87). In equation (5.88) for line contacts the origin $(i = 0)$ is a singular point since $B_{0j} = h_0 = 0$, but again equation (5.87) for the total load provides the missing equation.

It is unlikely in problems requiring numerical analysis that the shape or size of the contact area is known in advance. To start, therefore, a guess must be made of the shape of the contact surface and its size must be chosen to be sufficiently large to enclose the true area. Where the value of δ is specified or can be estimated, a first approximation to the contact area can be obtained from the 'interpenetration curve', that is the contour of separation $h(x, y) = \delta$. This is the area which is divided into an array of n pressure elements. After solving equation (5.88) or (5.89) for the unknown pressures, it will be found that the values of p_j near to the periphery are negative, which implies that a tensile traction is required at some mesh points to maintain contact over the whole of the assumed area. For the second iteration these mesh points are excluded from the assumed contact area and the pressures there put equal to zero. Experience confirms that repeated iterations converge to a set of values of p_j which are positive or zero and which satisfy equation (5.80a) within the region where $p_j > 0$ and equation (5.80b) in the region where $p_j = 0$. The boundary between the two regions defines the contact area to the accuracy of the mesh size.

In line contact the contact area is the strip $-b \leqslant x \leqslant a$ where, for a given load, a and b remain to be found. If the deformation is symmetrical about the origin, so that $b = a$, the pressure distribution can be found without iteration. It is preferable to take a as the independent load variable, to divide the contact strip into $2n$ segments and to use $2n - 1$ overlapping triangular pressure elements as shown in Fig. 5.17(c). This arrangement automatically ensures that the contact pressure falls to zero at $x = \pm a$. Equation (5.83) can then be inverted directly to find the values of p_j. This method has been used by Paul & Hashemi (1981) for normal contact and by Bentall & Johnson (1967) for problems in which tangential as well as normal tractions are present.

The procedure outlined above is appropriate for bodies whose profiles are both smooth and continuous. If one of the bodies has a sharp corner at the edge of the contact, the pressure at the edge of contact, instead of falling to zero, will rise to infinity according to $\rho^{-\lambda}$, where ρ is the distance from the edge and the value of λ depends upon the elasticities of the two bodies but is approximately 0.5 (see §1). This type of contact can be incorporated into the above method by using a boundary element in the segment adjacent to the edge of contact in which the pressure varies as $\rho^{-0.5}$ as shown in Fig. 5.19. The displacement at a distance x from the edge due to such an element of pressure is

Fig. 5.19. Singular pressure element.

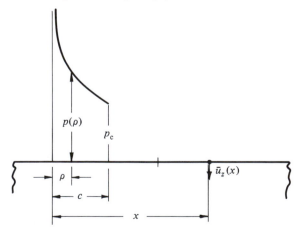

given by

$$\bar{u}_z(x) = -\frac{1-\nu^2}{\pi E} 2cp_c \left\{ \ln\left(\frac{x-c}{c}\right)^2 \right.$$

$$\left. + (x/c)^{1/2} \ln\left(\frac{x^{1/2}+c^{1/2}}{x^{1/2}-c^{1/2}}\right)^2 - 4 \right\} + c \qquad (5.90)$$

(b) Variational methods

Variational methods have been applied to non-Hertzian contact problems for two reasons: (i) to establish conditions which will determine the shape and size of the contact area and the contact stresses uniquely and (ii) to enable well-developed techniques of optimisation such as quadratic programming to be used in numerical solutions.

Fichera (1964) and Duvaut & Lions (1972) have investigated general principles which govern the existence and uniqueness of solution to contact problems. For two bodies having continuous profiles, pressed into contact by an overall displacement δ, Duvaut & Lions show that the true contact area and surface displacements are those which minimise the total strain energy U_E (with δ kept constant), provided that there is no interpenetration, i.e. provided

$$\bar{u}_{z1} + \bar{u}_{z2} + h(x,y) - \delta \geqslant 0$$

everywhere. An example of the application of this principle to a Hertz contact was given in §5.

For numerical solution of contact problems it is more convenient to work in terms of unknown tractions rather than displacements. Kalker (1977, 1978) has therefore proposed an alternative principle in which the true contact area and distribution of surface traction are those which minimise the total complementary energy (V^*), subject to the constraint that the contact pressure p is everywhere positive. Now the total complementary energy can be written:†

$$V^* = U_E^* + \int_S p(h-\delta)\,dS \qquad (5.91)$$

where S is the surface on which p acts and U_E^* is the internal complementary energy of the two stressed bodies. For linear elastic materials the complementary energy U_E^* is numerically equal to the elastic strain energy U_E, which can be expressed in terms of the surface tractions and displacements by

$$U_E^* = U_E = \tfrac{1}{2}\int_S p(\bar{u}_{z1} + \bar{u}_{z2})\,dS \qquad (5.92)$$

---

† For a discussion of the complementary energy principle see T. H. Richards, *Energy methods in stress analysis*, Ellis Horwood, 1977, p. 256.

To obtain a numerical solution the prospective contact area S is subdivided into a mesh on which elements of pressure act. Using equation (5.81) we have

$$U_{\mathrm{E}}^* = -\frac{(1-\nu^2)cA}{2E^*} \sum\{\sum C_{ij}p_j\}p_i \qquad (5.93)$$

and

$$\int_S p(h-\delta)\,\mathrm{d}S = A \sum p_i(h_i-\delta) \qquad (5.94)$$

where A is defined in equation (5.87). Thus substituting from (5.93) and (5.94) into (5.91) gives V^* as an object function quadratic in p_i. The values of p_i which minimise V^*, subject to $p_i > 0$, can be found by using a standard quadratic programming routine, e.g. that of Wolfe (1959) or Beale (1959). The contact is then defined, within the precision of the mesh size, by the boundary between the zero and non-zero pressures. This method has been applied to frictionless non-Hertzian contact problems by Kalker & van Randen (1972).

To find the subsurface stresses it is usually adequate to represent the surface tractions by an array of concentrated forces as in Fig. 5.17(*a*). The stress components at any subsurface point can then be found by superposition of the appropriate expressions for the stresses due to a concentrated force, normal or tangential, given in §§2.2 & 3 or §§3.2 & 6.

When the size of the contact region is comparable with the leading dimensions of one or both bodies, influence coefficients based on an elastic half-space are no longer appropriate. Bentall & Johnson (1968) have derived influence coefficients for thin layers and strips but, in general, a different approach is necessary. The finite-element method has been applied to contact problems, including frictional effects, notably by Fredriksson (1976). A more promising technique is the Boundary Element Method which has been applied to two-dimensional contact problems by Andersson *et al.* (1980).

6

Normal contact of inelastic solids

6.1 Onset of plastic yield

The load at which plastic yield begins in the complex stress field of two solids in contact is related to the yield point of the softer material in a simple tension or shear test through an appropriate yield criterion. The yield of most ductile materials is usually taken to be governed either by von Mises' shear strain-energy criterion:

$$J_2 \equiv \tfrac{1}{6}\{(\sigma_1 - \sigma_2)^2 + (\sigma_2 - \sigma_3)^2 + (\sigma_3 - \sigma_1)^2\} = k^2 = Y^2/3 \qquad (6.1)$$

or by Tresca's maximum shear stress criterion:

$$\max\{|\sigma_1 - \sigma_2|, |\sigma_2 - \sigma_3|, |\sigma_3 - \sigma_1|\} = 2k = Y \qquad (6.2)$$

in which σ_1, σ_2 and σ_3 are the principal stresses in the state of complex stress, and k and Y denote the values of the yield stress of the material in simple shear and simple tension (or compression) respectively. Refined experiments on metal specimens, carefully controlled to be isotropic, support the von Mises criterion of yielding. However the difference in the predictions of the two criteria is not large and is hardly significant when the variance in the values of k or Y and the lack of isotropy of most materials are taken into account. It is justifiable, therefore, to employ Tresca's criterion where its algebraic simplicity makes it easier to use. A third criterion of yield, known as the maximum reduced stress criterion, is expressed:

$$\max\{|\sigma_1 - \sigma|, |\sigma_2 - \sigma|, |\sigma_3 - \sigma|\} = k = \tfrac{2}{3}Y \qquad (6.3)$$

where $\sigma = (\sigma_1 + \sigma_2 + \sigma_3)/3$. It may be shown from conditions of invariance that, for a stable plastic material, the Tresca criterion and the reduced stress criterion provide limits between which any acceptable yield criterion must lie. We shall see that these limits are not very wide.

(a) Two-dimensional contact of cylinders

In two-dimensional contact the condition of plane strain generally ensures that the axial stress component σ_y is the intermediate principal stress, so that by the Tresca criterion yield is governed by the maximum principal stress difference (or maximum shear stress) in the plane of cross-section, i.e. the x–z plane. Contours of principal shear stress $\tau_1 = \frac{1}{2}|\sigma_1 - \sigma_2|$ are plotted in Fig. 4.5: they are also exhibited by the photo-elastic fringes in Fig. 4.1(d). The maximum shear stress is $0.30p_0$ at a point on the z-axis at a depth $0.78a$. Substituting in the Tresca criterion (6.2) gives

$$0.60p_0 = 2k = Y$$

whence yield begins at a point $0.78a$ below the surface when the maximum contact pressure reaches the value

$$(p_0)_Y = \frac{4}{\pi} p_m = 3.3k = 1.67Y \tag{6.4}$$

The von Mises and reduced stress criteria both depend upon the third principal stress and hence upon Poisson's ratio. Taking $\nu = 0.3$, the maximum value of the left-hand side of equation (6.1) is $0.104p_0^2$ at a depth $0.70a$, and the maximum value of the left-hand side of equation (6.3) is $0.37p_0$ at a depth $0.67a$. Thus by the von Mises criterion yield begins at a point $0.70a$ below the surface when

$$(p_0)_Y = 3.1k = 1.79Y \tag{6.5}$$

and by the reduced stress criterion yield first occurs when

$$(p_0)_Y = 2.7k = 1.80Y \tag{6.6}$$

We see from the three expressions (6.4), (6.5) and (6.6) that the value of the contact pressure to initiate yield is not influenced greatly by the yield criterion used. The value given by the von Mises criterion lies between the limits set by the Tresca and reduced stress criteria.

The load for initial yield is then given by substituting the critical value of p_0 in equation (4.45) to give

$$P_Y = \frac{\pi R}{E^*} (p_0)_Y^2 \tag{6.7}$$

where the suffix Y denotes the point of first yield and $1/R = 1/R_1 + 1/R_2$.

(b) Axi-symmetric contact of solids of revolution

The maximum shear stress in the contact stress field of two solids of revolution also occurs beneath the surface on the axis of symmetry. Along this axis σ_z, σ_r and σ_θ are principal stresses and $\sigma_r = \sigma_\theta$. Their values are given by equation (3.45). The maximum value of $|\sigma_z - \sigma_r|$, for $\nu = 0.3$, is $0.62p_0$ at a depth $0.48a$. Thus by the Tresca criterion the value of p_0 for yield is given by

$$p_0 = \tfrac{3}{2}p_m = 3.2k = 1.60Y \tag{6.8}$$

whilst by the von Mises criterion

$$p_0 = 2.8k = 1.60Y \tag{6.9}$$

The load to initiate yield is related to the maximum contact pressure by equation (4.24), which gives

$$P_Y = \frac{\pi^3 R^2}{6E^{*2}} (p_0)_Y^3 \tag{6.10}$$

It is clear from equations (6.7) and (6.10) that to carry a high load without yielding it is desirable to combine a high yield strength or hardness with a low elastic modulus.

(c) General smooth profiles

In the general case the contact area is an ellipse and the stresses are given by the equations (3.64)–(3.69). The stresses along the z-axis have been evaluated and the maximum principal stress difference is $|\sigma_y - \sigma_z|$ which lies in the plane containing the minor axis of the ellipse $(a > b)$. This stress difference and hence the maximum principal shear stress τ_1 maintain an almost constant value as the eccentricity of the ellipse of contact changes from zero to unity (see Table 4.1, p. 99). Thus there is little variation in the value of the maximum contact pressure to initiate yield, given by the Tresca criterion, as the contact geometry changes from axi-symmetrical (6.8) to two-dimensional (6.4). However the point of first yield moves progressively with a change in eccentricity from a depth of $0.48a$ in the axi-symmetrical case to $0.78b$ in the two-dimensional case. Similar conclusions, lying between the results of equation (6.9) for spheres and (6.5) for cylinders, are obtained if the von Mises criterion is used.

(d) Wedge and cone

The stresses due to the elastic contact of a blunt wedge or cone pressed into contact with a flat surface were found in §5.2, where it was shown that a theoretically infinite pressure exists at the apex. It might be expected that this would inevitably cause plastic yield at the lightest load, but this is not necessarily so. Let us first consider the case of an incompressible material. During indentation by a two-dimensional frictionless wedge the tangential stress $\bar{\sigma}_x$ at the interface is equal to the normal pressure p (see eq. (2.26)). If $\nu = 0.5$ then the axial stress $\bar{\sigma}_z$ to maintain plane strain is also equal to p. Thus the stresses are hydrostatic at the contact interface. The apex is a singular point.

By considering the variation in the principal stress difference $|\sigma_x - \sigma_z|$ along the z-axis, it may be shown that this difference has a maximum but finite value of $(2E^*/\pi)$ cot α at the apex. Then by the Tresca or von Mises criteria (which are identical for $\nu = 0.5$ when stated in terms of k) yield will initiate at the apex if the wedge angle α is such that

$$\cot \alpha \geqslant \pi k/E^* \tag{6.11}$$

Similar conclusions apply to indentation by a blunt cone when $\nu = 0.5$. An infinite hydrostatic pressure is exerted at the apex of the cone but the principal stress difference $|\sigma_r - \sigma_z|$ along the z-axis is finite and has a maximum value at the apex of E^* cot α. In this case two principal stresses are equal, so that the Tresca and von Mises criteria are identical if expressed in terms of Y. Thus yield will initiate at the apex if the cone angle is such that

$$\cot \alpha \geqslant Y/E^* \tag{6.12}$$

For compressible materials the results obtained above are no longer true. Instead of hydrostatic pressure combined with a finite shear, the infinite elastic pressure at the apex will give rise to theoretically infinite differences in principal stresses which will cause plastic flow however small the wedge or cone angle. Nevertheless the plastic deformation arising in this way will, in fact, be very small and confined to a small region close to the apex. In the case of the wedge the lateral stress σ_y is less than σ_x and σ_z, which are equal, so that a small amount of plastic flow will take place in the y–z plane. To maintain plane strain, this flow will give rise to a compressive residual stress in the y-direction until a state of hydrostatic pressure is established. Plastic flow will then cease. Similar behaviour is to be expected in the case of the cone.

It would seem to be reasonable, therefore, to neglect the small plastic deformation which arises in this way and to retain equations (6.11) and (6.12) to express the effective initiation of yield by a wedge and a cone respectively, even for compressible materials.

Even when the limits of elastic behaviour given by the above equations have been exceeded and plastic flow has begun, the plastic zone is fully contained by the surrounding material which is still elastic. This is clearly shown in the contours of principal shear stress given by the photo-elastic fringe patterns in Figs. 4.1 and 5.2. For bodies having smooth profiles, e.g. cylinders or spheres, the plastic enclave lies beneath the surface whilst for the wedge or cone it lies adjacent to the apex. Hence the plastic strains are confined to an elastic order of magnitude and an increase in load on the cylinders or spheres or an increase in wedge or cone angle gives rise only to a slow departure of the penetration, the contact area or the pressure distribution from the values given by elastic theory. For this reason Hertz' (1882b) original suggestion that the *initiation* of yield due to the impression of a hard ball could be used as a rational measure of the hardness of a material proved to be impracticable. The point of first yield is hidden beneath the surface and its effect upon measurable quantities such as mean contact pressure is virtually imperceptible. A refined attempt to detect by optical means the point of first yield during the impression of a hard ball on a flat surface has been made by Davies (1949).

We shall return to consider the growth of the plastic zone in more detail in §3, but meanwhile we shall turn to the other extreme: where the plastic deformation is so severe that elastic strains may be neglected in comparison with plastic strains. Analysis is then possible using the theory of rigid-perfectly-plastic solids.

6.2   Contact of rigid-perfectly-plastic solids

When the plastic deformation is severe so that the plastic strains are large compared with the elastic strains, the elastic deformation may be neglected. Then, provided the material does not strain-harden to a large extent, it may be idealised as a rigid-perfectly-plastic solid which flows plastically at a constant stress k in simple shear or Y in simple tension or compression. The theory of plane deformation of such materials is well developed: see, for example, Hill (1950a) or Ford & Alexander (1963).

A loaded body of rigid-perfectly-plastic material comprises regions in which plastic flow is taking place and regions in which, on account of the assumption of rigidity, there is no deformation. (It does not follow, however, that the stresses in the non-deforming regions are below the elastic limit.) The state of stress within the regions of flow can be represented by a *slip-line field*. The slip lines are drawn parallel to the directions of principal shearing stress at every point in the field, i.e. at 45° to the directions of principal direct stress. Thus they consist of a curvilinear net of 'α lines' and 'β lines' which are perpendicular to each other at all points. An element of such a slip-line field is shown in Fig. 6.1(a).

Since elastic compressibility is neglected, the principal stress acting perpendicular to the plane of deformation is given by

$$\sigma_3 = \tfrac{1}{2}(\sigma_1 + \sigma_2) \tag{6.13}$$

where σ_1 and σ_2 are the principal stresses acting in the plane of deformation. Under these conditions the Tresca and von Mises criteria of plastic flow both reduce to

$$|\sigma_1 - \sigma_2| = 2k \tag{6.14}$$

where $k = Y/2$ by Tresca or $Y/\sqrt{3}$ by von Mises. Thus the state of stress in the plastic zone comprises a variable hydrostatic stress $\tfrac{1}{2}(\sigma_1 + \sigma_2)$ denoted by σ,

Fig. 6.1. (*a*) Stresses acting on an element bounded by slip lines;
(*b*) Mohr's circle.

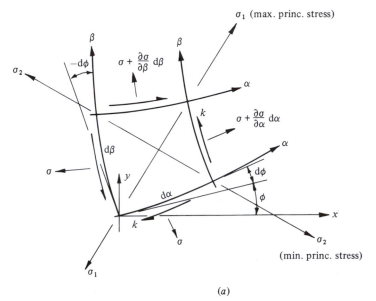

(*a*)

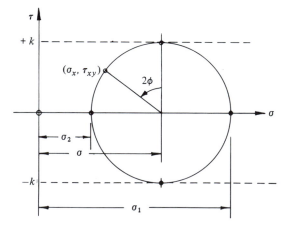

(*b*)

together with a constant simple shear k in the plane of deformation. This state of stress is represented by a Mohr's circle of constant radius k whose centre is located by the value of σ at the point in question, as shown in Fig. 6.1(*b*). The directions of the principal stresses relative to a fixed axis in the body are fixed by the directions of the slip lines. By considering the equilibrium of the element in Fig. 6.1(*a*) under the action of the direct stresses σ and the shear stress k, we obtain

$$\frac{\partial \sigma}{\partial \alpha} - 2k \frac{\partial \phi}{\partial \alpha} = 0 \tag{6.15a}$$

in the direction of an α line, and

$$\frac{\partial \sigma}{\partial \beta} + 2k \frac{\partial \phi}{\partial \beta} = 0 \tag{6.15b}$$

in the direction of a β line, which gives

$$\sigma - 2k\phi = \text{constant along an } \alpha \text{ line}\dagger \tag{6.16a}$$

$$\sigma + 2k\phi = \text{constant along a } \beta \text{ line}\dagger \tag{6.16b}$$

Thus, by starting at a point of known stress such as a free surface, equations (6.16) enable the variation in σ throughout the field to be found from the directions of the slip lines.

The constitutive relations for a plastically deforming solid relate the stresses to the small *increments* of strain. For convenience it is customary to think of the strain and displacement increments taking place in an interval of time dt and to work in terms of strain *rates* and *velocities* in place of increments of strain and displacement. The continuous deformation of the element of material shown in Fig. 6.1(*a*) consists of an extension along the direction of the maximum principal stress and a compression along the direction of the minimum principal stress. For constant volume $\dot{\epsilon}_2 = -\dot{\epsilon}_1$. There is no change in length along the direction of the slip lines, so that the deformation may be visualised as that of a 'net' in which the slip lines are inextensible strings. If a vector diagram is constructed by the velocities of particles in the deformation zone – a hodograph – the inextensibility of the slip lines requires that the 'velocity image' of a segment of a slip line is perpendicular to that line. A discontinuous mode of deformation is also possible in which an element such as that shown slides bodily relative to an adjacent element. It is clear that the line of discontinuity in particle velocity in such a deformation must coincide with a slip line.

---

† Some text-books take σ to be positive when *compressive*, which changes the signs in equations (6.16).

There is no progressive routine for constructing the slip-line field to solve a particular problem; it has to be found by trial. It must be self-consistent with a velocity field, and both must satisfy the boundary conditions of the problem. Finally it should be checked that the non-deforming (rigid) regions are capable of supporting the loads without violating the yield condition. When all these conditions are satisfied the slip-line field and the stresses found from it by equations (6.16) are unique, but the associated velocity field may not be.

We shall now proceed to discuss the slip-line fields associated with the rigid-plastic deformation of a wedge in contact with a plane surface. In the first instance we shall take the wedge to be appreciably harder than the flat, so that plastic deformation is confined to the flat. Secondly, we shall take the flat to be harder, so that it remains effectively rigid and crushes the apex of the wedge.

(a) Frictionless wedge indenting a rigid-plastic surface

The normal indentation of a rigid-perfectly-plastic half-space by a rigid wedge of semi-angle α is shown in Fig. 6.2(a). The material flows plastically in the two symmetrical regions lettered $ABCDE$. The material surrounding these regions, being assumed rigid, has not deformed at the current stage of indentation. The material displaced by the wedge is pushed up at the sides: for the volume to be conserved the areas of triangles AOF and FBC must be equal.

The slip-line field is shown in Fig. 6.2(a). Since the face of the wedge AB is frictionless it can sustain no shear stress. The normal pressure p_w on the wedge face is therefore a principal stress and the slip lines meet AB at 45°. Similarly the slip lines meet the free surface BC at 45°. The state of stress in the triangular region BCD is a uniform compression $2k$ acting parallel to the surface BC. It is represented by a Mohr's circle in Fig. 6.2(b). The hydrostatic component of stress in this region has the value $-k$ (shown by the centre of the circle). The state of stress in the triangular region ABE is also uniform and is represented by the other Mohr's circle. The distance between the two centres represents the difference in hydrostatic stress between the two regions, which is given by equation (6.16a) and has the value $2k\psi$, where ψ is the angle turned through by the α slip lines between the two regions. The state of stress in the fan BDE is represented by intermediate Mohr's circles whose centres are located by the inclination of the slip line at the point in question. The pressure on the wedge face is represented by point W on the circle: it has the uniform value given by

$$p_w = 2k(1 + \psi) \tag{6.17}$$

If the total normal load on the wedge is P and the projected area of contact is $2a$ per unit axial length, then the mean pressure acting normal to the original surface of the solid is given by

$$p_m = \frac{P}{2a} = p_w = 2k(1 + \psi) \tag{6.18}$$

The point B is a singular point in which the state of stress jumps from that at the free surface to that under the wedge face.

To locate the position of B and to determine the value of the angle ψ we must consider the mode of deformation. The velocity diagram (hodograph) for the right-hand deformation zone is shown in Fig. 6.2(c). The wedge is assumed to be penetrating the solid with a steady velocity V, represented by oa in the hodograph. $AEDC$, which separates the deforming region from the rigid region, is a line of discontinuity in velocity. The region ABE moves without distortion parallel to AE with the velocity oe and slides relative to the wedge face with velocity ae. The region BDC moves without distortion parallel to DC with velocity od. The velocity of the surface BC perpendicular to itself is represented by og. Now the state of stress and deformation shown in Fig. 6.2(a) should be independent of the depth of penetration: in other words geometrical similarity of the slip-line field should be maintained at all stages of the indentation. The condition of geometrical similarity controls the shape of the free surface BC. It requires that normal displacement from the origin O of any point on the

Fig. 6.2. Indentation by a rigid frictionless wedge.

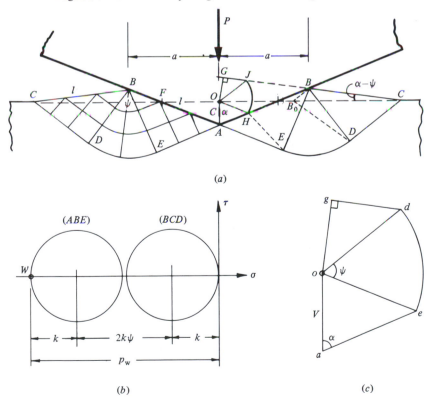

(a)

(b)

(c)

free surface *BC* and on the wedge face *AB* should grow in direct proportion to the component of velocity of that point normal to the surface. This condition may be visualised by superimposing the velocity diagram on the wedge so that *o* coincides with *O* and *a* with *A* as shown in Fig. 6.2(*a*). The above condition is then satisfied if the velocity image at a point on the free surface lies on the tangent to the surface at that point. In the example in Fig. 6.2, where the free surface is straight, all points on that surface have the same velocity *od*, so that *J*(*d*) and *G*(*g*) must lie on *CB* produced. The normal displacement of all points on *BC* is thus represented by *OG*. The angle ψ may now be found by geometry. Denoting *AB* = *BC* = *l*, the height of *B* above *A* is given by

$$l \cos \alpha = c + l \sin (\alpha - \psi) \tag{6.19}$$

The condition that *G* should lie on *CB* produced gives

$$l \cos \psi = c \sin \alpha + c \cos (\alpha - \psi) \tag{6.20}$$

Eliminating *l/c* from (6.19) and (6.20) gives

$$\cos (2\alpha - \psi) = \frac{\cos \psi}{1 + \sin \psi} \tag{6.21}$$

from which ψ can be found by trial for any value of α. Then substituting ψ into equation (6.18) gives the indentation pressure p_m. The variation of p_m with wedge angle α is shown by the curve marked 'frictionless' in Fig. 6.7.

Returning to examine the mode of deformation more closely, we observe that material particles originally lying along the internal line *OA* are displaced to lie along the wedge face in the segment *AH*. Thus the nose of the wedge acts like a cutting tool. Material lying within the triangle *OAE* has been displaced to *HAE*. Particles lying within the triangle *BDC* have moved in the direction *od* parallel to *DC*. Thus *B* originally lay on the surface at B_0 where $B_0 B$ is parallel to *DC*, and material in the deformed region *BDC* originally lay in the region $B_0 DC$. Material originally in the region $OB_0 DE$ has undergone a more complex deformation governed by the fan of slip lines *BDE*. The segment of the original surface OB_0 is folded into contact with the wedge face along *BH*. The distortion of a square grid calculated from the hodograph is shown in Fig. 6.3(*a*). Details of the method of calculation are given in the original paper by Hill *et al.* (1947).

(b) Influence of friction on wedge indentation

Friction between the face of the wedge and the indented material influences the mode of deformation significantly. The relative motion of the material up the face of the wedge now introduces an opposing shear stress $\tau_w = \mu p_w$ so that the slip lines intersect the face of the wedge at an angle $ABE = \lambda$ which is less than 45°. The modified slip-line field, and Mohr's circle for the stress at the wedge face, are shown in Fig. 6.4. The pressure on the wedge face is now given by

$$p_w = k(1 + 2\psi + \sin 2\lambda) \tag{6.22}$$

and the shear stress is

$$\tau_w = k \cos 2\lambda \tag{6.23}$$

hence λ is related to the coefficient of friction by

$$\cos 2\lambda = \mu(1 + 2\psi + \sin 2\lambda) \tag{6.24}$$

Equilibrium of the wedge gives the indentation pressure

$$p_m = \frac{P}{2a} = p_w(1 + \mu \cot \alpha) \tag{6.25}$$

Fig. 6.3. Deformation by a blunt wedge: (*a*) frictionless; (*b*) no slip.

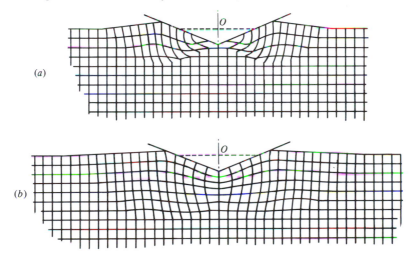

Fig. 6.4. Indentation by a rough wedge with slip at the interface.

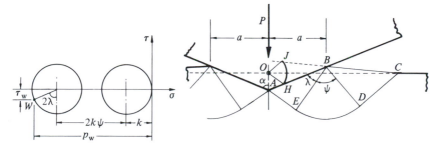

When the hodograph is superimposed on the diagram of the wedge, as before, the condition of similarity requires that J should again lie on the line of the free surface BC produced. The determination of the angles ψ and λ for given values of α and μ must be carried out by a process of trial. Such calculations were made by Grunsweig, Longman & Petch (1954) and the results are plotted in Fig. 6.7.

We note that friction causes the point H to move towards the apex of the wedge which reduces the cutting action. With increasing friction a limit is reached when slip at the wedge face ceases: the surface of the solid adheres to the wedge and shear takes place in the body of the material. For a wedge of semi-angle less than $45°$, the limiting slip-line field is shown in Fig. 6.5. A slip line coincides with the wedge face and concentrated shearing takes place just within the material at a shear stress $\tau_w = k$. The critical value of μ and the corresponding pressures are found by putting $\lambda = 0$ in equations (6.24), (6.22) and (6.25).

With a blunt wedge ($\alpha > 45°$) this solution is incorrect since the slip lines at the apex cannot meet at an angle less than $90°$. In this case a cap of undeforming material adheres to the wedge and shear takes place on slip line BE, as shown in Fig. 6.6. The stress state along the slip line BE is indicated by point S on the Mohr's circle. Equilibrium of the cap of metal ABE results in the stresses on the face of the wedge being represented by W, i.e. the normal stress is $k(1 + 2\psi - \cos 2\alpha)$ and the shear stress is $k \sin 2\alpha$. Hence for the cap to adhere to the wedge

$$\mu \geqslant \frac{\sin 2\alpha}{1 + 2\psi - \cos 2\alpha} \tag{6.26}$$

whereupon the indentation pressure is given by

$$p_m = \frac{P}{2a} = 2k(1 + \psi) \tag{6.27}$$

Fig. 6.5. Indentation by a sharp wedge with no slip.

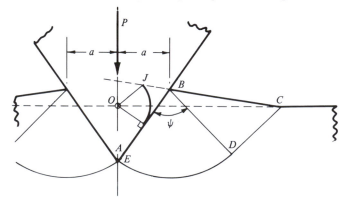

As before, ψ is determined from the hodograph and the condition of geometrical similarity. Values have been found by Haddow (1967) and are shown in Fig. 6.7.

The distortion of a square grid in the mode of deformation involving a built-up nose on the wedge has been constructed and is shown in Fig. 6.3(*b*) for direct comparison with the frictionless mode. The difference is striking. Deformation occurs beneath the apex of the wedge and the displacements of grid points from their undeformed positions are approximately radial from *O*.

The influence of strain hardening on the slip-line field for an indenting wedge has been investigated by Bhasin *et al.* (1980).

(c) Crushing of a plastic wedge by a rigid flat

If the wedge is appreciably softer than the flat surface the wedge will deform plastically and the flat will remain rigid. The slip-line field suggested by Hill (1950*a*) for frictionless contact between the wedge and the plane is shown in Fig. 6.8(*a*). *AEDC* is a line of velocity discontinuity. The triangular region *ABE* slides outwards relative to the face of the flat and, in the absence of fric-tin, the slip lines meet the interface at 45°. The pressure on the interface is uniform and given by

$$p_{\mathrm{m}} = \frac{P}{2a} = 2k(1 + \psi) \qquad (6.28)$$

where the angle ψ is determined by the condition of geometrical similarity, that *J* should lie on *CB* produced.

Fig. 6.6. Indentation by a blunt wedge with no slip.

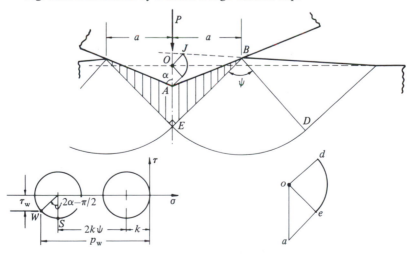

An alternative mode of deformation and associated slip-line field is shown in Fig. 6.8(*b*). The triangular region *ABE* adheres to the flat and moves vertically with it and there is intense shear on the slip line *BE*. The pressure on the interface is still given by equation (6.28) but, in this case, the angle is slightly larger than before so that the pressure required to produce this mode of deformation is higher, as shown in Fig. 6.9. The difference in pressure is due to the difference in inclination on the surface *BC* of the shoulders.

For an ideally frictionless flat it is to be presumed that the deformation would follow the pattern in Fig. 6.8(*a*) since this requires the lowest pressure to cause deformation.† Friction at the interface will oppose the sliding motion and cause the slip lines to meet the surface at *B* and *B'* at less than 45°. This would result in an unacceptable state of stress at *A* unless a cap of undeforming material adheres to the contacting flat in the vicinity of *A*, in the manner found when a rectangular block is compressed between two flat rigid plates (see Ford & Alexander, 1963). The slip-line field and the profile of the free surfaces *BC*

---

† Asymmetrical modes of deformation are also possible but, since they require higher pressures than the symmetrical mode, they will not be considered.

Fig. 6.7. Mean contact pressure for a rigid wedge indenting a rigid-perfectly-plastic half-space.

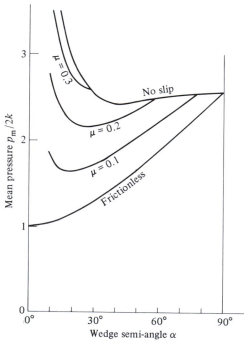

Fig. 6.8. Crushing of a plastic wedge by a rigid flat: (*a*) frictionless;
(*b*) no slip.

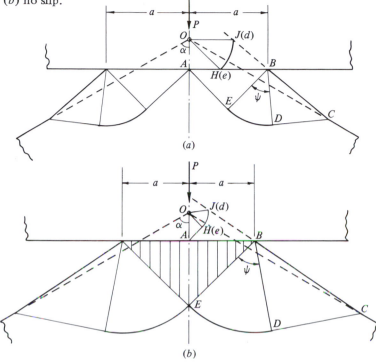

Fig. 6.9. Contact pressure on a flat crushing a plastic wedge.

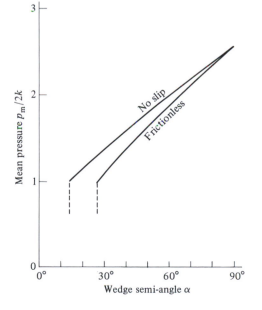

will then be curved. This problem has not yet been solved quantitatively, so that it is not possible to say exactly how much interfacial friction is necessary to eliminate sliding and ensure that the adhesive mode of Fig. 6.8(b) is obtained.†
This is not a serious shortcoming, however, since the difference in indentation pressure between the frictionless and adhesive modes is not large. It is somewhat paradoxical that although friction is necessary to ensure that the adhesive mode takes place, when it does so no friction forces are transmitted at the interface.

The paradox arises (a) through the neglect of *elastic* deformation which will influence the frictional conditions at the interface in the 'adhesive' mode and (b) through the assumption of perfect plasticity which leads in some cases to a lack of uniqueness in the mode of deformation. Strain hardening, which is a feature of real materials, will favour the mode of deformation in which the plastic strains are least and are most uniformly distributed through the deforming region.

The solutions given in Fig. 6.8 cease to apply when the angle ψ vanishes. In the limit ($\alpha = 26.6°$ and $\alpha = 14°$ respectively) the sides of the shoulders of displaced material are parallel and the deforming region of the wedge is in simple compression.

If the wedge and the flat are of comparable hardness both will deform. This state of affairs has been examined by Johnson *et al.* (1964) who have determined the limiting values of the yield stress ratio for the deformation to be restricted to either the wedge or the flat.

(d) Conical indenters

Problems of axi-symmetrical plastic flow cannot, in general, be solved by the method of characteristics (slip lines) as in plane strain. However Shield (1955) has shown that, in certain cases, for material which flows plastically according to the Tresca criterion, a slip-line field can be constructed which specifies the state of stress. Such a field must be consistent with an associated velocity field of axi-symmetrical deformation. As an example Shield found the stresses in a rigid-plastic semi-infinite solid under the action of a frictionless, flat-ended, cylindrical punch. Following Shield's method, Lockett (1963) was able to construct the fields due to a rigid frictionless cone penetrating a flat

---

† An estimate of the *average* friction force necessary to produce the adhesive mode may be obtained by applying the principle of virtual work to the frictionless mode with friction forces introduced at the interface. The calculation gives the critical coefficient of friction as

$$\mu_c \approx 1 - (p_m)_s/(p_m)_a$$

where $(p_m)_s$ and $(p_m)_a$ are the values of p_m in the frictionless and adhesive modes respectively. It is apparent from Fig. 6.9 that the value of μ_c is generally small.

surface, provided the semi-cone-angle α exceeded $52.5°$. The slip-line field is similar to that for a two-dimensional wedge shown in Fig. 6.8, but the slip lines and the profile of the deformed surface are no longer straight. The pressure on the surface of the cone is not uniform but rises to a peak at the apex. The limiting case, where $\alpha = 90°$, corresponds to the cylindrical punch studied by Shield. The pressure distributions and the mean indentation pressures are shown in Fig. 6.10. The mean pressure is slightly higher than that of a frictionless wedge of the same angle.

Indentation by a cylindrical punch in which the material does not slip relative to the flat face of the punch has been analysed by Eason & Shield (1960). A cone of undeforming material adheres to the surface of the punch. The mean contact pressure is $6.05k$ compared with $5.69k$ for a frictionless punch. Strictly speaking the distribution of stress in the cone of material adhering to the punch is indeterminate, but an indication may be obtained by continuing the slip-line field into this region, which results in the pressure distribution on the punch face shown by the broken line in Fig. 6.10. From the ratio of shear stress to normal pressure it transpires that a coefficient of friction >0.14 is required to prevent slip.

Fig. 6.10. Indentation of a rigid-plastic half-space by a rigid cone. Solid line – smooth, Lockett (1963); broken line – adhesive, Eason & Shield (1960).

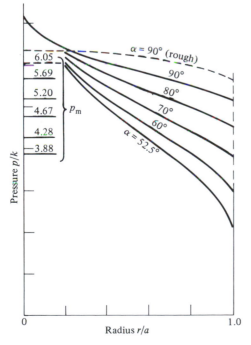

(e) Curved indenters

When an indenter having a convex profile is pressed into a surface geometrical similarity is not maintained. The intensity of strain increases with increasing penetration. The slip-line field is curvilinear and difficulties of analysis arise with the changing shape of the free surface. By assuming that the free surface remains flat, Ishlinsky (1944) constructed a field for the indentation of a rigid-plastic half-space by a rigid frictionless sphere.† In an indentation for which the ratio of contact radius to sphere radius $a/R = 0.376$, the mean indentation pressure was found to be $5.32k$.

An exact solution has been found by Richmond *et al.* (1974) for indentation of a perfectly-plastic half-space by a rigid sphere with no slip at the interface. The mean contact pressure is found to be almost independent of the penetration, varying from $6.04k$ when $a/R = 0.07$ to $5.91k$ when $a/R = 0.30$ (cf. $6.05k$ for the flat punch). This result is not surprising when it is remembered that the indenter is covered by a nose of undeforming material; the profile of the indenter can then only influence the contact pressure through small changes in the profile of the free surface outside the contact.

In the slip-line fields considered in this section the boundary which separates the plastically deforming region from the region below it should not be confused with the elastic-plastic boundary which exists in any real material having some elasticity. Rigid-plastic theory is not able to locate the elastic-plastic boundary, but a rough estimate of its position in a plane-strain indentation may be obtained from the elastic stress distribution under a uniform pressure acting on a half-space given by equations (2.30) *et seq*. Contours of constant principal shear stress τ_1 are circles through the edges of the contact as shown in Fig. 2.7(a). Taking the indentation pressure for a blunt wedge to be $5.1k$, τ_1 will have the value k along the circular contour of radius $1.6a$. If this circle is taken as a first approximation to the elastic-plastic boundary, the plastic zone extends to a depth of $2.9a$ beneath the point of first contact. This is well below the boundaries of the deforming region shown in Figs. 6.2, 6.3, 6.4 or 6.5. Another simple approach to finding the position of the elastic-plastic boundary is given in §3.

A method for extending the slip-line field into the rigid region has been described by Bishop (1953). He has extended the field around a two-dimensional flat-ended punch (wedge angle $\alpha = 90°$) until a boundary of zero stress is reached. Provided these boundaries lie wholly within the actual free surfaces of the solid body which is being indented, it can be concluded that the rigid

---

† It has been shown subsequently that the slip-line field proposed by Ishlinsky is not compatible with the associated velocity field, since there are some regions in which the plastic work is non-positive. His result must be regarded, therefore, as approximate. No exact solutions have been found for *frictionless* curved indenters.

region can support the stresses given by the slip-line field in the deforming region without plastic failure occurring elsewhere. From this construction the minimum dimensions of a rectangular block can be found to ensure that the indentation is not influenced by the size of the block. The minimum depth is $8.8a$ and the minimum width from the centre-line of the impression to the side of the block is $8.7a$, where a is the half-width of the punch. Since the flat punch is the most severe case, a block of the above dimensions would contain the indentation made by any other profile. For further discussion of this question see Hill (1950*b*).

An extension of the field under a flat-ended cylindrical punch by Shield (1955) shows that the indented solid should have a minimum depth of $3.4a$ and a minimum radius from the axis of the indenter of $3.2a$.

Strain hardening, discussed in more detail in §3, has the effect of pushing the elastic-plastic boundary, where the yield stress is lower than in the more severely strained region close to the indenter, further into the solid than perfectly plastic theory would predict. Thus a block of strain-hardening material should be somewhat larger than the critical dimensions given above to ensure that the impression is not influenced by the size of the block (see experiments by Dugdale, 1953, 1954).

6.3   Elastic-plastic indentation

The elasticity of real materials plays an important part in the plastic indentation process. When the yield point is first exceeded the plastic zone is small and fully contained by material which remains elastic so that the plastic strains are of the same order of magnitude as the surrounding elastic strains. In these circumstances the material displaced by the indenter is accommodated by an elastic expansion of the surrounding solid. As the indentation becomes more severe, either by increasing the load on a curved indenter or by using a more acute-angled wedge or cone, an increasing pressure is required beneath the indenter to produce the necessary expansion. Eventually the plastic zone breaks out to the free surface and the displaced material is free to escape by plastic flow to the sides of the indenter. This is the 'uncontained' mode of deformation analysed by the theory of rigid-plastic solids in the previous section. We would expect the plastic zone to break out to the surface and the uncontained mode to become possible when the pressure beneath the indenter reaches the value given by rigid-plastic theory. From the results of the previous section we can write this pressure:

$$p_{\mathrm{m}} = cY \qquad (6.29)$$

where c has a value about 3.0 depending on the geometry of the indenter and friction at the interface. From the results of §6.1, first yield is also given by equation (6.29), where the constant c has a value about unity. There is a transitional range of contact pressures, lying between Y and $3Y$, where the plastic flow is contained by elastic material and the mode of deformation is one of roughly radial expansion. The three ranges of loading: purely elastic, elastic-plastic (contained) and fully plastic (uncontained) are a common feature of most engineering structures.

The deformation of an elastic-perfectly-plastic material is governed by the stress–strain relations of Reuss (see Hill, 1950a or Ford & Alexander, 1963). In principle the contact stresses due to an elastic-plastic indentation in which the strains remain small can be calculated. In practice this is very difficult because the shape and size of the elastic-plastic boundary is not known *a priori*. The technique whereby the solid continuum is replaced by a mesh of 'finite elements' shows promise, but the high stress concentration makes it difficult to obtain a refined picture of the stress field in the contact zone. First studies by the Finite Element method of the two-dimensional stresses beneath a cylindrical indenter were made by Akyuz & Merwin (1968). More complete computations for the indentation of an elastic-perfectly-plastic half-space by a cylinder and sphere have been presented by Hardy *et al.* (1971), Dumas & Baronet (1971), Lee *et al.* (1972) and Skalski (1979). They follow the development of the plastic zone as the load is increased and more elements of the mesh reach the elastic limit. Computational difficulties arose when the fully plastic state was approached and the calculations were restricted to a load $P < 100P_Y$, where P_Y is the load for first yield. These difficulties have been overcome by Follansbee & Sinclair (1984) for ball indentation of a strain-hardening solid well into the fully plastic state.

The pressure distributions found by Hardy *et al.* are shown in Fig. 6.11. As expected, plastic flow leads to a flattening of the pressure distribution and at high loads may peak slightly towards the edge. The development of the plastic zone is also shown. It roughly follows the contours of J_2 (defined by eq. (6.1)) and, for the range of loads investigated, is almost completely contained beneath the contact area.

An alternative approach to the analysis of an elastic-plastic indentation, which avoids the numerical complexities of finite elements, follows an early suggestion of Bishop, Hill & Mott (1945), which was developed by Marsh (1964) and Johnson (1970a). It is based on the observations of Samuels & Mulhearn (1956) and Mulhearn (1959) that the subsurface displacements produced by any blunt indenter (cone, sphere or pyramid) are approximately radial from the point of first contact, with roughly hemi-spherical contours of equal strain (Fig. 6.12).

In this simplified model of an elastic-plastic indentation we think of the contact surface of the indenter being encased in a hemi-spherical 'core' of radius a (Fig. 6.13). Within the core there is assumed to be a hydrostatic component of stress $\bar{p}$. Outside the core it is assumed that the stresses and displacements have radial symmetry and are the same as in an infinite elastic, perfectly-plastic body which contains a spherical cavity under a pressure $\bar{p}$. The elastic-plastic boundary lies at a radius c, where $c > a$. At the interface between core and the plastic zone (*a*) the hydrostatic stress in the core is just equal to the radial component of stress in the external zone, and (*b*) the radial

Fig. 6.11. Indentation of an elastic-plastic half-space by a rigid sphere, Hardy *et al.* (1971): development of the plastic zone. Broken line – contours of J_2.

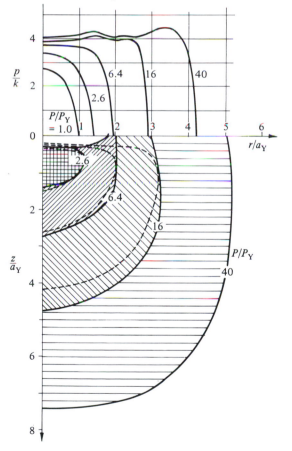

displacement of particles lying on the boundary $r = a$ during an increment of penetration dh must accommodate the volume of material displaced by the indenter (neglecting compressibility of the core).

The stresses in the plastic zone $a \leqslant r \leqslant c$ are given by (Hill, 1950a, p. 99):

$$\sigma_r/Y = -2 \ln (c/r) - 2/3 \qquad (6.30a)$$

$$\sigma_\theta/Y = -2 \ln (c/r) + 1/3 \qquad (6.30b)$$

In the elastic zone $r \geqslant c$

$$\sigma_r/Y = -\tfrac{2}{3}(c/r)^3, \qquad \sigma_\theta/Y = \tfrac{1}{3}(c/r)^3 \qquad (6.31)$$

Fig. 6.12. Experimental contours of plastic strain produced by (*a*) ball indentation ($a/R = 0.51$) and (*b*) Vicker's hardness pyramid indenter.

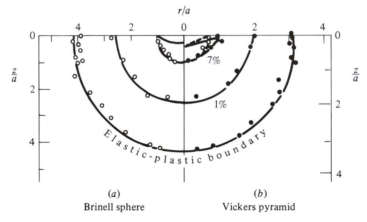

(*a*)
Brinell sphere

(*b*)
Vickers pyramid

Fig. 6.13. Cavity model of an elastic-plastic indentation by a cone.

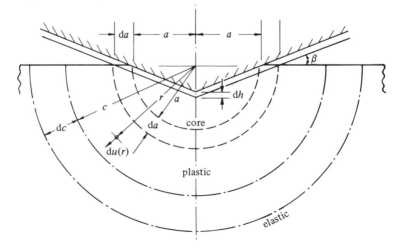

At the boundary of the core, the core pressure is given by

$$\bar{p}/Y = -[\sigma_r/Y]_{r=a} = 2/3 + 2 \ln (c/a) \tag{6.32}$$

The radial displacements are given by (Hill, 1950, p. 101)

$$\frac{du(r)}{dc} = \frac{Y}{E} \{3(1-v)(c^2/r^2) - 2(1-2v)(r/c)\} \tag{6.33}$$

Conservation of volume of the core requires

$$2\pi a^2 \, du(a) = \pi a^2 \, dh = \pi a^2 \tan \beta \, da \tag{6.34}$$

where β is the inclination of the face of the cone to the surface ($\beta = \pi/2 - \alpha$). If we put $r = a$ in (6.33) and note that for a conical indenter geometrical similarity of the strain field with continued penetration requires that $dc/da = c/a = $ constant, then equations (6.33) and (6.34) locate the elastic-plastic boundary by

$$E \tan \beta/Y = 6(1-v)(c/a)^3 - 4(1-2v) \tag{6.35}$$

Substitution for (c/a) in (6.32) gives the pressure in the core. For an incompressible material a simple expression is obtained:

$$\frac{\bar{p}}{Y} = \frac{2}{3} \left\{ 1 + \ln \left(\frac{1}{3} \frac{E \tan \beta}{Y} \right) \right\} \tag{6.36}$$

Of course the stress in the material immediately below an indenter is not purely hydrostatic. If $\bar{p}$ denotes the hydrostatic component, the normal stress will have a value $\sigma_z \approx -(\bar{p} + 2Y/3)$ and the radial stress $\sigma_r \approx -(\bar{p} - Y/3)$. A best estimate of the indentation pressure p_m for the spherical cavity model would therefore be $\bar{p} + 2Y/3$. A similar analysis may be made for two-dimensional indentation by a rigid wedge (Johnson, 1970a).

It appears from equation (6.36) that the pressure in the hydrostatic core beneath the indenter is a function of the single non-dimensional variable $E \tan \beta/Y$, which may be interpreted as the ratio of the strain imposed by the indenter ($\tan \beta$) to the elastic strain capacity of the material (Y/E). Elasticity of the indenter can be taken into account by replacing E by E^* (as defined in §4.2(a)).

The indentation pressure under elastic, elastic-plastic and fully plastic conditions may be correlated on a non-dimensional graph of p_m/Y as a function of ($E^* \tan \beta/Y$) where β is the (small) angle of the indenter at the edge of the contact (Fig. 6.14). With a spherical indenter we put $\tan \beta \approx \sin \beta = a/R$ which varies during the indentation. Integration of equations (6.33) and (6.34), with $c/a = 1$ at the point of first yield ($p_m = 1.1Y$), leads to equation (6.36) with an additional constant (≈ 0.19) on the right-hand side. For a Vickers diamond pyramid, β is taken to be the angle of the cone (19.7°) which displaces the same volume. First yield for a spherical indenter occurs at $p_m \approx 1.1Y$ and for a conical indenter at $p_m \approx 0.5Y$. Fully plastic deformation sets an upper limit of $\sim 3Y$ for the indentation pressure which is reached at a value of $(E^* \tan \beta/Y) \approx 30$ for a cone and $(E^*a/YR) \approx 40$ for a sphere.

The discussion above has been restricted to elastic-perfectly-plastic solids having a constant yield stress Y in simple compression. Tabor (1951) has shown that the results for a perfectly plastic solid may be applied with good approximation to a strain-hardening solid if Y is replaced by a 'representative' flow stress Y_R, measured in simple compression at a representative strain ϵ_R, where

$$\epsilon_R \approx 0.2 \tan \beta \qquad (6.37)$$

For a Vickers hardness pyramid, therefore, $\epsilon_R \approx 0.07$ (Tabor suggested 0.08) and for a spherical indenter, $\epsilon_R \approx 0.2a/R$. Matthews (see §6) has considered materials which strain-harden according to a power law of index n, with the result $\epsilon_R = 0.28(1 + 1/n)^{-n}(a/R)$, which varies slightly with n from $0.17a/R$ to $0.19a/R$. In this way good agreement is found between elastic-perfectly-

Fig. 6.14. Indentation of an elastic-plastic half-space by spheres and cones. Small-dashed line – elastic: A cone, B sphere. Solid line – finite elements: C Hardy *et al.* (1971), D Follansbee & Sinclair (1984). Chain line – cavity model: F cone, G sphere. Large-dashed line – rigid-plastic (sphere), E Richmond *et al.* (1974). Experiments: cross – pyramids, Marsh (1964), circle – spheres, Tabor (1951).

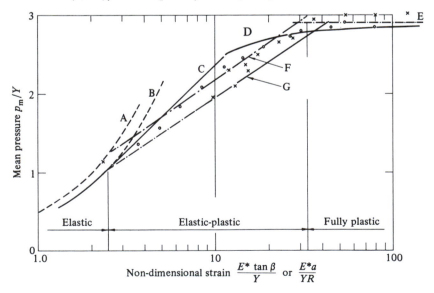

plastic theory and experiments on strain-hardening materials. Thus Fig. 6.14 gives a measure of the mean indentation pressure by an axi-symmetrical indenter of arbitrary profile pressed into any elastic-plastic solid whose stress–strain curve in simple compression is known. In particular it gives a relationship between hardness and the flow stress in simple compression. Thus the Vickers diamond pyramid hardness H_V is given by

$$H_V = 0.93p_m \approx 2.8Y_R \qquad (6.38)$$

where Y_R is the flow stress in simple compression at a strain of about 0.08.

Experimental data for spherical indentation of various materials have been examined by Francis (1976) and been shown to correlate well using the variables defined above.

For a fully plastic spherical indentation $((E^*a/YR) = 170)$, the stresses along the z-axis and close to the surface found by the finite element method (Follansbee & Sinclair, 1984) (1) are compared in Figs. 6.15 and 6.16 with (2) rigid-plastic theory for a rigid punch and (3) the spherical cavity model. The general agreement between the rigid-plastic theory and the finite element analysis theory is good, particularly when it is remembered that the latter was carried out for a strain-hardening material. Adhesion at the face of the punch causes the maximum values of σ_r and σ_z to be located beneath the surface, an effect which is

Fig. 6.15. Plastic indentation by a rigid sphere, subsurface stresses, (A) frictionless, and (B) adhesive. (1) Finite elements, Follansbee & Sinclair (1984). (2) Rigid-plastic punch, Shield (1955), Eason & Shield (1960). (3) Cavity model, eqs. (6.30) and (6.31).

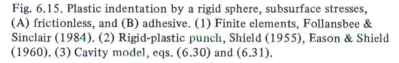

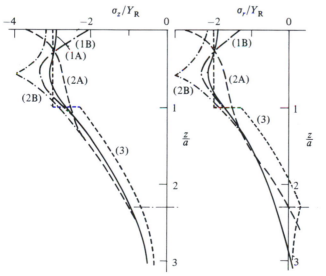

also apparent in the finite elements results. For the spherical cavity model, in the fully plastic state $p_m/Y = 3$; $E^* \tan \beta/Y = 40$ whereupon equation (6.35) gives the elastic-plastic boundary at $c/a \approx 2.3$. The stresses are then given by equations (6.30) and (6.31). Along the z-axis they show the same trend as the finite elements but underestimate their magnitude. On the surface, the cavity model predicts circumferential *tension* and radial *compression* for $r > 2a$. This is the opposite of the elastic state of stress in which the radial stress is tensile. The finite element calculations also show circumferential tension but of rather smaller magnitude. Rigid-plastic theory gives $\sigma_\theta = \sigma_z = 0$ (the Haar–von Kármán condition) as a result of using the Tresca yield criterion. The change from radial tension under purely elastic conditions to circumferential tension under elastic-plastic conditions is largely responsible for the change in the mode of indentation fracture from a ring crack with very brittle materials such as glass to a radial crack in semi-brittle materials such as perspex (Puttick *et al.*, 1977). Porous materials behave differently. They respond to indentation by crushing and the indentation pressure is of order Y, where Y is the crushing strength in uni-axial compression (see Wilsea *et al.*, 1975).

In addition to the contact pressure, the depth of penetration of the indenter is also of interest. It is a difficult quantity to determine theoretically, because of the uncertain 'pile-up' at the edge of the indentation. With a rigid-plastic solid the material displaced by the indenter appears in the piled-up shoulder, but with an elastic-plastic solid this is not the case. Most, if not all, of the displaced material is accommodated by radial expansion of the elastic hinterland.

Fig. 6.16. Plastic indentation by a rigid sphere, surface stresses. Legend as for Fig. 6.15.

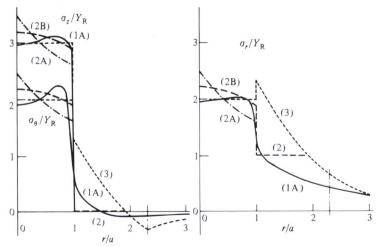

It reappears in an imperceptible increase in the external dimensions of the indented body. Pile-up is also influenced by the strain-hardening properties of the material. A large capacity for strain hardening pushes the plastic zone further into the material and thereby decreases the pile-up adjacent to the indenter.

A plot of penetration δ against load P is usually referred to as a compliance curve. Suitable non-dimensional variables for a spherical indenter are:

$$\delta/\delta_Y \equiv 0.148(\delta E^{*2}/RY^2) \qquad (6.39a)$$

and

$$P/P_Y \equiv 0.043(PE^{*2}/R^2 Y^3) \qquad (6.39b)$$

where the load at first yield P_Y is given by equation (6.10) and the corresponding displacement δ_Y is related to P_Y by the elastic equation (4.23). The fully plastic condition is reached when $E^*a/RY \approx 40$, i.e. when $P/P_Y \approx 400$. If, in the fully plastic regime, it is assumed that the edges of the impression neither pile up nor sink in, then the penetration is given approximately by†

$$\delta = a^2/2R \qquad (6.40)$$

Taking the fully plastic contact pressure to be $3.0Y$ and constant, this gives

$$P/P_Y = 0.81(\delta E^{*2}/RY^2) = 5.5(\delta/\delta_Y) \qquad (6.41)$$

Accurate measurements of penetration are not so easy to obtain as those of contact pressure. Measurements by Foss & Brumfield (1922) of the penetration under load and the depth of the residual crater are plotted non-dimensionally‡ in Fig. 6.17. On the basis of the non-dimensional variables (6.39) the results for materials of different hardness and elastic modulus lie on a common curve which approaches the elastic line at light loads and equation (6.41) in the fully plastic regime.

6.4 Unloading of a plastic indentation, cyclic loading and residual stresses

The contact of *elastic* solids loaded by a normal force, discussed in detail in Chapters 4 and 5, is generally regarded as a reversible process. The stresses and deformation produced by a specific contact load are then independent of the history of loading. Small departures from perfect reversibility, however, can arise in two ways; by slip and friction at the contact interface, or by internal hysteresis of the materials under the action of cyclic stress.

If the materials of two contacting bodies are dissimilar, we have seen in §5.4 that some slip occurs at the periphery of the contact area. During unloading

---

† An expression for δ which takes account of strain hardening is given by equation (6.82).

‡ Since Foss & Brumfield quote hardness H rather than yield stress of their specimens, Y has been taken as $H/2.8$.

the direction of slip will reverse and the tangential surface traction will differ from that during loading. The contact force to produce a given contact area will be slightly greater during loading than during unloading. In a complete load cycle a small amount of energy is dissipated through interfacial slip. Although precise calculations have not been made it is clear that this energy dissipation is very small. The difference in the bulk stresses between loading and unloading will be negligible although in a situation of rapid cyclic loading the interfacial slip itself and the heat generated may be responsible for progressive surface damage.

Fig. 6.17. Penetration of a spherical indenter into an elastic-plastic half-space. Solid line – penetration under load. Broken line – depth of unloaded crater.

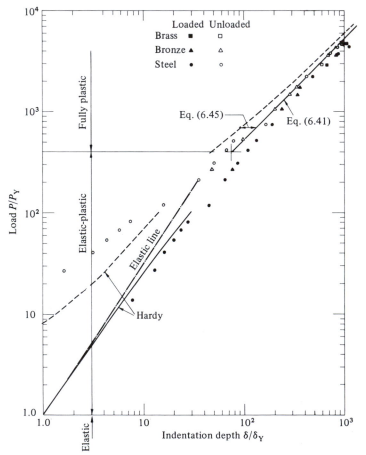

Real materials, even metals below their yield point, are not perfectly elastic, but exhibit some hysteresis during a cycle of stress. Such 'elastic hysteresis' or 'internal damping' gives rise to a slight irreversibility in a contact stress cycle. Provided that the departure from perfect elasticity is small the effect upon the overall distribution of contact stress will be small. An estimate of the energy dissipated in one cycle of the load may be made. The usual way of expressing the internal hysteresis of a material is by the ratio of the energy dissipated per cycle ΔW to the maximum elastic strain energy in the cycle W. This ratio $\alpha(= \Delta W/W)$ is known as the *hysteresis-loss factor* or the *specific damping capacity*. Representative values for a wide variety of materials are quoted by Lazan (1968). The damping capacity of most materials depends upon the amplitude of cyclic strain; the values tend to rise as the elastic limit is approached, but are roughly constant at low and moderate strains.

The elastic strain energy when two bodies are in contact can be calculated from the relationship between load and compression, thus

$$W = \int P \, d\delta$$

For two spherical surfaces the relationship between P and δ is given by equation (4.23) for which

$$W = \tfrac{2}{5} \left(\frac{9E^{*2}P^5}{16R} \right)^{1/3} \tag{6.42}$$

The energy loss in a load cycle from zero to a maximum and back to zero is then given approximately by

$$\Delta W = \alpha W$$

where α is a representative value of the hysteresis-loss factor. Direct measurements by the author of the energy loss in the cyclic normal contact of spheres over a range of loads yielded a fairly constant value of α equal to 0.4% for a hard bearing steel. This value is not inconsistent with internal hysteresis measurements at high stress.

When the initial loading takes the material well into the plastic range the above approach is no longer appropriate, since the differences between loading and unloading will no longer be small. Even though large plastic deformations occur during loading, however, it is intuitive to expect the unloading process to be perfectly elastic. A simple check of this hypothesis was carried out by Tabor (1948) from observations of the permanent indentations made by a hard steel ball of radius R in the flat surface of a softer metal (see Fig. 6.18). The indentation under load has a radius R', which is slightly greater than R due to elastic compression of the ball (Fig. 6.18(b)). When the load is removed the plastic indentation shallows to some extent due to elastic recovery, so that its permanent radius ρ is slightly greater than R' (Fig. 6.18(c)). If the unloading process is elastic, and hence reversible, a second loading of the plastic indentation will follow the elastic process in which a ball of radius R is pressed into contact with a concave spherical cup of radius ρ. Provided that the indentation is not too deep, so that the assumptions of the Hertz theory still apply, the permanent radius of the indentation can be related to the radius of the ball by equation (4.22). Remembering that ρ, being concave, is negative,

$$4a^3\left(\frac{1}{R}-\frac{1}{\rho}\right) = 3P/E^* \tag{6.43}$$

Tabor's measurements of ρ were consistent with this equation to the accuracy of the observations.

The elastic deflexion which is recovered when the load is removed can be estimated in the same way. By eliminating R from the elastic equations (4.22) and (4.23), the elastic deflexion δ' can be expressed in terms of the mean contact pressure p_m by

$$\delta'^2 = \frac{9\pi}{16}\frac{Pp_m}{E^{*2}} \tag{6.44}$$

In the fully plastic state $p_m \approx 3.0Y$, so that equation (6.44) can then be written, in terms of the non-dimensional variables of equation (6.41), as

$$P/P_Y = 8.1 \times 10^{-3}(\delta'E^{*2}/RY^2)^2 = 0.38(\delta'/\delta_Y)^2 \tag{6.45}$$

The residual depth of the indentation after the load is removed is therefore $(\delta - \delta')$. In the fully plastic range it may be estimated from equations (6.41) and (6.45) resulting in the line at the right-hand side of Fig. 6.17. The residual depth calculated by the finite element analysis is plotted at the left-hand side of Fig. 6.17. It appears that the elastic recovery given by equation (6.45) is in good agreement with Foss & Brumfield's measurements.

Fig. 6.18. Unloading a spherical indenter. (a) before loading, (b) under load, (c) after unloading.

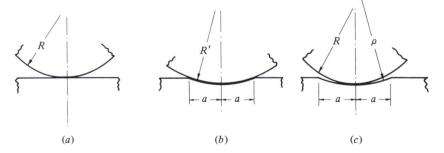

(a) (b) (c)

A similar investigation of the unloading of conical indenters (Stilwell & Tabor, 1961)) showed that the shallowing of the indentation could be ascribed to elastic recovery and calculated from the elastic theory of cone indentation (§5.2).

This treatment of the unloading process is only approximate, however, since it tacitly assumes that the pressure distribution before unloading is Hertzian and hence that the recovered profile is a circular arc. The actual pressure distribution is flatter than that of Hertz as shown in Fig. 6.11. This pressure distribution, when released by unloading, will give rise to an impression whose profile is not exactly circular, but whose shape is related to the pressure by the elastic displacement equations. Hence, accurate measurement of the recovered profile enables the actual pressure distribution before unloading to be deduced. This has been done by Johnson (1968b) for copper spheres and cylinders and by Hirst & Howse (1969) for perspex indented by a hard metal wedge.

After unloading from a plastically deformed state the solid is left in a state of residual stress. To find the residual stress it is first necessary to know the stresses at the end of the plastic loading. Then, assuming unloading to be elastic, the residual stresses can be found by superposing the elastic stress system due to a distribution of surface normal traction equal and opposite to the distribution of contact pressure. The contact surface is left free of traction and the internal residual system is self-equilibrating. Such calculations have been made in detail by the finite element method (Hardy *et al.*, 1971; Follansbee & Sinclair, 1984). They show that the material beneath the indenter is left in a state of residual compression and the surface outside the impression contains radial compression and circumferential tension.

The residual stresses left in the solid after a fully plastic indentation may be estimated using the slip-line field solutions or the spherical cavity model. A rough idea what to expect can be gained by simple reasoning: during a plastic indentation the material beneath the indenter experiences permanent compression in the direction perpendicular to the surface and radial expansion parallel to the surface. During the recovery, the stress normal to the surface is relieved, but the permanent radial expansion of the plastically deformed material induces a radial compressive stress exerted by the surrounding elastic material. The shot-peening process, which peppers a metal surface with a large number of plastic indentations, gives rise to a residual bi-axial compressive stress, acting parallel to the surface, whose intensity is greatest in the layers just beneath the surface. The aim of the process is to use the residual compression in the surface layers to inhibit the propagation of fatigue cracks.

Along the axis of symmetry, during plastic loading

$$|\sigma_z - \sigma_r| = Y \tag{6.46}$$

During *elastic* unloading,

$$|\sigma_r - \sigma_z| = Kp_m = KcY \tag{6.47}$$

where K depends upon the pressure distribution at the end of loading and upon the depth below the surface. The residual stress difference is then given by the superposition of (6.46) and (6.47), i.e.

$$|\sigma_r - \sigma_z|_r = (Kc - 1)Y \tag{6.48}$$

Since $(\sigma_z)_r$ is zero at the surface, its value beneath the surface is likely to be small compared with $(\sigma_r)_r$. In a fully plastic indentation $c \approx 3.0$ and the pressure distribution is approximately uniform which, by equation (3.33), gives $K = 0.65$ at $z = 0.64a$. Hence $(Kc - 1) \approx 0.95$ so that, from equation (6.48), reversed yielding on unloading is not to be expected except as a consequence of the Bauchinger effect. Even if some reversed yielding does take place it will be fully contained and its influence on the surface profile will be imperceptible.

At the contact surface the situation is different, taking the pressure p to be uniform, $\sigma_z = -p$, $\sigma_r = \sigma_\theta = -(p - Y)$. Elastic unloading superposes stresses $\sigma_z = p$, $\sigma_r = \sigma_\theta = \frac{1}{2}(1 + 2\nu)p \approx 0.8p$, leaving residual stresses

$$(\sigma_z)_r = 0, \quad (\sigma_r)_r = (\sigma_\theta)_r = Y - 0.2p \tag{6.49}$$

Putting $p = 3Y$ for a fully plastic indentation gives $(\sigma_r)_r = (\sigma_\theta)_r \approx 0.4Y$ which is *tensile*.

At the surface in the plastic zone outside the contact area the stresses due to loading are given approximately by the cavity model (eqs. (6.30) and (6.31)). The radial stress is compressive and the circumferential stress though small is tensile. Elastic unloading would add

$$\sigma_r = -\sigma_\theta = -\tfrac{1}{2}(1 - 2\nu)p_m a^2/r^2$$

but additional radial compression and circumferential tension are not possible. Instead slight additional plastic deformation will occur whilst the stresses remain roughly constant. Subsequent loading and unloading will then be entirely elastic.

6.5   Linear viscoelastic materials

Many materials, notably polymers, exhibit time-dependent behaviour in their relationships between stress and strain which is described as viscoelastic. The common features of viscoelastic behaviour are illustrated in Fig. 6.19 which shows the variation of strain $\epsilon(t)$ in a specimen of material under the action of a constant stress σ_0 applied for a period t_1. The strain shows an *initial elastic response* OA to the applied stress; a further *delayed elastic strain* AB is acquired in time. If the material is capable of flow or creep it will also acquire a steadily increasing *creep strain* BC. When the stress is removed there is an immediate elastic response CD $(= -OA)$ and a delayed elastic response DE. The specimen is left with a permanent strain at E which it acquired through the action of creep.

This material behaviour can be incorporated into a rigorous theory of contact stresses provided that the viscoelastic stress–strain relationships of the material can be taken to be *linear*. For this requirement to be met, the strains must remain small (as in the linear theory of elasticity) and the principle of super-position must apply. Thus, for linearity, an increase in the stress in Fig. 6.19 by a constant factor must produce an increase in the strain response by the same factor; further, the strain response to different stress histories, acting simultaneously, must be identical to the sum of the responses to the stress histories applied separately. The stress–strain relations for a linear viscoelastic material can be expressed in various ways, but the most common is to make use of the *creep compliance function* which expresses the strain response to a step change in stress or, alternatively, the *relaxation function* which expresses the stress response to a step change in strain. An isotropic material in a state of complex stress requires two independent functions to express its response to shear and volumetric deformation respectively. These functions correspond to the shear modulus and bulk modulus of purely elastic solids. For simplicity, in what follows, we shall restrict our discussion to an incompressible material so that its stress–strain relations can be expressed in terms of a single function describing its behaviour in shear.† The approximation is a reasonable one for polymers, whose values of Poisson's ratio usually exceed 0.4. It is convenient to write the stress–strain relations in terms of the deviatoric stress components $s = (\sigma - \bar{\sigma})$ and the deviatoric strains $e = (\epsilon - \bar{\epsilon})$ where $\bar{\sigma} = \frac{1}{3}(\sigma_1 + \sigma_2 + \sigma_3)$ and

---

† An alternative simplification is to assume that Poisson's ratio remains constant with time in which case the relaxation functions in response to volumetric and shear deformations are in the fixed ratio of $2(1 + v)/3(1 - 2v)$.

Fig. 6.19

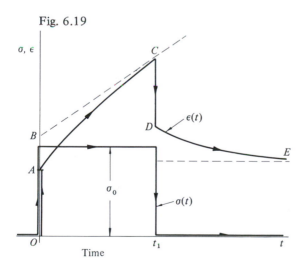

$\bar{\epsilon} = \frac{1}{3}(\epsilon_1 + \epsilon_2 + \epsilon_3)$. For an incompressible elastic solid $\bar{\epsilon} = 0$, so that

$$s = 2Ge = 2G\epsilon \tag{6.50}$$

where G is the shear modulus. The corresponding relationship for an incompressible viscoelastic material may be written as either

$$s(t) = \int_0^t \Psi(t - t') \frac{\partial e(t')}{\partial t'} \, dt' \tag{6.51}$$

or

$$e(t) = \int_0^t \Phi(t - t') \frac{\partial s(t')}{\partial t'} \, dt' \tag{6.52}$$

The function $\Psi(t)$ is the relaxation function, which specifies the stress response to a unit step change of strain; the function $\Phi(t)$ is the creep compliance, which specifies the strain response to a unit step change in stress. For particular materials, they may be deduced from appropriate spring and dashpot models or obtained by experiment (see Lee & Rogers, 1963). Equation (6.51), expressed in terms of the relaxation function $\Psi(t)$, can be regarded as the superposition of the stress responses to a sequence of small changes of strain $de(t')$ at times t'. Similarly equation (6.52) expresses the total strain response to a sequence of step changes in stress.

By way of example we shall make use of two idealised viscoelastic materials which demonstrate separately the effects of delayed elasticity and steady creep. The first material is represented in Fig. 6.20(a) by two springs of modulus g_1 and g_2 together with a dashpot of viscosity η connected as shown. For this material the creep response to a step change is stress s_0 is given by

$$e(t) = \Phi(t)s_0 = \left[\frac{1}{g_1} + \frac{1}{g_2} \{1 - \exp(-t/T_1)\} \right] s_0 \tag{6.53}$$

where $T_1 = \eta/g_2$. The response to a step change of strain e_0 is given by

$$s(t) = \Psi(t)e_0 = \frac{g_1}{g_1 + g_2} \{g_2 + g_1 \exp(-t/T_2)\} e_0 \tag{6.54}$$

where $T_2 = \eta/(g_1 + g_2)$.

The second material - a Maxwell body - is represented in Fig. 6.20(b) by a spring of modulus g in series with a dashpot of viscosity η. The creep response is

$$e(t) = \Phi(t)s_0 = \left\{ \frac{1}{g} + \frac{1}{\eta} t \right\} s_0 \tag{6.55}$$

and the relaxation response is

$$s(t) = \Psi(t)e_0 = ge^{-t/T}e_0 \tag{6.56}$$

where $T = \eta/g$ is the relaxation time of the material.

The first material exhibits delayed elasticity but the ultimate strain is limited to a finite value. The second material shows a steady creep under constant stress, so that the strains increase continuously with time. Of course this model is only valid during time intervals when the strains remain small. For the Maxwell model to be representative of a 'solid' rather than a 'fluid', the viscosity η must be large, comparable in magnitude with the modulus of elasticity g. The simplest model of a material which exhibits both delayed elasticity and steady creep is made up of four elements, by adding a second dashpot in series with the model shown in Fig. 6.20(a).

We shall now examine the behaviour when a rigid spherical indenter is pressed into contact with a viscoelastic solid. Under the action of a constant normal force the penetration of the indenter and the contact area will both grow with time and the distribution of contact pressure will change. In principle we wish to find, for a given material, the variation with time of the contact area and pressure distribution resulting from any prescribed programme of loading or penetration. The simplest approach to this problem follows a suggestion by Radok (1957) for

Fig. 6.20. Simple viscoelastic materials which display (*a*) delayed elasticity, (*b*) steady creep (Maxwell).

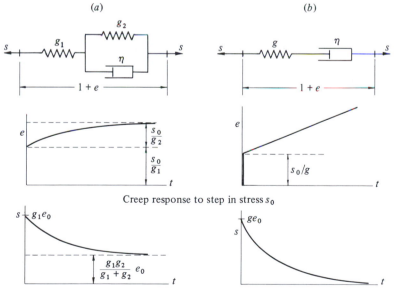

Creep response to step in stress s_0

Stress relaxation due to step in strain e_0

finding the stresses and deformations in cases where the corresponding solution for a purely elastic material is known.

It consists of replacing the elastic constant in the elastic solution by the corresponding integral operator from the viscoelastic stress–strain relations. If the deformation history is known, the stresses are found by replacing $2G$ in the elastic solution by the integral operator, expressed in terms of the relaxation function $\Psi(t)$ (eq. (6.51)). On the other hand, if the load or stress history is known, the variation in deformation is found from the elastic solution by replacing the constant $(1/2G)$ by the integral operator involving the creep compliance $\Phi(t)$ (eq. (6.52)). Lee & Radok (1960) show that this approach can be applied to the contact problem provided that the loading programme is such that the contact area is increasing throughout.

When two purely elastic spherical bodies are pressed into contact by a force P, the radius of the contact circle a, the penetration δ and the contact pressure p are given by equations (4.22), (4.23) and (4.24). If one sphere is rigid and the other is incompressible with a shear modulus G, these expressions for a and δ may be written:

$$a^3 = (R\delta)^{3/2} = \tfrac{3}{8}\left(\frac{1}{2G}\right)RP \tag{6.57}$$

and for the pressure distribution:

$$p(r) = \frac{4}{\pi R}\,2G(a^2 - r^2)^{1/2} \tag{6.58}$$

where $1/R$ is the relative curvature of the two surfaces $(1/R_1 + 1/R_2)$. When the material is viscoelastic a and p vary with time so, following Radok's suggestion, we rewrite equation (6.58) for the pressure, replacing $2G$ by the relaxation operator for the material, thus for $r < a(t')$

$$p(r, t) = \frac{4}{\pi R}\int_0^t \Psi(t - t')\,\frac{\mathrm{d}}{\mathrm{d}t'}\,\{a^2(t') - r^2\}^{1/2}\,\mathrm{d}t' \tag{6.59}$$

Similarly the contact force is given by

$$P(t) = \frac{8}{3R}\int_0^t \Psi(t - t')\,\frac{\mathrm{d}}{\mathrm{d}t'}\,a^3(t')\,\mathrm{d}t' \tag{6.60}$$

If the variation of penetration $\delta(t)$ is prescribed, then the variation in contact radius $a(t)$ is given directly by the first of (6.57), i.e.

$$a^2(t) = R\delta(t)$$

which can be substituted in (6.59) to find the variation in pressure distribution. It is more common however, for the load variation $P(t)$ to be prescribed. In this case we replace $(1/2G)$ in (6.57) by the creep compliance operator to obtain

$$a^3(t) = \tfrac{3}{8}R \int_0^t \Phi(t-t')\,\frac{d}{dt'}\,P(t')\,dt' \tag{6.61}$$

The integral form of equations (6.59) and (6.61) may be interpreted as the linear superposition of small changes in $p(r)$ brought about by a sequence of infinitesimal step changes in δ or P. Lee & Radok (1960) show that the pressure distribution given in equation (6.59) produces normal displacements of the surface of the solid $\bar{u}_z(r,t)$ which conform to the profile of the rigid sphere within the contact area $(r \leqslant a)$ at all times, i.e.

$$\bar{u}_z(r,t) = \delta(t) - r^2/2R$$

for all t.

Some of the significant features of viscoelastic contact will now be illustrated by applying the above method of analysis to two particular cases: the response of each of the two idealised viscoelastic materials shown in Fig. 6.20 to a step load applied to a rigid sphere. Thus the variation in load is prescribed:

$$P(t) = 0, \quad t < 0; \quad P(t) = P_0, \quad t > 0.$$

For a single step, equation (6.61) becomes

$$a^3(t) = \tfrac{3}{8}RP_0\Phi(t), \quad t > 0 \tag{6.62}$$

(a) Material with delayed elasticity

The material characterised in Fig. 6.20(a) has the creep compliance $\Phi(t)$ given by equation (6.53), which can be substituted in equation (6.62) to give the variation in contact radius

$$a^3(t) = \tfrac{3}{8}RP_0\left\{\frac{1}{g_1} + \frac{1}{g_2}(1 - e^{-t/T})\right\} \tag{6.63}$$

Immediately the load is applied there is an instantaneous elastic response to give a contact radius $a_0 = (3RP_0/8g_1)^{1/3}$. The contact size then grows with time as shown by curve A in Fig. 6.21, and eventually approaches

$$a_1 = \{3RP_0(1/g_1 + 1/g_2)/8\}^{1/3}.$$

Initially the contact pressure follows the elastic distribution of the Hertz theory given by (6.58) with $2G = g_1$ and $a = a_0$. Finally the pressure distribution again approaches the elastic form with $2G = g_1g_2/(g_1 + g_2)$ and $a = a_1$. At intermediate times the pressure distribution can be found by substituting the relaxation function $\Psi(t)$ from (6.54) and $a(t)$ from (6.63) into equation (6.59) and performing the integrations. These computations have been carried out by Yang (1966) for $g_2 = g_1$; the results are plotted in Fig. 6.22. It is apparent that the pressure distribution is not very different from the Hertz distribution at any

Fig. 6.21. Growth of contact radius $a(t)$ due to a step load P_{β} applied to a rigid sphere of radius R. (A) Three parameter solid (Fig. 6.20(a)) with $g_1 = g_2 = g$ and $T = \eta/2g$, (B) Maxwell solid with $T = \eta/g$, and (C) viscous solid, $T = 2\eta/g$.

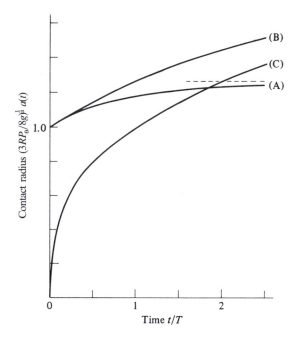

Fig. 6.22. Variation of pressure distribution when a step load is applied to a sphere indenting the 3-parameter solid of Fig. 6.20(A).

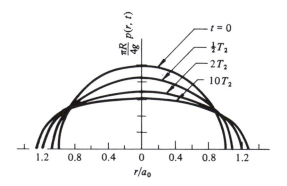

stage in the deformation. The effect of delayed elasticity, therefore, would seem to comprise a growth in the contact from its initial to its final size; the stresses at any instant in this process being distributed approximately according to elastic theory.

(b) Material with steady creep

The simplest material which exhibits steady creep is the Maxwell solid depicted in Fig. 6.20(*b*). The growth of contact size produced by a step load is found by substituting the creep compliance in (6.55) into equation (6.62) with the result

$$a^3(t) = \tfrac{3}{8}RP_0\left(\frac{1}{g} + \frac{1}{\eta}t\right) \qquad (6.64)$$

Once again, the initial elastic deformation will give $a_0 = (3RP_0/8g)^{1/3}$ immediately the load is applied. The value of a will then grow continuously according to equation (6.64), as shown in Fig. 6.21, although it must be remembered that the theory breaks down when a becomes no longer small compared with R. Substituting the relaxation function $\Psi(t) = g\,e^{-t/T}$ from (6.56) together with (6.64) into equation (6.59) enables the variation in pressure distribution to be found. Numerical evaluation of the integral results in the pressure distributions shown in Fig. 6.23. The initial elastic response gives a Hertzian distribution of stress. As the material creeps the pressure distribution changes markedly. The growth of the contact area brings new material into the deformed region which responds elastically, so that, at the periphery of the contact circle, the pressure distribution continues to follow the Hertzian 'elastic' curve. In the centre of the contact the deformation does not change greatly, so that the stress relaxes which results in a region of low contact pressure. Thus, as time progresses, we see that the effect of continuous creep is to change the pressure distribution from the elastic form, in which the maximum pressure is in the centre of the contact area, to one where the pressure is concentrated towards the edge.

(c) Purely viscous material

It is interesting to observe that the phenomenon of concentration of pressure at the edge of contact arises in an extreme form when the material has no initial elastic response. A purely viscous material such as pitch, for example, may be thought of as a Maxwell material (Fig. 6.20(*b*)) in which the elastic modulus g becomes infinitely high. For such a material the stress response to a step change of strain – the relaxation function – involves a theoretically infinite stress exerted for an infinitesimally short interval of time. This difficulty can be avoided by rewriting the viscoelastic stress–strain relations (eq. (6.51) and (6.52)) in terms of differential rather than integral operators. Thus a purely viscous material, with viscosity η (as usually defined), has the stress–strain relationship

$$s(t) = 2\eta \mathbf{D}e(t), \quad \text{where } \mathbf{D} \equiv d/dt. \tag{6.65}$$

Following Radok's method we can now replace G in the elastic solution by the differential operator $\mathbf{D}$. From the elastic equation (6.57) we get

$$a^3 = (R\delta)^{3/2} = \frac{3}{16\eta} \frac{1}{\mathbf{D}} RP$$

For the case of a step load, in which P has a constant value P_0 ($t > 0$),

$$a^3 = (R\delta)^{3/2} = \frac{3}{16} \frac{R}{\eta} P_0 t \tag{6.66}$$

The pressure distribution is obtained by replacing $2G$ by $2\eta\mathbf{D}$ in the elastic equation (6.58)

$$p(r, t) = \frac{8\eta}{\pi R} \mathbf{D}(a^2 - r^2)^{1/2}$$

$$= \frac{P_0}{2\pi a} (a^2 - r^2)^{-1/2} \tag{6.67}$$

Fig. 6.23. Variation of pressure distribution when a step load is applied to a sphere. Solid line – Maxwell fluid, $T = \eta/g$. Chain line – purely viscous fluid, $a^3 = 3a_0^3$. Broken line – elastic solid (Hertz).

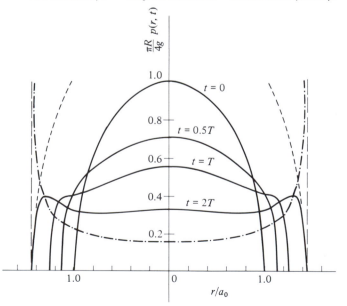

This pressure distribution maintains the same shape as the contact size grows; it rises to a theoretically infinite value at the periphery of the contact circle, as shown in Fig. 6.23. This result is not so surprising if it is remembered that, when the moving boundary of the contact circle passes an element of material in the surface, the element experiences a sudden jump in shear strain, which gives rise to the theoretically infinite stress. Other idealised materials which have no initial elastic response, for example a Kelvin solid (represented by the model in Fig. 6.20(a) in which the spring g_1 is infinitely stiff) would also give rise to an infinite pressure at the edge of the contact. Real solid-like materials, of course, will have sufficient elasticity to impose some limit to the edge pressures.

So far we have considered a rigid sphere indenting a viscoelastic solid: Yang (1966) has investigated the contact of two viscoelastic bodies of arbitrary profile. He shows that the contact region is elliptical and that the eccentricity of the ellipse is determined solely by the profiles of the two surfaces, as in the contact of elastic bodies, i.e. by equation (4.28). Further, the approach of the two bodies at any instant $\delta(t)$ is related to the size of the contact region, $a(t)$ and $b(t)$, at that instant by the elastic equations. When the material of both bodies is viscoelastic, the deformation of each surface varies with time in such a way that each body exerts an identical contact pressure on the other. In these circumstances equations (6.59) and (6.61) still yield the variations of contact pressure and contact size with time, provided that the relaxation and creep compliance functions $\Psi(t)$ and $\Phi(t)$ are taken to refer to a fictitious material whose elements may be thought of as a *series* combination of elements of the two separate materials. This procedure is equivalent to the use of the combined modulus E^* for elastic materials.

The method of analysis used in this section is based upon Radok's technique of replacing the elastic constants in the elastic solution by the corresponding integral or differential operators which appear in the stress–strain relations for linear viscoelastic materials. Unfortunately this simple technique breaks down when the loading history is such as to cause the contact area to *decrease* in size. Lee & Radok (1960) explain the reason for this breakdown. They show that, when their method is applied to the case of a shrinking contact area, negative contact pressures are predicted in the contact area. In reality, of course, the contact area will shrink at a rate which is different from their prediction such that the pressure will remain positive everywhere.

This complication has been studied by Ting (1966, 1968) and Graham (1967), with rather surprising conclusions. If, at time t, the contact size $a(t)$ is decreasing, a time t_1 is identified as that instant previously when the contact size $a(t_1)$ was increasing and equal to $a(t)$. It then transpires that the contact pressure $p(r, t)$

depends only upon the variation of contact size prior to t_1 during which it is less than $a(t)$. Hence equation (6.59) can still be used to find the contact pressure at time t, by making the limits of integration 0 and t_1. Equation (6.61) can be used to obtain $a(t')$, since the contact is increasing in the range $0 \leqslant t' \leqslant t_1$. The penetration $\delta(t)$, on the other hand, exhibits the opposite characteristics. During the period $0 \leqslant t' \leqslant t_1$, whilst $a(t')$ is increasing to $a(t_1)$, the penetration $\delta(t')$ is related to $a(t')$ by the elastic equation (6.57) and is not dependent upon the rate of loading. But when $a(t)$ is decreasing, the penetration $\delta(t)$ depends upon the time history of the variation of contact size during the interval from t_1 to t. For the relationship governing the variation of penetration with time when the contact area is shrinking, the reader is referred to the paper by Ting (1966).

If the loading history $P(t)$ is prescribed, the variation in contact size $a(t)$ may be found without much difficulty. As an example, we shall investigate the case of a rigid sphere pressed against a Maxwell material (Fig. 6.20(b)) by a force which increases from zero to a maximum P_0 and decreases again to zero, according to

$$P(t) = P_0 \sin{(t/T)} \tag{6.68}$$

as shown in Fig. 6.24. The material has an elastic modulus g and time constant $\eta/g = T$. Whilst the contact area is increasing its size is given by substituting the creep compliance of the material from (6.55) and the load history of (6.68) into equation (6.61), to give

$$a^3(t) = \tfrac{3}{8}R \int_0^t \frac{1}{g} \{1 + (t - t')/T\} \frac{\partial P_0 \sin{(t'/T)}}{\partial t'} \, dt'$$

$$= \frac{3RP_0}{8g} \{\sin{(t/T)} - \cos{(t/T)} + 1\} \tag{6.69}$$

This relationship only holds up to the maximum value of $a(t)$, which occurs at $t = t_m = 3\pi T/4$. When $a(t)$ begins to decrease we make use of the result that the contact pressure $p(r, t)$ and hence the total load $P(t)$ depend upon the contact stress history up to time t_1 only, where $t_1 \leqslant t_m$, and t_1 is given by

$$a(t_1) = a(t). \tag{6.70}$$

To find t_1, we use equation (6.60) for the load; and since the range of this integral lies within the period during which the contact size is increasing, we may substitute for $a^3(t')$ from equation (6.69) with the result that

$$P(t) = P_0 e^{-t/T} \int_0^{t_1/T} e^{t'/T} \{\cos{(t'/T)} + \sin{(t'/T)}\} \, d(t'/T)$$

$$= P_0 e^{-t/T} e^{t_1/T} \sin{(t_1/T)} \tag{6.71}$$

Since the load variation is known from (6.68), equation (6.71) reduces to

$$e^{t/T} \sin (t/T) = e^{t_1/T} \sin (t_1/T) \tag{6.72}$$

This equation determines t_1 corresponding to any given t. It is then used in conjunction with the relationship (6.69) to find $a(t)$ during the period when the contact is decreasing $(t > t_m)$. The process is illustrated in Fig. 6.24 by plotting the function $e^{t/T} \sin (t/T)$.

From the figure we see that the maximum contact area is not coincident with the maximum load; the contact continues to grow by creep even when the load has begun to decrease. Only at a late stage in the loading cycle does the contact rapidly shrink to zero as the load is finally removed. The penetration of the sphere $\delta(t)$ also reaches a maximum at t_m. During the period of increasing indentation $(0 \leqslant t \leqslant t_m)$, the penetration is related to the contact size by the elastic equation (6.57). During the period when the contact size is decreasing the penetration is greater than the 'elastic' value by an amount which depends upon

Fig. 6.24. Contact of a sphere with a Maxwell solid under the action of a sinusoidally varying force $P = P_0 \sin (t/T)$.

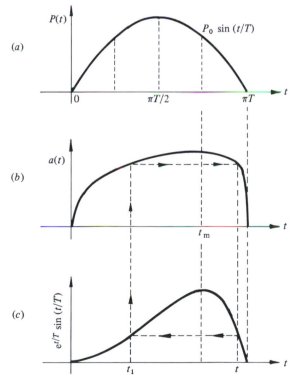

the detailed variation of $a(t)$ in this period. Thus an indentation remains at the time when the load and contact area have vanished.

The example we have just discussed is related to the problem of *impact* of a viscoelastic body by a rigid sphere. During impact, however, the force variation will be only approximately sinusoidal; it will in fact be related to the penetration, through the momentum equation for the impinging sphere. Nevertheless it is clear from our example that the maximum penetration will lag behind the maximum force, so that energy will be absorbed by the viscoelastic body and the coefficient of restitution will be less than unity. This problem has been studied theoretically by Hunter (1960) and will be discussed further in §11.5(*c*).

6.6   Nonlinear elasticity and creep

Many materials, particularly at elevated temperatures, exhibit non-linear relationships between stress, strain and strain rate. Rigorous theories of nonlinear viscoelasticity do not extend to the complex stress fields at a non-conforming contact, but some simplified analytical models have proved useful. Two related cases have received attention: (i) a nonlinear elastic material with the power law stress–strain relationship

$$\epsilon = \epsilon_0(\sigma/\sigma_0)^n \tag{6.73}$$

and (ii) a material which creeps according to the power law:

$$\dot{\epsilon} = \dot{\epsilon}_0(\sigma/\sigma_0)^n = B\sigma^n \tag{6.74}$$

where σ_0, ϵ_0 and $\dot{\epsilon}_0$ are representative values of stress, strain and strain rate which, together with the index n, may be found by fitting the above relationships to uniaxial test data. The most frequent use of this nonlinear elastic model is to describe the *plastic* deformation of an annealed metal which severely strain-hardens. The model only applies, of course, while the material is being loaded, i.e. when the principal strain increments satisfy the condition

$$\{(d\epsilon_1 - d\epsilon_2)^2 + (d\epsilon_2 - d\epsilon_3)^2 + (d\epsilon_3 - d\epsilon_1)^2\} > 0$$

By taking values of n from 1 to ∞ a range of material behaviour can be modelled from linear elastic, in which Young's modulus $E = \sigma_0/\epsilon_0$, to rigid-perfectly-plastic in which yield stress $Y = \sigma_0$ (Fig. 6.25). The power law creep relationship (6.74) applies to the steady-state or 'secondary' creep of metals at elevated temperatures, provided the strain rate is less than about 10^{-2} s^{-1}.

Concentrated line load

The stresses and deformation produced by a concentrated normal line load acting on a half-space which deforms according to either of the constitutive laws (6.73) or (6.74) can be found exactly. This is the nonlinear analogue of the linear elastic problem considered in §2.2. For a nonlinear elastic material the stress system was shown by Sokolovskii (1969) to be (in the notation of Fig. 2.2)

$$\sigma_r = -\frac{P}{D}(\cosh k\theta)^{1/n} \qquad (6.75)$$

$$\sigma_\theta = \tau_{r\theta} = 0$$

where $D = 2\int_0^{\pi/2}(\cosh k\theta)^{1/n}\cos\theta \, d\theta$, and $k = n(n-2)$ for conditions of plane strain.† The stress field is a simple radial one as in the linear case. In that case $n = 1, k = \sqrt{(-1)}, (\cosh k\theta)^{1/n} = \cos\theta, D = \pi/2$ whereupon equation (6.75) reduces to equation (2.15). Note that the stress is zero at the free surface where $\theta = \pm\pi/2$. A special case arises when $n = 2; k = 0, D = 2$ and $\sigma_r = -P/2r$, which is independent of θ. With materials which strain-harden only to a moderate extent, $n > 2$, so that $k > 0$; the stress σ_r at constant radius r increases with θ from a minimum directly below the load ($\theta = 0$) to a maximum at the surface ($\theta = \pm\pi/2$). Expressions for the displacements $u_r(r, \theta)$ and $u_\theta(r, \theta)$ are given by Arutiunian (1959) for plane strain and by Venkatraman (1964) for plane stress.

The stresses, strains and displacements at any point in the solid are directly proportional to the applied load P; so also are increments of stress and strain acquired in a time increment dt. This implies that the above solution, derived

† The problem has been solved by Venkatraman (1964) for 'plane stress', for which $k = n(n-3)/2$.

Fig. 6.25. Nonlinear power-law material: $\epsilon = \epsilon_0(\sigma/\sigma_0)^n$. Linear elastic: $n = 1, E = \sigma_0/\epsilon_0$. Perfectly-plastic, $n = \infty, Y = \sigma_0$.

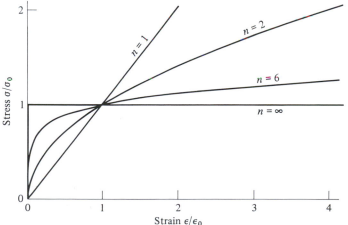

for a nonlinear elastic material specified by equation (6.73), applies equally to a creeping material specified by (6.74) in which strains are replaced by strain rates and displacements by velocities.

The stresses and displacements due to a point force acting on a nonlinear half-space – the nonlinear equivalent of the problem considered in §3.2 – have been analysed by Kuznetsov (1962).

Contact of nonlinear solids

For linear materials, whether elastic or viscoelastic, the stresses and displacements caused by a concentrated force can be superposed to find the stresses and displacements caused by a distributed load or by the contact of bodies with known profiles. With nonlinear materials the principle of super-position is no longer applicable, but Arutiunian (1959) argues that the surface displacement produced by a distributed load acting on a small segment of the boundary of a nonlinear half-space can be expressed by a series expansion whose dominant term is that obtained by the superposition of the displacements given by the solution for a concentrated force described above. On the basis of this approximation, expressions are developed from which the contact size and pressure distribution at any time can be found numerically when the value of the index n in equation (6.73) or (6.74) is specified.

In the special case of a rigid die with a flat base indenting a nonlinear half-space, the boundary conditions at the contact surface for a nonlinear elastic body (e.g. (6.73)) and a power law creeping solid (e.g. (6.74)) are analogous. In the first case the displacement $\bar{u}_z = $ constant $= \delta$, and in the second the surface velocity $\dot{\bar{u}}_z = $ constant $= \dot{\delta}$. The situation is similar, therefore, to loading by a concentrated force discussed above: the pressure on the face of the die is the same for both nonlinear elasticity and nonlinear creep. For a two-dimensional punch Arutiunian (1959) finds

$$p(x,t) = \frac{\Gamma\left(\dfrac{3n-1}{2n}\right)\Gamma\left(\dfrac{1}{2n}\right)\sin(\pi/2n)}{a\pi^{1/2}}\frac{P(t)}{\pi(1-x^2/a^2)^{1/2n}} \tag{6.76}$$

where Γ denotes a Gamma function, and for the axi-symmetric punch Kuznetsov (1962) shows that

$$p(r,t) = \frac{(2n-1)}{2\pi na^2}\frac{P(t)}{(1-r^2/a^2)^{1/2n}} \tag{6.77}$$

For a linear elastic material, where $n = 1$, both expressions reduce as expected to the elastic pressure on the base of a rigid punch, i.e. to equations (2.64) and (3.34). In the other extreme of a perfectly plastic material, $n \to \infty$, and the contact pressures given by equations (6.76) and (6.77) become uniform. This agrees exactly with the slip-line field solution for a two-dimensional punch in which the pressure is $2.97Y$, where Y is the yield stress in tension or compression. For a punch of circular plan-form the slip-line field solution gives a pressure which peaks slightly towards the centre of the punch (see Fig. 6.10), with an average value $p_m = 2.85Y$.

With an indenter whose profile is curved rather than flat the behaviour in steady creep is different from that of nonlinear elasticity because (a) the *displacements* imparted to the surface vary over the face of the indenter whereas the *velocity* of the indenter is uniform and (b) the contact area grows during the indentation so that material elements do not experience proportional loading. In these circumstances the analysis by Arutiunian's method becomes very involved while remaining approximate through the use of superposition. An alternative approximate treatment of the contact of spheres which is attractive by its simplicity has been suggested by Matthews (1980). We shall look first at a nonlinear material which deforms according to equation (6.73), as shown in Fig. 6.25. Guided by Kuznetsov's result for a rigid punch (eq. (6.77)), the pressure distribution is assumed to be given by:

$$p(r) = \frac{2n + 1}{2n} \, p_m (1 - r^2/a^2)^{1/2n} \tag{6.78}$$

For a linear elastic incompressible material ($n = 1$) this distribution reduces to that of Hertz for which $p_m = 16Ea/9\pi R$ (eqs. (4.21), (4.22) and (4.24)). Putting $n = \infty$ in equation (6.78) gives a uniform pressure. For a perfectly plastic material the pressure distribution given by the slip-line field solution, shown in Fig. 6.10, is roughly uniform with $p_m \approx 3Y$. Both of these values of p_m, corresponding to the extreme values of n (1 and ∞), are realised if we write

$$\frac{P}{\pi a^2} = p_m = \frac{6n\sigma_0}{2n + 1} \left(\frac{8a}{9\pi R\epsilon_0} \right)^{1/n} \tag{6.79}$$

since, in the constitutive equation (6.73), $\sigma_0/\epsilon_0^{1/n} = E$ when $n = 1$ and $\sigma_0 = Y$ when $n \to \infty$. It is reasonable to suppose that equation (6.79) then gives a good approximation to the relationship between load P and contact size a for intermediate values of n. In his study of hardness, Tabor (1951) suggested the empirical relationship

$$p_m \approx 3Y_R \tag{6.80}$$

where Y_R is the stress at a representative strain ϵ_R in a simple compression test. For a material with power law hardening, $Y_R = \sigma_0(\epsilon_R/\epsilon_0)^{1/n}$. Substituting for Y_R in (6.80) enables ϵ_R to be found from equation (6.79), thus:

$$\epsilon_R = \frac{8}{9\pi} \left(\frac{2n}{2n + 1} \right)^n \frac{a}{R} \tag{6.81}$$

which varies from $0.188a/R$ to $0.171a/R$ as n varies from 1 to ∞. The variation with n is small and the values are reasonably consistent with Tabor's empirical result: $\epsilon_R \approx 0.2(a/R)$, independent of the precise shape of the stress–strain curve.

The penetration δ of the indenter into the half-space is of interest. Based on the experimental data of Norbury & Samnuel (1928) Matthews proposes the expression:

$$\delta = \left(\frac{2n}{2n+1}\right)^{2(n-1)} \frac{a^2}{R} \tag{6.82}$$

Thus δ varies from the elastic value a^2/R when $n = 1$ to $0.368a^2/R$ when $n = \infty$, which is in good agreement with the perfectly plastic analysis of Richmond *et al.* (1974). When $\delta > a^2/2R$ the periphery of the indentation 'sinks in' below the surface of the solid, as described in §3. This occurs for low values of n, i.e. with annealed materials. When n exceeds 3.8, $\delta < a^2/2R$ and 'piling up' occurs outside the edge of the contact.

We turn now to penetration by a spherical indenter under conditions of power law creep governed by equation (6.74). Matthews assumes that the pressure distribution is the same as that found by Kuznetsov for a flat-faced punch (eq. (6.77)), i.e.

$$p(r, t) = \frac{2n-1}{2n} p_m(t)(1 - r^2/a^2)^{-1/2n} \tag{6.83}$$

For $n = 1$, equation (6.74) describes a linear viscous material of viscosity $\eta = \sigma_0/3\dot{\epsilon}_0 = 1/3B$. Spherical indentation of such a material was analysed in §5, where it was shown (eq. (6.67)) that

$$p(r, t) = \frac{8\eta\dot{a}}{\pi R}(1 - r^2/a^2)^{-1/2} \tag{6.84}$$

By differentiating equation (6.66) with respect to time we get

$$p_m(t) = \frac{P(t)}{\pi a^2} = \frac{16\eta}{\pi R}\dot{a} = \frac{8\eta\dot{\delta}}{\pi(R\delta)^{1/2}} \tag{6.85}$$

For $n \to \infty$, the nonlinear viscous material also behaves like a perfectly plastic solid of yield stress $Y = \sigma_0$, so that $p_m \approx 3Y$.

If, for the nonlinear material, we now write

$$p_m(t) = \frac{P(t)}{\pi a^2} = \frac{6n\sigma_0}{2n-1}\left(\frac{8\dot{a}}{9\pi R\dot{\epsilon}_0}\right)^{1/n} \tag{6.86}$$

Equations (6.83) and (6.86) reduce to (6.84) and (6.85) when $n = 1$ and $\sigma_0/3\dot{\epsilon}_0 = \eta$, and reduce to $p_m = \text{constant} = 3Y$ when $n = \infty$. Making use of the relationship (6.82) the velocity of penetration $\dot{\delta}$ can be written

$$\dot{\delta}(t) = 2(\delta/R)^{1/2} \left(\frac{2n}{2n + 1} \right)^{n-1} \dot{a}(t) \tag{6.87}$$

where $\dot{a}$ is related to the load by equation (6.86). In a given situation either the load history $P(t)$ or the penetration history $\delta(t)$ would be specified, whereupon equations (6.86) and (6.83) enable the variations in contact size $a(t)$ and contact pressure $p(r, t)$ to be found if the material parameters σ_0, $\dot{\varepsilon}_0$ and n are known.

7

Tangential loading and sliding contact

7.1 Sliding of non-conforming elastic bodies

In our preliminary discussion in Chapter 1 of the relative motion and forces which can arise at the point of contact of non-conforming bodies we distinguished between the motion described as *sliding* and that described as *rolling*. Sliding consists of a relative peripheral velocity of the surfaces at their point of contact, whilst rolling involves a relative angular velocity of the two bodies about axes parallel to their tangent plane. Clearly rolling and sliding can take place simultaneously, but in this chapter we shall exclude rolling and restrict our discussion to the contact stresses in simple rectilinear sliding. The system is shown in Fig. 7.1. A slider, having a curved profile, moves from right to left over a flat surface. Following the approach given in Chapter 1, we regard the point of initial contact as a fixed origin and imagine the material of the lower surface flowing through the contact region from left to right with a steady velocity V. For convenience we choose the x-axis parallel to the direction of sliding.

Fig. 7.1. Sliding contact.

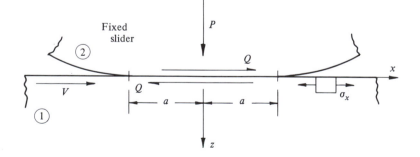

A normal force P pressing the bodies together gives rise to an area of contact which, in the absence of friction forces, would have dimensions given by the Hertz theory. Thus in a frictionless contact the contact stresses would be unaffected by the sliding motion. However a sliding motion, or any tendency to slide, of real surfaces introduces a tangential force of friction Q, acting on each surface, in a direction which opposes the motion. We are concerned here with the influence of the tangential force Q upon the contact stresses. In this section we shall imagine that the bodies have a steady sliding motion so that the force Q represents the force of 'kinetic friction' between the surfaces. In the next section we shall investigate the situation of two bodies, nominally with no relative velocity, but subjected to a tangential force tending to cause them to slide. The force Q then arises from 'static friction'; it may take any value which does not exceed the force of 'limiting friction' when sliding is just about to occur.

The first question to consider is whether the tangential traction due to friction at the contact surface influences the size and shape of the contact area or the distribution of normal pressure. For the purposes of calculating the elastic stresses and displacements due to the tangential tractions, we retain the basic premise of the Hertz theory that the two bodies can each be regarded as an elastic half-space in the proximity of the contact region. The methods of analysis given in Chapter 2 and 3 are then appropriate. From equation (2.22) in two dimensions and from equation (3.75) in three dimensions we note that the *normal* component of displacement at the surface $\bar{u}_z$ due to a concentrated *tangential* force Q is proportional to the elastic constant $(1 - 2\nu)/G$. The tangential tractions acting on each surface at the interface are equal in magnitude and opposite in direction, viz.:

$$q_1(x, y) = -q_2(x, y) \tag{7.1}$$

Hence the normal displacements due to these tractions are proportional to the respective values of $(1 - 2\nu)/G$ of each body and are of opposite sign:

$$\frac{G_1}{1 - 2\nu_1} \bar{u}_{z1}(x, y) = -\frac{G_2}{1 - 2\nu_2} \bar{u}_{z2}(x, y) \tag{7.2}$$

It follows from equation (7.2) that, if the two solids have the same elastic constants, any tangential traction transmitted between them gives rise to equal and opposite normal displacements of any point on the interface. Thus the warping of one surface conforms exactly with that of the other and does not disturb the distribution of normal pressure. The shape and size of the contact area are then fixed by the profiles of the two surfaces and the normal force, and are independent of the tangential force. With solids of different elastic properties this is no longer the case and the tangential tractions do interact with the normal pressure. The effect is entirely analogous to the interaction between normal and tangential tractions in normal contact of dissimilar solids discussed in §5.4. However, as we shall see later, it transpires that the influence of tangential tractions upon the normal pressure and the contact area is generally small, particularly when the coefficient of limiting friction is appreciably less than unity. In our analysis of problems involving tangential tractions, therefore, we shall neglect this interaction and assume that the stresses and deformation due to (a) the normal pressure and (b) the tangential traction are independent of each other, and that they can be superposed to find the resultant stress.

We must now prescribe the relationship between the tangential traction and the normal pressure in sliding contact. It is usual to assume that Amontons' law of sliding friction applies at each elementary area of the interface, so that

$$\frac{|q(x,y)|}{p(x,y)} = \frac{|Q|}{P} = \mu \tag{7.3}$$

where μ is a constant coefficient of kinetic friction whose value is determined by the materials and the physical conditions of the interface. Some indication of the circumstances which account for the validity of Amonton's law with dry surfaces can be found in Chapter 13. It is also found to be approximately valid when *non-conforming* sliding surfaces are separated by thin lubricating films. Experimental confirmation that the tangential traction is distributed in direct proportion to normal pressure is provided by photo-elastic work of Ollerton & Haines (1963) using large models of araldite epoxy resin. For a full discussion of the physics of friction and the conditions which determine the value of the coefficient of friction, the reader is referred to Bowden & Tabor (1951, 1964).

We are now in a position to examine the elastic stress distributions in sliding contact. The two-dimensional case of a cylinder sliding in a direction perpendicular to its axis has been studied in more detail than the corresponding three-dimensional case. We will look at the two-dimensional problem first:

(a) Cylinder sliding perpendicular to its axis

If the cylinder and the plane on which it slides have the same elastic properties, we have seen that the width of the contact strip $2a$ and the normal pressure distribution are given by the Hertz theory (equations (4.43)–(4.45)), i.e.

$$p(x) = \frac{2P}{\pi a^2} (a^2 - x^2)^{1/2}$$

where P is the normal force per unit axial length pressing the cylinder into contact with the plane. Then, assuming Amonton's law of friction (equation

(7.3)) the tangential traction is

$$q(x) = \mp \frac{2\mu P}{\pi a^2} (a^2 - x^2)^{1/2} \tag{7.4}$$

where the negative sign is associated with a positive velocity V as shown in Fig. 7.1. The stress components within both the cylinder and the plane are now given by equations (2.23). These integrals have been evaluated by several workers (McEwen, 1949; Poritsky, 1950; Smith & Liu, 1953; Sackfield & Hills, 1983c). A short table of values is given in Appendix 4. From equations (2.23), since $q(x)$ and $p(x)$ are in proportion, it may be seen that there are some analogies between the stresses due to the tangential traction and those due to the normal traction, which may be expressed by:

$$\frac{(\sigma_z)_q}{q_0} = \frac{(\tau_{xz})_p}{p_0} \tag{7.5a}$$

and

$$\frac{(\tau_{xz})_q}{q_0} = \frac{(\sigma_x)_p}{p_0} \tag{7.5b}$$

where $q_0 = \mu p_0$ is the tangential traction at $x = 0$, and the suffixes p and q refer to stress components due to the normal pressure and tangential traction acting separately. Hence $(\sigma_z)_q$ and $(\tau_{xz})_q$ can be found directly from the expressions for $(\tau_{xz})_p$ and $(\sigma_x)_p$ given by equations (4.49). The direct stress parallel to the surface $(\sigma_x)_q$, however, must be evaluated independently. In the notation of equations (4.49) it may be shown that

$$(\sigma_x)_q = \frac{q_0}{a} \left\{ n \left(2 - \frac{z^2 - m^2}{m^2 + n^2} \right) - 2x \right\}^\dagger \tag{7.6}$$

At the surface $z = 0$ the expression for the direct stress reduces to:

$$(\bar{\sigma}_x)_q = \begin{cases} -2q_0 x/a, & |x| \leqslant a \tag{7.7a} \\ -2q_0 \left\{ \frac{x}{a} \mp \left(\frac{x^2}{a^2} - 1 \right)^{1/2} \right\}, & |x| > a \tag{7.7b} \end{cases}$$

The surface stresses in the moving plane (Fig. 7.1) are shown in Fig. 7.2. The tangential traction acting on the moving plane is negative so that the direct stress reaches a maximum compressive stress $-2q_0$ at the leading edge of the contact area $(x = -a)$ and a maximum tension $2q_0$ at the trailing edge $(x = a)$. We recall that the normal pressure gives rise to an equal compressive stress at the surface, $(\sigma_x)_p = -p(x)$, within the contact region and no stress outside. Hence, whatever the coefficient of friction, the maximum resultant tensile stress in sliding contact occurs at the trailing edge with the value $2\mu p_0$.

† A short table of values is given in Appendix 4 (p. 430).

The onset of plastic yield in sliding contact will be governed (using the Tresca yield criterion) by the maximum value of the principal shear stress throughout the field. Contours of τ_1 in the absence of friction are shown in Fig. 4.5. The maximum value is $0.30p_0$ on the z-axis at a depth $0.78a$. Contours of τ_1 due to combined normal pressure and tangential traction, taking $\mu = 0.2$, are plotted in Fig. 7.3. The maximum value now occurs at a point closer to the surface. The position and magnitude of the maximum principal shear stress may be computed and equated to the yield stress k in simple shear to find the contact pressure p_0 for first yield (by the Tresca yield criterion). This is shown for increasing values of the coefficient of friction in Fig. 7.4. The frictional traction also introduces shear stresses into the contact surface which can reach yield if the coefficient of friction is sufficiently high. The stresses in the contact surface

Fig. 7.2. Surface stresses due to frictional traction $q = q_0'(1 - x^2/a^2)^{1/2}$.

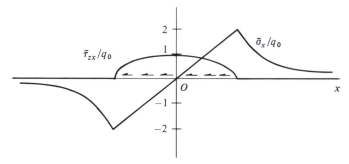

Fig. 7.3. Contours of the principal shear stress τ_1 beneath a sliding contact, $Q_x = 0.2P$.

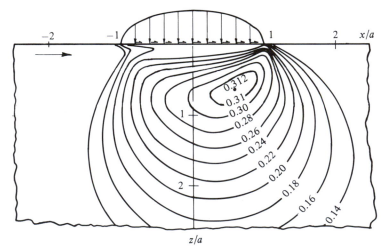

due to both pressure and frictional tractions are

$$\bar{\sigma}_x = -p_0\{(1-x^2/a^2)^{1/2} + 2\mu x/a\} \tag{7.8a}$$

$$\bar{\sigma}_z = -p_0(1-x^2/a^2)^{1/2} \tag{7.8b}$$

$$\bar{\sigma}_y = -2\nu p_0\{(1-x^2/a^2)^{1/2} + \mu x/a\} \tag{7.8c}$$

$$\bar{\tau}_{xz} = -\mu p_0(1-x^2/a^2)^{1/2} \tag{7.8d}$$

The principal shear stress in the plane of the deformation is

$$\tau_1 = \tfrac{1}{2}\{(\sigma_x - \sigma_z)^2 + 4\tau_{xz}^2\}^{1/2} = \mu p_0 \tag{7.9}$$

This result shows that the material throughout the width of the contact surface will reach yield when

$$p_0/k = 1/\mu \tag{7.10}$$

Yield may also occur by 'spread' of the material in the axial direction although such flow must of necessity be small by the restriction of plane strain. Calculations of the contact pressure for the onset of yield in sliding contact have been made by Johnson & Jefferis (1963) using both the Tresca and von Mises yield criteria; the results are shown in Fig. 7.4. For low values of the coefficient of friction ($\mu < 0.25$ by Tresca and $\mu < 0.30$ by von Mises) the yield point is first reached at a point in the material beneath the contact surface. For larger values of μ yield first occurs at the contact surface. The Tresca criterion predicts lateral yield for $0.25 < \mu < 0.44$; but when $\mu > 0.44$ the onset of yield is given by equation (7.9).

In the above discussion the tangential traction has been assumed to have no effect upon the normal pressure. This is strictly true only when the elastic constants of the two bodies are the same. The influence of a difference in elastic constants has been analysed by Bufler (1959) using the methods of §2.7. The boundary condition $q(x) = \mu p(x)$ is of class IV, which leads to a singular integral equation of the second kind (2.53). Solving the integral equation by (2.55) and (2.56), the surface traction within the contact area is found to be:

$$q(x) = \frac{\mu E^*}{2R(1+\beta^2\mu^2)^{1/2}}\left(\frac{a+x}{a-x}\right)^{\gamma}(a^2-x^2)^{1/2} \tag{7.11}$$

where β is a measure of the difference in the elastic constants defined in equation (5.3) and

$$\gamma = -(1/\pi)\tan^{-1}(\beta\mu) \approx -\beta\mu/\pi \tag{7.12}$$

provided that $\beta\mu$ is small. The semi-width of the contact strip is given by

$$a^2 = \frac{1}{1-4\gamma^2}\frac{4PR}{E^*} \tag{7.13}$$

The contact strip is no longer symmetrically placed; its centre is displaced from the axis of symmetry by a distance

$$x_0 = 2\gamma a \tag{7.14}$$

Bufler also finds the distribution of direct stress σ_x in the surface. When the elastic constants of the two bodies are equal, β is zero and hence γ vanishes. The contact area is then seen to be symmetrical and of a size given by the Hertz theory; the pressure distribution also reduces to that of Hertz. Values of β for various combinations of materials are given in Table 5.1. The values do not exceed 0.21. Since coefficients of friction rarely exceed 1.0, the maximum likely value of $|\gamma|$ is about 0.06. The distribution of surface traction and contact area have been calculated from equations (7.11)–(7.13), taking $\gamma = 0.06$, for the arrangement in Fig. 7.1 in which it has been assumed that the lower surface

Fig. 7.4. Effect of sliding friction on the contact pressure for first yield and shakedown (see §9.2). Large-dashed line – line contact, first yield (Tresca). Chain line – line contact, first yield (von Mises). Solid line – line contact, shakedown (Tresca). Small-dashed line – point contact, first yield (von Mises).

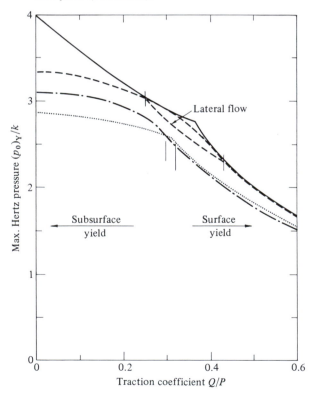

is the more elastic (i.e. β + ve and μ − ve). The results are shown in Fig. 7.5. The effect of the tangential traction is to shift the centre of the contact region by a distance $x_0 = 0.12a$ towards the trailing edge; the contact width is increased by 0.8% and the centre of pressure moves towards the trailing edge. However the comparison with the Hertz pressure distribution shows that the effect is small even for an extreme value of the product $\beta\mu$. For more representative values of $\beta\mu$ the influence of frictional traction upon the contact area and pressure distribution is negligible.

(b) Sliding sphere

We now consider a sphere, carrying a normal load P, which slides over a plane surface in a direction chosen parallel to the x-axis. Neglecting any interaction between normal pressure and tangential traction arising from a difference in elastic constants of the two solids, the size of the circular contact area and the pressure distribution are given by the Hertz theory (equations (3.39), (4.22) and (4.24)). Amontons' law of friction specifies the tangential traction to be

$$q(r) = \frac{3\mu P}{2\pi a^3} (a^2 - r^2)^{1/2} \tag{7.15}$$

acting parallel to the x-axis everywhere in the contact area.

We wish to find the stress components in the solid produced by the surface traction. In principle they may be found by using the stress components due to a concentrated force, given by equations (3.76), weighted by the distribution

Fig. 7.5. Influence of a difference in elastic constants on the pressure and traction distribution in sliding contact, for $\beta = 0.2$, $\mu = 1$ ($\gamma = 0.06$).

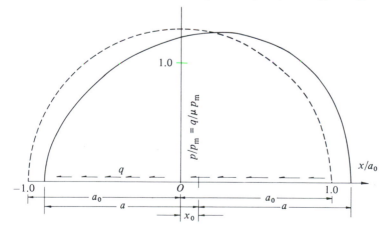

(7.15) and integrated throughout the contact area. However such integrations can only be performed numerically. A different approach has been taken by Hamilton & Goodman (1966), by extending a method introduced by Green (1949) for the stress analysis of a normally loaded half-space. They computed stresses in the x-z plane and at the surface (x-y plane) for values of $\mu = 0.25$ and 0.50, ($\nu = 0.3$). Explicit equations for calculating the stress components at any point in the solid have since been given by Hamilton (1983) and by Sackfield & Hills (1983c).

The von Mises criterion has been used to calculate the point of first yield. As for a two-dimensional contact, the point of first yield moves towards the surface as the coefficient of friction is increased; yield occurs at the surface when μ exceeds 0.3. The values of the maximum contact pressure $(p_0)_Y$ to initiate yield have been added to Fig. 7.4 from which it will be seen that they are not significantly different from the two-dimensional case.

The normal contact of elastic spheres introduces a radial tension at $r = a$ of magnitude $(1 - 2\nu)p_0/3 \approx 0.13p_0$. The effect of the tangential traction is to add to the tension on one side of the contact and to subtract from it at the other. The maximum tension, which occurs at the surface point $(-a, 0)$ rises to $0.5p_0$ and $1.0p_0$ for $\mu = 0.25$ and 0.5 respectively. This result is again comparable with the two-dimensional case.

The analysis has been extended to elliptical contacts by Bryant & Keer (1982) and by Sackfield & Hills (1983b) who show that the contact pressure for first yield $(p_0)_Y$ is almost independent of the shape of the contact ellipse.

7.2 Incipient sliding of elastic bodies

A tangential force whose magnitude is less than the force of limiting friction, when applied to two bodies pressed into contact, will not give rise to a sliding motion but, nevertheless, will induce frictional tractions at the contact interface. In this section we shall examine the tangential surface tractions which arise from a combination of normal and tangential forces which does not cause the bodies to slide relative to each other.

The problem is illustrated in Fig. 7.6. The normal force P gives rise to a contact area and pressure distribution which we assume to be uninfluenced by the existence of the tangential force Q, and hence to be given by the Hertz theory. The effect of the tangential force Q is to cause the bodies to deform in shear, as indicated by the distorted centre-line in Fig. 7.6. Points on the contact surface will undergo tangential displacements $\bar{u}_x$ and $\bar{u}_y$ relative to distant points T_1 and T_2 in the undeformed region of each body. Clearly, if there is no sliding motion between the two bodies as a whole, there must be at least one point at the interface where the surfaces deform without relative motion; but it does not follow that there is no slip anywhere within the contact area. In fact it will be shown that the effect of a tangential force less than the limiting friction force $(Q < \mu P)$ is to cause a small relative motion, referred to as 'slip' or 'micro-slip', over part of the interface. The remainder of the interface deforms without relative motion and in such regions the surfaces are said to adhere or to 'stick'.

To proceed with an analysis we must consider the conditions governing 'stick' and 'slip'. In Fig. 7.6, A_1 and A_2 denote two points on the interface which were coincident before the application of the tangential force. Under the action of the force, points in the body such as T_1 and T_2, distant from the interface, move through effectively rigid displacements δ_{x1}, δ_{y1} and δ_{x2}, δ_{y2} while A_1 and A_2 experience tangential elastic displacements $\bar{u}_{x1}, \bar{u}_{y1}$ and $\bar{u}_{x2}, \bar{u}_{y2}$ relative to T_1 and T_2. If the absolute displacements of A_1 and A_2 (i.e. relative to O) are denoted by s_{x1}, s_{y1} and s_{x2}, s_{y2}, the components of slip between A_1 and A_2 may be written

$$s_x \equiv s_{x1} - s_{x2} = (\bar{u}_{x1} - \delta_{x1}) - (\bar{u}_{x2} - \delta_{x2})$$
$$= (\bar{u}_{x1} - \bar{u}_{x2}) - (\delta_{x1} - \delta_{x2}) \qquad (7.16)$$

A similar relation governs the tangential displacements in the y-direction. If the points A_1 and A_2 are located in a 'stick' region the slip s_x and s_y will be zero so

Fig. 7.6

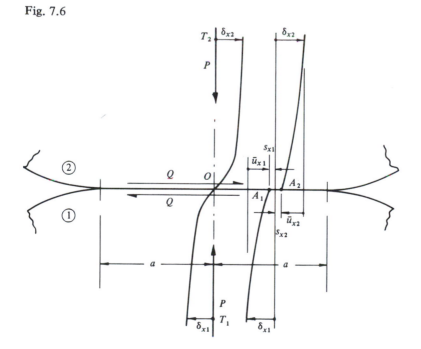

that

$$\bar{u}_{x1} - \bar{u}_{x2} = (\delta_{x1} - \delta_{x2}) \equiv \delta_x \qquad (7.17a)$$

$$\bar{u}_{y1} - \bar{u}_{y2} = (\delta_{y1} - \delta_{y2}) \equiv \delta_y \qquad (7.17b)$$

We note that the right-hand sides of equations (7.17) denote relative tangential displacements between the two bodies as a whole under the action of the tangential force. Thus δ_x and δ_y are constant, independent of the position of A_1 and A_2, within the 'stick' region. Further, if the two bodies have the same elastic moduli, since they are subjected to mutually equal and opposite surface tractions, we can say at once that $\bar{u}_{x2} = -\bar{u}_{x1}$ and $\bar{u}_{y2} = -\bar{u}_{y1}$. The condition of no slip embodied in equations (7.17) can then be stated: *all surface points within a 'stick' region undergo the same tangential displacement.* The statement is also true when the elastic constants are different but the overall relative displacements δ_x and δ_y are then divided unequally between the two bodies according to equation (7.2).

At points within a stick region the resultant tangential traction cannot exceed its limiting value. Assuming Amonton's law of friction with a constant coefficient μ, this restriction may be stated:

$$|q(x,y)| \leqslant \mu |p(x,y)| \qquad (7.18)$$

In a region where the surfaces slip, the conditions of equations (7.17) are violated, but the tangential and normal tractions are related by

$$|q(x,y)| = \mu |p(x,y)| \qquad (7.19)$$

In addition, the direction of the frictional traction q must oppose the direction of slip. Thus

$$\frac{q(x,y)}{|q(x,y)|} = -\frac{s(x,y)}{|s(x,y)|} \qquad (7.20)$$

Equations (7.17)–(7.20) provide boundary conditions which must be satisfied by the surface tractions and surface displacements at the contact interface. Equations (7.17) and (7.18) apply in a stick region and equations (7.19) and (7.20) apply in a slip region. Difficulty arises in the solution of such problems because the division of the contact area into stick and slip regions is not known in advance and must be found by trial. In these circumstances a useful first step is to assume that no slip occurs anywhere in the contact area. Slip is then likely to occur in those regions where the tangential traction, so found, exceeds its limiting value.

A few particular cases will now be examined in detail.

(a) Two-dimensional contact of cylinders – no slip

We shall first consider two cylinders in contact with their axes parallel to the y-axis, compressed by a normal force P per unit axial length, to which a tangential force Q per unit length ($<\mu P$) is subsequently applied (see Fig. 7.7). The contact width and the pressure distribution due to P are given by the Hertz theory. These quantities are assumed to be unaffected by the subsequent application of Q. In view of the difficulty of knowing whether Q causes any micro-slip and, if so, where it occurs, we start by assuming that the coefficient of friction is sufficiently high to prevent slip throughout the whole contact area. Thus the complete strip $-a \leqslant x \leqslant a$ is a 'stick' area in which the condition of no-slip (eq. (7.17)) applies, i.e.

$$\bar{u}_{x1} - \bar{u}_{x2} = \text{constant} = \delta_x, \quad -a \leqslant x \leqslant a \tag{7.21}$$

The distribution of tangential traction at the interface is thus one which will give rise to a constant tangential displacement of the contact strip. For the purpose of finding the unknown traction each cylinder is regarded as a half-space to which the results of Chapter 2 apply. The analogous problem of finding the distribution of *pressure* which gives rise to a constant *normal* displacement, i.e.

Fig. 7.7. Contact of cylinders with their axes parallel. Surface tractions and displacements due to a tangential force $Q < \mu P$. Curve A – no slip, eq. (7.22); curve B – partial slip, eq. (7.28).

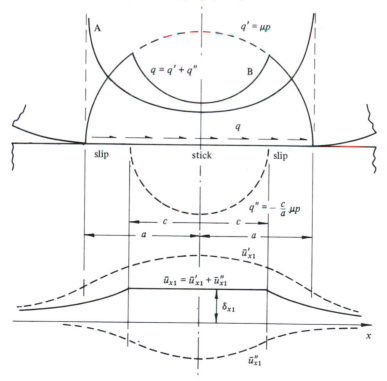

the pressure on the face of a flat frictionless punch, has been discussed in §2.8. The pressure is given by equation (2.64). Using the analogy between tangential and normal loading of an elastic half-space in plane strain we can immediately write down the required distribution of tangential traction, viz.

$$q(x) = \frac{Q}{\pi(a^2 - x^2)^{1/2}} \tag{7.22}$$

This traction acts on the surface of each body in opposing directions so that $\bar{u}_{x1}$ and $\bar{u}_{x2}$ will be of opposite sign and therefore additive in equation (7.21). The actual values of $\bar{u}_{x1}$ and $\bar{u}_{x2}$ and hence the value of δ_x, as in all two-dimensional problems, depend upon the choice of the reference points T_1 and T_2.

The traction given by (7.22) is plotted in Fig. 7.7 (curve A). It rises to a theoretically infinite value at the edges of the contact. This result is not surprising when it is remembered that the original assumption that there should be no slip at the interface effectively requires that the two bodies should behave as one. The points $x = \pm a$ then appear as the tips of two sharp deep cracks in the sides of a large solid block, where singularities in stress would be expected. It is clear that these high tangential tractions at the edge of the contact area cannot be sustained, since they would require an infinite coefficient of friction. There must be some micro-slip, and the result we have just obtained suggests that it occurs at both edges of the contact strip. We might expect a 'stick' region in the centre of the strip where the tangential traction is low and the pressure high. This possibility will now be investigated.

(b) Contact of cylinders – partial slip

The method of solution to the problem of partial slip was first presented by Cattaneo (1938) and independently by Mindlin (1949).

If the tangential force Q is increased to its limiting value μP, so that the bodies are on the point of sliding, the tangential traction is given by equation (7.4), viz.

$$q'(x) = \mu p_0(1 - x^2/a^2)^{1/2} \tag{7.23}$$

where $p_0 = 2P/\pi a$.

The tangential displacements within the contact surface due to the traction can be found. By analogy with the normal displacements produced by a Hertzian distribution of normal pressure, we conclude that the surface displacements are distributed parabolically within the contact strip. If no slip occurs at the mid-point $x = 0$, then we can write

$$\bar{u}'_{x1} = \delta'_{x1} - (1 - \nu_1^2)\mu p_0 x^2/aE_1 \tag{7.24}$$

and a similar expression of opposite sign for the second surface. These distributions of tangential displacement satisfy equation (7.21) at the origin only; elsewhere in the contact region the surfaces must slip.

We now consider an additional distribution of traction given by

$$q''(x) = -\frac{c}{a}\mu p_0 (1 - x^2/c^2)^{1/2} \tag{7.25}$$

acting over the strip $-c \leqslant x \leqslant c$ $(c < a)$, as shown in Fig. 7.7. The tangential displacements produced by this traction within the surface $-c \leqslant x \leqslant c$ follow by analogy with equation (7.24), viz.:

$$\bar{u}''_{x1} = -\delta''_{x1} + \frac{c}{a}(1 - \nu_1^2)\mu p_0 x^2/cE_1 \tag{7.26}$$

If we now superpose the two tractions q' and q'', the resultant displacements within the central strip $-c \leqslant x \leqslant c$ are constant, as shown in Fig. 7.7.

$$\bar{u}_{x1} = \bar{u}'_{x1} + \bar{u}''_{x1} = \delta'_{x1} + \delta''_{x1} = \delta_{x1} \tag{7.27a}$$

and for the second surface to which an equal and opposite traction is applied

$$\bar{u}_{x2} = -\delta_{x2} \tag{7.27b}$$

Substitution for $\bar{u}_{x1}$ and $\bar{u}_{x2}$ in equation (7.21) shows that the condition of no-slip is satisfied in the strip $-c \leqslant x \leqslant c$. Furthermore in this region the resultant traction is given by

$$q(x) = q'(x) + q''(x) = \mu p_0 \{(a^2 - x^2)^{1/2} - (c^2 - x^2)^{1/2}\}/a \tag{7.28}$$

which is everywhere less than μp. Thus the two necessary conditions that the central strip should be a 'stick' region are satisfied. At the edges of the contact, $c \leqslant |x| \leqslant a$, $q(x) = \mu p(x)$, as required in a slip region. It remains to prove that the direction of the traction opposes the direction of slip in these regions as required by equation (7.20). To do this we require the surface displacements in these regions. The surface displacements due to an elliptical distribution of tangential traction have been evaluated by Poritsky (1950), from which $\bar{u}'_{x1}$ and $\bar{u}''_{x1}$ throughout the surface are plotted (dotted) in Fig. 7.7. From equation (7.16) the slip s_x is given by

$$s_x = (\bar{u}_{x1} - \bar{u}_{x2}) - \delta_x$$

From the figure it is clear that $(\bar{u}_{x1} - \bar{u}_{x2})$ is less than δ_x in each slip region, so that s_x is negative in each region. This is consistent with the positive traction q acting on body (1). We have shown, therefore, that the resultant distribution of tangential traction shown in Fig. 7.7 produces surface displacements which satisfy the necessary conditions in a central stick region $-c \leqslant x \leqslant c$ and two peripheral slip regions $c \leqslant |x| \leqslant a$.

The size of the stick region is determined by the magnitude of the tangential force

$$Q = \int_{-a}^{a} q(x) \, dx = \int_{-a}^{a} q'(x) \, dx + \int_{-c}^{c} q''(x) \, dx$$

$$= \mu P - \frac{c^2}{a^2} \mu P$$

so that

$$\frac{c}{a} = \left(1 - \frac{Q}{\mu P}\right)^{1/2} \tag{7.29}$$

The physical behaviour is now clear. If, keeping P constant, Q is increased steadily from zero, micro-slip begins immediately at the two edges of the contact area and spreads inwards according to equation (7.29). As Q approaches μP, c approaches zero and the stick region shrinks to a line at $x = 0$. Any attempt to increase Q in excess of μP causes the contact to slide.

The stresses within either solid due to the elliptical distribution of traction $q'(x)$ have been discussed in the last section. The stresses due to a force Q less than μP can be found by superposing a distribution of stress due to $q''(x)$ which is similar in form but is reduced in scale.

(c) Contact of spheres – no slip

Two spherical bodies pressed into contact by a normal force P have a circular area of contact whose radius is given by equation (4.22) and an ellipsoidal pressure distribution given by equation (3.39). If a tangential force Q, applied subsequently, causes elastic deformation without slip at the interface, then it follows from equation (7.17) that the tangential displacement of all points in the contact area is the same. If the force Q is taken to act parallel to the x-axis, then it follows from symmetry that this tangential displacement must also be parallel to the x-axis. The distribution of tangential traction which produces a uniform tangential displacement of a circular region on the surface of an elastic half-space has been found in Chapter 3. The traction (equation (3.82)) is radially symmetrical in magnitude and everywhere parallel to the x-axis:

$$q_x(r) = q_0(1 - r^2/a^2)^{-1/2} \tag{7.30}$$

from which $q_0 = Q_x/2\pi a^2$. The corresponding displacement, which, in this case, can be precisely defined, is given by equation (3.86a) i.e.

$$\bar{u}_x = \frac{\pi(2 - \nu)}{4G} q_0 a \tag{7.31}$$

Substituting into equation (7.17) gives the relative tangential displacement between distant points T_1 and T_2 in the two bodies:

$$\delta_x = \bar{u}_{x1} - \bar{u}_{x2} = \frac{Q_x}{8a}\left(\frac{2-\nu_1}{G_1} + \frac{2-\nu_2}{G_2}\right) \tag{7.32}$$

This relationship is shown by the broken line in Fig. 7.8; the tangential displacement is directly proportional to the tangential force. This is unlike the normal approach of two elastic bodies which varies in a nonlinear way with normal load because the contact area grows as the load is increased.

The tangential traction necessary for no slip rises to a theoretically infinite value at the periphery of the contact circle so that some micro-slip is inevitable at the edge of contact.

(d) Contact of spheres – partial slip

Cattaneo's technique can also be applied to the case of spheres in contact. The axial symmetry of the tangential traction given by equation (7.30) suggests that the 'stick' region in this case might be circular and concentric with the contact circle. On the point of sliding, when only the two points in contact at the origin are 'stuck', the distribution of traction is

$$q'(x,y) = \mu p(x,y) = \mu p_0(1 - r^2/a^2)^{1/2} \tag{7.33}$$

The tangential displacements within the contact circle, $r \leqslant a$, are then given by

Fig. 7.8. Tangential displacement δ_x of a circular contact by a tangential force Q_x; (A) with no slip, (B) with slip at the periphery of the contact.

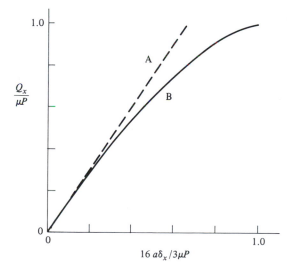

$\frac{Q_x}{\mu P}$

$16\,a\delta_x/3\mu P$

equations (3.91), viz.:

$$\bar{u}_x' = \frac{\pi\mu p_0}{32Ga} \{4(2-v)a^2 + (4-v)x^2 + (4-3v)y^2\} \tag{7.34a}$$

and

$$\bar{u}_y' = \frac{\pi\mu p_0}{32Ga} 2vxy \tag{7.34b}$$

If we now consider a distribution of traction

$$q''(x,y) = -\frac{c}{a}p_0(1-r^2/c^2)^{1/2} \tag{7.35}$$

acting over the circular area $r \leqslant c$, by analogy the tangential displacements within that circle are

$$\bar{u}_x'' = -\frac{c}{a}\frac{\pi\mu p_0}{32Gc} \{4(2-v)c^2 + (4-v)x^2 + (4-3v)y^2\} \tag{7.36a}$$

$$\bar{u}_y'' = -\frac{c}{a}\frac{\pi\mu p_0}{32Gc} 2vxy \tag{7.36b}$$

The resultant displacements in the circle, $r \leqslant c$, are given by adding equations (7.34) and (7.36), with the result:

$$\bar{u}_x = \frac{\pi\mu p_0}{8Ga} (2-v)(a^2 - c^2) \tag{7.37a}$$

$$\bar{u}_y = 0 \tag{7.37b}$$

We see that these displacements satisfy the condition for no-slip (7.17) within the circle $r \leqslant c$, with the result that

$$\delta_x = \frac{3\mu P}{16} \left(\frac{2-v_1}{G_1} + \frac{2-v_2}{G_2} \right) \frac{a^2 - c^2}{a^3} \tag{7.38}$$

Thus the stick region is the circle of radius c whose value can be found from the magnitude of the tangential force.

$$Q_x = \int_0^a 2\pi q' r \, dr - \int_0^c 2\pi q'' r \, dr = \mu P(1 - c^3/a^3)$$

whence

$$\frac{c}{a} = (1 - Q/\mu P)^{1/3} \tag{7.39}$$

The tangential traction acts parallel to the x-axis at all points; it is given by q' (eq. (7.33)) in the annulus $c \leqslant r \leqslant a$ and by the resultant of q' and q'' (which is less than μp) in the central circle $r \leqslant c$. To confirm that the slip conditions are satisfied in the annulus $c \leqslant r \leqslant a$, the displacements in that annulus due to q'' are required. These are given in equations (3.92). The relative slip at any point in the annulus is then found from equation (7.16) which, in this case, becomes

$$s_x = (\bar{u}'_{x1} + \bar{u}''_{x1}) + (\ddot{\bar{u}}_{x2} + \bar{u}''_{x2}) - \delta_x \tag{7.40a}$$

and

$$s_y = (\bar{u}'_{y1} + \bar{u}''_{y1}) + (\bar{u}'_{y2} + \bar{u}''_{y2}) \tag{7.40b}$$

where δ_x is given by (7.38), $\bar{u}'_x$ and $\bar{u}'_y$ are given by (7.34), $\bar{u}''_x$ and $\bar{u}''_y$ are given by (3.92). It is clear from the form of equations (3.92b) that, when substituted into equation (7.40b), the slip in the y-direction will not vanish, i.e. $s_y \neq 0$. On the other hand the frictional traction in the slip annulus, q', is assumed to be everywhere parallel to the x-axis, so that the condition that the slip must be in the direction of the frictional traction is not precisely satisfied. However, we note that the ratio of s_y to s_x is of the order $\nu/(4 - 2\nu) \approx 0.09$ so that the inclination of the resultant slip direction to the x-axis will not be more than a few degrees. We conclude, therefore, that the distribution of tangential traction which has been postulated, acting everywhere parallel to the tangential force, is a good approximation to the exact solution.

As the tangential force is increased from zero, keeping the normal force constant, the stick region decreases in size according to equation (7.39). An annulus of slip penetrates from the edge of the contact area until, when $Q_x = \mu P$, the stick region has dwindled to a single point at the origin and the bodies are on the point of sliding.

The magnitude of the slip at a radius r within the annulus $c \leqslant r \leqslant a$ is found from equation (7.40). Neglecting terms of order $\nu/(4 - 2\nu)$ this turns out to be

$$s_x \approx \frac{3\mu P}{16Ga}(2 - \nu)$$

$$\times \left\{ \left(1 - \frac{2}{\pi}\sin^{-1}\frac{c}{r}\right)\left(1 - 2\frac{c^2}{r^2}\right) + \frac{2}{\pi}\frac{c}{r}\left(1 - \frac{c^2}{r^2}\right)^{1/2} \right\} \tag{7.41a}$$

$$s_y \approx 0 \tag{7.41b}$$

The maximum value of this micro-slip occurs at the edge of the contact.

The relative tangential displacement of the two bodies is found by substituting equation (7.39) into (7.38)

$$\delta_x = \frac{3\mu P}{16a}\left(\frac{2 - \nu_1}{G_1} + \frac{2 - \nu_2}{G_2}\right)\left\{1 - \left(1 - \frac{Q_x}{\mu P}\right)^{2/3}\right\} \tag{7.42}$$

This nonlinear expression is also plotted in Fig. 7.8. For very small values of tangential force, when the slip annulus is very thin, it follows the linear relationship for no-slip (eq. (7.32)). As Q approaches μP the tangential displacement departs further from the no-slip solution until the point of sliding is reached. On the point of sliding, the overall displacement δ_x is just twice the relative slip s_x at the edge of the contact circle. Experimental measurements of the displacement by Johnson (1955) substantiate equation (7.42).

It is instructive to compare the compliance of two spherical bodies to tangential force with the compliance to normal force found from the Hertz theory (eq. (4.23)). Since the normal displacement δ_z is nonlinear with load, it is most meaningful to compare the *rates* of change of displacement with load. For bodies having the same elastic constants, differentiating equation (4.23) gives a normal compliance

$$\frac{d\delta_z}{dP} = \frac{2}{3}\left\{\frac{9}{4}\left(\frac{1-\nu^2}{E}\right)^2\left(\frac{1}{R_1}+\frac{1}{R_2}\right)\frac{1}{P}\right\}^{1/3}$$

$$= \frac{(1-\nu)}{2Ga} \tag{7.43}$$

The tangential compliance for small values of Q_x is given by equation (7.32):

$$\frac{d\delta_x}{dQ_x} = \frac{(2-\nu)}{4Ga} \tag{7.44}$$

So that the ratio of the tangential to normal compliance is $(2-\nu)/2(1-\nu)$, which varies from 1.17 to 1.5 as Poisson's ratio varies from 0.25 to 0.5 and is independent of the normal load. Thus the tangential and normal compliances are roughly similar in magnitude.

Non-conforming surfaces of general profile will have an elliptical contact area under normal load. Their behaviour under the action of a subsequently applied tangential force is qualitatively the same as for spherical bodies. Micro-slip occurs at the edge of the contact area and a stick area is found which is elliptical in shape and which has the same eccentricity as the contact ellipse. Expressions for the tangential displacement have been found by Mindlin (1949) for the case of no-slip and by Deresiewicz (1957) for partial slip.

7.3 Simultaneous variation of normal and tangential forces

In the previous section we discussed the contact stresses introduced by a steadily increasing tangential force into two bodies which were pressed into contact by a normal force which was maintained constant. We saw that the tangential force, however small, causes some slip to occur over part of the contact area. The 'irreversibility' implied by frictional slip suggests that the final state of contact stress will depend upon the history of loading and not solely upon the final values of the normal and tangential forces. Two examples demonstrate that this is indeed the case.

In the problem of the last section, in which the normal force was kept constant and the tangential force was increased, the annulus of slip spread inwards from its inner boundary. If, on the other hand, the tangential force were subsequently *decreased* this process would not simply reverse. Instead micro-slip *in the opposite direction* would begin at the edge of the contact. Hence the state of stress during unloading is different from that during loading, showing that the process is irreversible. We shall return to this problem in the next section.

As a second example, consider the case where tangential force, having been applied, is kept constant while the normal force is varied. Increasing the normal force increases the area of contact, but leaves the tangential traction unchanged, so that a traction-free annulus grows at the edge of the contact area. Decreasing the normal force causes a reduction in the contact area and thereby releases some of the tangential traction. In order to maintain equilibrium with the constant tangential force the inner boundary of the annulus of slip must contract until eventually, when P reaches the value Q/μ, the contact will slide. Clearly, in this example also, the behaviour in normal unloading is different from that during loading.

It is evident from the foregoing discussion that the state of contact stress between two bodies subjected to variations in normal and tangential load is dependent upon the sequence of application of the loads, so that the surface tractions can only be determined with certainty by following, in incremental steps, the complete loading history. In a paper of considerable complexity Mindlin & Deresiewicz (1953) have investigated the changes in surface traction and compliance between spherical bodies in contact arising from the various possible combinations of incremental change in loads: P increasing, Q increasing; P decreasing, Q increasing; P increasing, Q decreasing; etc. In this way it is possible to build up the stress and displacement variation throughout any prescribed sequence of loading.

In this section we shall consider just one example of practical interest. Two spherical bodies, pressed together initially by a normal force P_0, are subsequently compressed by an increasing oblique force F, which is inclined at a constant angle α to the common normal. This loading is equivalent to increasing the tangential load Q and the normal load P by increments in the constant proportion $\tan \alpha$ (see Fig. 7.9).

The contact radius is determined by the current value of the total normal load according to the Hertz theory (eq. (4.22)) which may be written

$$a^3 = KP \tag{7.45}$$

Due to the initial normal load

$$a_0{}^3 = KP_0 \tag{7.46}$$

During the subsequent application of the oblique force F, the increments in tangential and normal force are $dQ = dF \sin \alpha$ and $dP = dF \cos \alpha$ respectively. The incremental growth of the contact radius is given by differentiating (7.45), thus

$$3a^2\,da = K\,dP = K\,dQ/\tan \alpha \tag{7.47}$$

We will assume first that the increment of tangential force does not give rise to any slip. The consequent increment in tangential traction is then given by equation (7.30), viz.:

$$dq(r) = \frac{dQ}{2\pi a}(a^2 - r^2)^{-1/2} \tag{7.48}$$

To find the resultant distribution of tangential traction when the contact radius has grown to the value a, we substitute for dQ in (7.48) from (7.47) and integrate with respect to a.

For points originally within the contact circle the lower limit of integration is a_0, but points lying outside the original contact only start to acquire tangential traction when the contact circle has grown to include them, so that the lower limit of integration is then r.

Fig. 7.9. Circular contact subjected to a steady normal load P_0 and an oblique force F. A – $\tan \alpha < \mu$ (no micro-slip); B – $\tan \alpha > \mu$ (slip in annulus $c \leqslant r \leqslant a$).

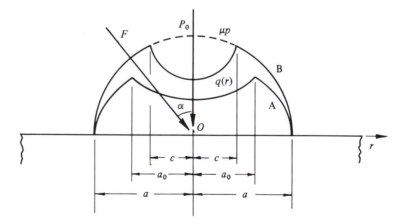

$$q(r) = \frac{3 \tan \alpha}{2\pi K} \int_{a_0}^{a_1} a(a^2 - r^2)^{-1/2} \, da$$

$$= \frac{3 \tan \alpha}{2\pi K} \{(a_1^2 - r^2)^{1/2} - (a_0^2 - r^2)^{1/2}\}, \quad 0 \leqslant r \leqslant a_0 \qquad (7.49a)$$

and

$$q(r) = \frac{3 \tan \alpha}{2\pi K} \int_{r}^{a_1} a(a^2 - r^2)^{-1/2} \, da$$

$$= \frac{3 \tan \alpha}{2\pi K} (a_1^2 - r^2)^{1/2}, \quad a_0 \leqslant r \leqslant a_1 \qquad (7.49b)$$

The normal pressure at this stage of loading is given by the Hertz theory:

$$p(r) = \frac{3(P_0 + F_1 \cos \alpha)}{2\pi a_1^3} (a_1^2 - r^2)^{1/2}$$

$$= \frac{3}{2\pi K} (a_1^2 - r^2)^{1/2} \qquad (7.50)$$

For our original assumption of no slip to be valid, $q(r)$ must not exceed $\mu p(r)$ at any point. This condition is satisfied provided that

$$\tan \alpha \leqslant \mu \qquad (7.51)$$

Thus an oblique force inclined to the normal axis at an angle less than the angle of friction produces no micro-slip within the contact area. The consequent distribution of traction is shown in Fig. 7.9, curve A; it is everywhere less than the limiting value $\mu p(r)$.

On the other hand, if the inclination of the force F exceeds the angle of friction, some slip must occur and the above analysis breaks down. An annulus of slip must develop at the edge of the contact circle. Within this annulus the tangential traction will maintain its limiting value $\mu p(r)$ at all stages of the oblique loading. The inner boundary of the annulus will lie within the original contact circle, its value being determined by the usual condition of equilibrium with the tangential force, whence

$$1 - \frac{c^3}{a^3} = \frac{F \sin \alpha}{\mu(P_0 + F \cos \alpha)} \qquad (7.52)$$

This state of affairs is shown in Fig. 7.9, curve B. The stick region will vanish $(c = 0)$ and sliding will begin when

$$F = \frac{\mu P_0}{\cos \alpha (\tan \alpha - \mu)} \qquad (7.53)$$

In the case where there is no initial compression the above results reduce to the elementary rule of dry friction: if the inclination of the oblique force is less than the angle of friction no slip will occur and, moreover, the distribution of frictional stress at the interface is everywhere proportional to the normal contact pressure, $(q = p \tan \alpha)$; if the inclination of the force exceeds the angle of friction, sliding begins at once and the frictional traction is everywhere equal to its limiting value $(q = \mu p)$.

7.4 Oscillating forces

In this section we examine contacts which are compressed by a steady mean normal load P_0 while being subjected to an oscillating force of prescribed amplitude. It will be taken for granted that the magnitude of the oscillating force is insufficient to cause the two surfaces in contact to separate or to slide at any instant during the loading cycle.

We shall consider first an oscillating *tangential* force of amplitude $\pm Q_*$ applied to spherical surfaces in contact. Since the normal force remains constant at P_0, the contact area and the normal pressure will remain constant and as given by Hertz. The first application of Q in a positive direction will cause micro-slip in the annulus $c \leqslant r \leqslant a$, where

$$c/a = (1 - Q/\mu P_0)^{1/3}$$

in the manner discussed in §2. The distribution of tangential traction is shown by curve A in Fig. 7.10(a); it reaches its limiting value in the positive sense in the annulus of slip. The tangential displacement of one body relative to the other is given by equation (7.42), and shown by OA in Fig. 7.10(b). At point A on this curve $Q = +Q_*$. The tangential force now begins to decrease, which is equivalent to the application of a *negative* increment in Q. If there were no slip during this reduction, the increment in tangential traction would be negative and infinite at the edge of the contact area. Hence there must be some negative slip immediately unloading starts and the tangential traction near to the edge must take the value $q(r) = -\mu p(r)$. During the unloading the reversed slip penetrates to a radius c' and, within this radius, there is no reversed slip. The increment in tangential traction due to unloading, by Cattaneo's technique, is therefore

$$\Delta q = -2 \frac{3\mu P_0}{2\pi a^3} (a^2 - r^2)^{1/2}, \quad c' \leqslant r \leqslant a \tag{7.54a}$$

and

$$\Delta q = -2 \frac{3\mu P_0}{2\pi a^3} \{(a^2 - r^2)^{1/2} - (c'^2 - r^2)^{1/2}\}, \quad r \leqslant c' \tag{7.54b}$$

Fig. 7.10. Circular contact subjected to a steady normal load P_0 and an oscillating tangential load of amplitude Q_*. (*a*) Traction distributions at $A\ (Q = Q_*)$; $B\ (Q = 0)$ and $C\ (Q = -Q_*)$. (*b*) Load–displacement cycle.

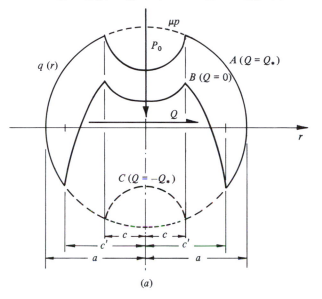

(*a*)

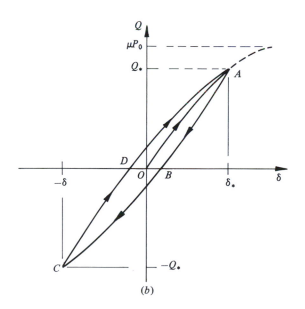

(*b*)

The resultant traction at a point on the unloading curve is then given by adding this increment to the traction at A with the result:

$$
q = \begin{cases}
-\dfrac{3\mu P_0}{2\pi a^3}(a^2 - r^2)^{1/2}, & c' \leqslant r \leqslant a & (7.55a) \\[2mm]
-\dfrac{3\mu P_0}{2\pi a^3}\{(a^2 - r^2)^{1/2} - 2(c'^2 - r^2)^{1/2}\}, & c \leqslant r \leqslant c' & (7.55b) \\[2mm]
-\dfrac{3\mu P_0}{2\pi a^3}\{(a^2 - r^2)^{1/2} - 2(c'^2 - r^2)^{1/2} + (c^2 - r^2)^{1/2}\}, & r \leqslant c & (7.55c)
\end{cases}
$$

as shown by curve B in Fig. 7.10(a). The radii of the stick regions are found from equilibrium of the traction distribution given above with the applied force. At point A

$$
\frac{Q_*}{\mu P_0} = 1 - c_*^3/a^3 \tag{7.56}
$$

During unloading

$$
\frac{Q}{\mu P_0} = \frac{Q_*}{\mu P_0} - \frac{\Delta Q}{\mu P_0} = (1 - c_*^3/a^3) - 2(1 - c'^3/a^3) \tag{7.57}
$$

which fixes the extent of reversed slip c'/a. At point B, when the tangential load is removed, $Q = 0$, so that

$$
c'^3/a^3 = \tfrac{1}{2}(1 + c_*^3/a^3) \tag{7.58}
$$

The tangential displacement during unloading is found using equation (7.38), viz.:

$$
\begin{aligned}
\delta &= \delta_* - \Delta\delta \\[1mm]
&= \frac{3\mu P_0}{16a^3}\left(\frac{2 - \nu_1}{G_1} + \frac{2 - \nu_2}{G_2}\right)\{(a^2 - c_*^2) - 2(a^2 - c'^2)\} \\[1mm]
&= \frac{3\mu P_0}{16a}\left(\frac{2 - \nu_1}{G_1} + \frac{2 - \nu_2}{G_2}\right)\left\{2\left(1 - \frac{Q_* - Q}{2\mu P_0}\right)^{2/3}\right. \\[1mm]
&\quad \left. - \left(1 - \frac{Q_*}{\mu P_0}\right)^{2/3} - 1\right\}
\end{aligned} \tag{7.59}
$$

This expression is shown in Fig. 7.10(b) by the curve ABC. At point C, when the tangential force is completely reversed, substituting $Q = -Q_*$ in equations (7.57) and (7.59) gives

$$
c = c_* \quad \text{and} \quad \delta = -\delta_*
$$

Thus the reversed slip has covered the original slip annulus and the distribution of tangential traction is equal to that at A, but of opposite sign. The conditions at C are a complete reversal of those at A, so that a further reversal of Q produces a sequence of events which is similar to unloading from A, but of opposite sign. The displacement curve CDA completes a symmetrical hysteresis loop.

The work done by the tangential force during a complete cycle, represented by the area of the loop, is dissipated by a reversal of micro-slip in the annulus $c \leqslant r \leqslant a$. This problem was first studied by Mindlin *et al.* (1952), who derived an expression for the energy dissipated per cycle, viz.:

$$\Delta W = \frac{9\mu^2 P_0^2}{10a} \left(\frac{2-\nu_1}{G_1} + \frac{2-\nu_2}{G_2} \right)$$

$$\times \left[1 - \left(1 - \frac{Q_*}{\mu P_0} \right)^{5/3} - \frac{5Q_*}{6\mu P_0} \left\{ 1 - \left(1 - \frac{Q_*}{\mu P_0} \right)^{2/3} \right\} \right] \tag{7.60}$$

During repeated oscillation a tangential force might be expected to produce some attrition of interface in the annulus where oscillating slip is taking place. Measurements by Goodman & Brown (1962) of the energy dissipated in micro-slip compare favourably with equation (7.60).

We turn now to the case where the line of action of the oscillating force $\pm F_*$ is not tangential to the surface, but makes a constant angle α to the z-axis. If the inclination of the force is less than the angle of friction, we saw in the last section that the first application of the oblique force F causes no slip anywhere in the contact area. The derivation of this result is equally applicable when F is decreasing, so no slip would be expected and hence no energy would be dissipated in a cycle of oscillation.

When the inclination of the force exceeds the angle of friction, slip arises on first loading and on unloading. Mindlin & Deresiewicz (1953) have traced the variations in slip, traction and tangential compliance when two spherical bodies, compressed by a steady normal load P_0, are subjected to a cyclic oblique force F which oscillates between the extreme values F_* and $-F_*$. They show that the first loading from $F = 0$ to $F = +F_*$ and the first unloading from $F = +F_*$ to $F = -F_*$ are unique. Subsequently a steady cycle is repeated. The contact radius varies from a_* to a_{-*} determined by the maximum and minimum values of the normal load, i.e. $P_0 \pm F_* \cos \alpha$. Oscillating slip occurs in the annulus $c_* \leqslant r \leqslant a_*$, where

$$\left(\frac{c_*}{a_*} \right)^3 = \frac{P_0 - F_* (\sin \alpha)/\mu}{P_0 + F_* \cos \alpha} \tag{7.61}$$

The energy dissipated per cycle in oscillating slip is shown to be:

$$\Delta W = \frac{9(\mu P_0)^2}{10a_0} \left(\frac{2-\nu_1}{G_1} + \frac{2-\nu_2}{G_2} \right) \left[\frac{1}{4\lambda} \left\{ \frac{1+\lambda}{1-\lambda} (1 - \lambda L_*)^{5/3} \right. \right.$$

$$\left. \left. - \frac{1-\lambda}{1+\lambda} (1 + \lambda L_*)^{5/3} \right\} - \frac{6 - L_* - 5\lambda^2 L_*}{6(1-\lambda^2)} (1 - L_*)^{2/3} \right] \tag{7.62}$$

where $L_* = F_* \sin \alpha / \mu P_0$, $\lambda = \mu / \tan \alpha$ and a_0 is the contact radius due to P_0. If the force acts in a tangential direction, $\alpha = \pi/2$ and $\lambda = 0$. Equation (7.62) then reduces to equation (7.60). When the angle α diminishes to the value $\tan^{-1} \mu$, $\lambda = 1$ and the energy loss given by (7.62) vanishes.

This interesting result – that oscillating forces, however small their amplitude compared with the steady compressive load, produce oscillating slip and consequent energy dissipation if their inclination to the normal exceeds the angle of friction – has been subjected to experimental scrutiny by Johnson (1961) using a hard steel sphere in contact with a hard flat surface. The angle of friction was approximately $29°$ ($\mu \approx 0.56$). Photographs of the surface attrition due to repeated cycles of oscillating force are shown in Fig. 7.11. Measurements of the energy dissipated per cycle at various amplitudes of force F_* and angles of obliquity α are plotted in Fig. 7.12. Serious surface damage is seen to begin

Fig. 7.11. Annuli of slip and fretting at the contact of a steel sphere and flat produced by an oscillating oblique force at an angle α to the normal.

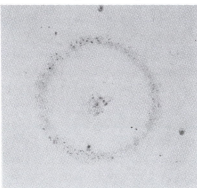

(a) $\alpha = 20°$.

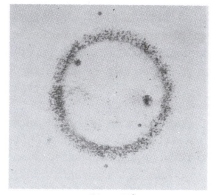

(b) $\alpha = 30°$.

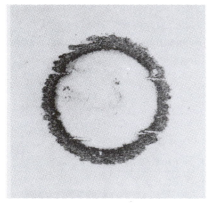

(c) $\alpha = 60°$.

(d) $\alpha = 90°$.

at values of α in excess of $29°$, when the theory would predict the onset of slip, and the severity of attrition is much increased as α approaches $90°$. This is consistent with the large increase in energy dissipated as the angle α is increased. There is generally reasonable agreement between the measured energy dissipation and that predicted by equation (7.62), taking $\mu = 0.56$. The small energy loss measured at $\alpha = 0$ is due to elastic hysteresis.

It is evident from Fig. 7.11 that some slight surface damage occurred at angles at which no slip would be expected. More severe damage has been observed by Tyler *et al.* (1963), within the annulus $a_* < r < a_{-*}$, under the action of a purely normal load. The difference in curvature between the sphere and the mating flat surface must lead to tangential friction and possible slip, but this effect is very much of second order and cannot be analysed using small-strain elastic theory. It is more likely that the damage is associated with plastic deformation of the surface asperities.

The contact problems involving oscillating forces discussed in this section are relevant to various situations of engineering interest. Oscillating micro-slip at

Fig. 7.12. Energy dissipated in micro-slip when a circular contact is subjected to a steady load P_0 and an oscillating oblique force of amplitude F_* at an angle α to the normal. Eq. (7.62) compared with experimental results (Johnson, 1961).

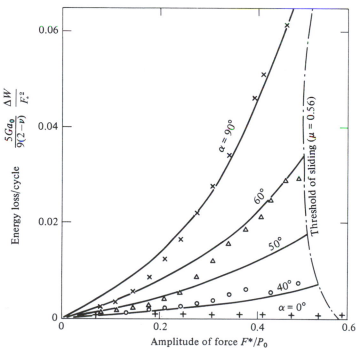

the interface between two surfaces which are subjected to vibration, often combined with corrosion, produces the characteristic surface damage described as 'fretting'. In components which also carry a high steady stress, the presence of fretting can lead to premature failure of the component by fatigue. An ideal solution to this problem is to eliminate the possibility of micro-slip. From the examples discussed in this section, two lessons can be learnt. Firstly, the design should be arranged so that the line of action of the oscillating force is close to the direction of the common normal to the two mating surfaces. Secondly, the profiles of the two contacting surfaces should be designed so that, when they are in contact and under load, high concentrations of tangential traction at the edge of the contact area are avoided. This means that the 'sharp notch' which arises at the edge of the contact of non-conforming surfaces must be avoided. Such a modification to prevent micro-slip is illustrated in Fig. 7.13. These questions have been discussed in a paper by Johnson & O'Connor (1964).

The energy dissipated in micro-slip at surfaces in contact provides one source of vibration damping in a built-up mechanical structure. In particular we have seen how slip and frictional damping may be expected even though the amplitude of the oscillating force transmitted by the surfaces in contact is only a small fraction of the force necessary to cause bulk sliding at the interface. If the amplitude of oscillation is small, i.e. if $Q_*/\mu P_0 \ll 1.0$, equation (7.60) for the energy loss per cycle reduces to

$$\Delta W = \frac{1}{36a\mu P_0} \left(\frac{2 - \nu_1}{G_1} + \frac{2 - \nu_2}{G_2} \right) Q_*^{\,3} \tag{7.63}$$

i.e. the energy loss is proportional to the *cube* of the amplitude of the oscillating force. The same is true for an oblique force (eq. (7.62)). In a review of the damping in built-up structures arising from interfacial slip Goodman (1960) shows that the variation of energy loss with the cube of the amplitude is a general rule which applies to clamped joints having a variety of geometric forms. These cases all have one feature in common: they develop a region of micro-slip which

Fig. 7.13. Influence of the profiles of contacting bodies on micro-slip and fretting.

(a)   (b)

grows in size in direct proportion to amplitude of the force. Experimental measurements of slip damping however tend to result in energy losses which are more nearly proportional to the *square* of the amplitude at small amplitudes. In part this discrepancy between theory and experiments is due to inelastic effects within the body of the solids (internal hysteresis damping) which provide a large proportion of the measured damping at small amplitudes. But variations in the coefficient of friction and effects of roughness of the experimental surface are also influential (see Johnson, 1961).

A third practical application of the contact stress theory discussed in this section lies in the mechanics of granular media. Mindlin and his colleagues used the compliance of elastic spheres in contact to calculate the speed of propagation of elastic waves through an idealised granular 'solid' made up of elastic spheres packed in a regular array. This work is summarised by Mindlin (1954) and by Deresiewicz (1958).

7.5 Torsion of elastic spheres in contact

A situation which is qualitatively similar to those discussed in the previous sections of this chapter arises when two elastic solids are pressed together by a constant normal force and are then subjected to a varying twisting or 'spinning' moment about the axis of their common normal. The physical behaviour is easy to visualise. The normal force produces an area of contact and distribution of normal pressure given by the Hertz theory. The twisting moment causes one body to rotate about the z-axis through a small angle β relative to the other. Slip at the interface is resisted by frictional traction. Each body is regarded as an elastic half-space from the point of view of calculating its elastic deformation. Under the action of a purely twisting couple M_z the state of stress in each body is purely torsional, i.e. all the direct stress components vanish, as discussed in §3.9. In the case of spheres in contact the system is axi-symmetrical; $\tau_{r\theta}$ and $\tau_{z\theta}$ are the only non-zero stress components and u_θ is the only non-zero displacement.

If there were to be no slip at the interface, it follows that the contact surface must undergo a rigid rotation relative to distant points in each body. Thus

$$\bar{u}_{\theta 1} = \beta_1 r, \quad \bar{u}_{\theta 2} = -\beta_2 r$$

The distribution of tangential traction to produce a rigid rotation of a circular region on the surface of an elastic half-space is shown to be (equations (3.109) and (3.111))

$$q(r) = \frac{3M_z r}{4\pi a^3} (a^2 - r^2)^{-1/2} \tag{7.64}$$

where $q(r)$ acts in a circumferential direction at all points in the contact circle

$r \leqslant a$. The rotation is given by

$$\beta = \beta_1 + \beta_2 = \tfrac{3}{16}\left(\frac{1}{G_1} + \frac{1}{G_2}\right)\frac{M_z}{a^3} \qquad (7.65)$$

A similar result for bodies of general profile, whose area of contact is elliptical, follows from equations (3.114) and (3.115).

As we might expect, the surface traction to prevent slip entirely, given by equation (7.64), rises to infinity at the edge of the contact circle so that an annular region of slip will develop. This slip will be in a circumferential direction and the surface traction in the slip annulus $c \leqslant r \leqslant a$ will take its limiting value

$$q(r) = \mu p(r) = \frac{3\mu P}{2\pi a^3}(a^2 - r^2)^{1/2} \qquad (7.66)$$

Fig. 7.14. A circular contact area subjected to a twisting moment M_z. (a) Shear traction $q(r/a)$: A – no slip, eq. (7.64); B – partial slip, taking $c = a/\sqrt{2}$. (b) Angle of twist β: A – no slip; B with slip.

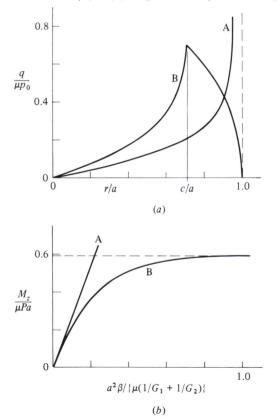

(a)

(b)

The radius c of the stick region is given by

$$\frac{3}{4\pi}(1 - c^2/a^2)\left(\frac{1}{G_1} + \frac{1}{G_2}\right)\mathbf{D}(k) = a^2\beta/\mu P \tag{7.67}$$

where $\mathbf{D}(k) = \mathbf{K}(k) - \mathbf{E}(k); k = (1 - c^2/a^2)^{1/2}$, and $\mathbf{K}(k)$ and $\mathbf{E}(k)$ are complete elliptic integrals of the first and second kind respectively. The distribution of traction in the stick circle and the relation between the twisting moment M_z and the angle of twist β are shown in Fig. 7.14(a) and 7.14(b) respectively. These results are due to Lubkin (1951).

As the twisting moment is increased the radius of the stick region decreases and eventually shrinks to a point in the centre. One body is then free to 'spin' relative to the other resisted by a constant moment

$$M_z = 3\pi\mu Pa/16 \tag{7.68}$$

The shear stress reaches its limiting value at all points. The state of stress within the solid under the combined action of this tangential traction and the Hertz pressure has been investigated by Hetenyi & McDonald (see §3.9).

Spherical bodies in contact which are subjected to an oscillating twisting moment $\pm M_z^*$ have been studied by Deresiewicz (1954). Unloading after the application of a moment M_z^* leads to a reversal of micro-slip at the edge of the contact circle with consequent hysteresis in the angle of twist. In a complete oscillation of the moment a hysteresis loop is traced out, similar to the loop arising from an oscillating tangential force shown in Fig. 7.10. Deresiewicz shows that the energy dissipated per cycle, represented by the area of the hysteresis loop, is given by

$$\Delta W = \frac{2\mu^2 P^2}{Ga}\left[\frac{8}{9}\left\{1 - \left(1 - \frac{3}{2}\frac{M_z^*}{\mu Pa}\right)^{2/3}\right\}\right.$$
$$\left. - \frac{M_z^*}{\mu Pa}\left\{1 + \left(1 - \frac{3}{2}\frac{M_z^*}{\mu Pa}\right)^{1/2}\right\}\right] \tag{7.69}$$

For small amplitudes $(M_z^* \ll \mu Pa)$ this expression reduces to

$$\Delta W \approx 3M_z^{*3}/16Ga^4\mu P \tag{7.70}$$

in which the energy loss per cycle is again proportional to the cube of the amplitude of the twisting moment.

7.6   Sliding of rigid-perfectly-plastic bodies

In §6.2 we considered the contact of perfectly plastic bodies under the action of normal compression. The plastic deformation was regarded as sufficiently large to justify neglecting the elastic strains and applying the theory of rigid-perfectly-plastic solids. We shall now consider such contacts to which tangential as well as normal forces are applied so that sliding motion or, at least, incipient sliding occurs. The analyses presented in this section have been applied mainly to the interaction between the irregularities on the sliding surfaces of ductile solids and thereby relate to theories of friction and wear (see Bowden & Tabor, 1951). The simplest example which has been solved completely is that of a plastic wedge which is compressed and subsequently sheared (Johnson, 1968a). We shall take this example first.

(a) Combined shear and pressure on a plastic wedge

A rigid-perfectly-plastic wedge, of semi-angle α, is deformed by a rigid flat die. If the interface were frictionless no tangential force could be realised, so that sliding could not contribute to the deformation. We shall assume the opposite extreme: that there is no slip at the interface.

Under an initial compressive load P the wedge is crushed as described in §6.2(c) and shown in Fig. 6.8(b). The pressure on the die face is given by equation (6.28) viz.:

$$p_0 = P/l_0 = 2k(1 + \psi_0)$$

The normal force is now kept constant and a steadily increasing tangential force Q is applied, which introduces a shear traction q at the interface and causes the slip lines to meet the die face at angles $\pi/4 \pm \phi$. The slip-line field for the combined action of P and Q is shown in Fig. 7.15(a). The triangle ABC adheres to the face of the die and moves with it both normally and tangentially. Triangles BDE and EHJ also move as rigid blocks. The left-hand shoulder of the wedge (apex A) is unloaded by Q and does not deform further. At the stage of the process shown in Fig. 7.15(a) the die pressure is

$$p = k(1 + 2\psi + \cos 2\phi) = p_0 - 2k(\phi - \sin^2 \phi) \tag{7.71}$$

and the shear stress on the die face is

$$q = 2k \sin \phi \cos \phi \tag{7.72}$$

Hence

$$\frac{Q}{P} = \frac{q}{p} = \frac{\sin \phi \cos \phi}{(1 + \psi_0) - (\phi + \sin^2 \phi)} \tag{7.73}$$

As the tangential force is increased, ϕ increases and, in due course, $\phi = \pi/4$. A second stage of deformation is then reached, illustrated in Fig. 7.15(b), in which

$$p = k(1 + 2\psi) \tag{7.74}$$

and

$$q = k = \text{constant} \tag{7.75}$$

hence

$$\frac{Q}{P} = \frac{1}{1 + 2\psi} \tag{7.76}$$

Further increases in Q causes the block BHJ to rotate clockwise so that ψ steadily decreases to zero and Q/P approaches unity. At this point sliding might be expected by shear of the interface AB. With real materials, which strain-harden, it seems likely that further bulk deformation could occur by shearing virgin material along a slip line from A to J.

It is apparent from the analysis that the tangential force Q causes an increase in contact area even though the normal load P remains constant. This process has been called 'junction growth' by Tabor (1959). In plane deformation the

Fig. 7.15. Combined shear and pressure on a rigid-perfectly-plastic wedge ($\alpha = 60°$): (a) first stage ($Q/P < 0.72$); (b) second stage ($0.72 < Q/P < 1.0$).

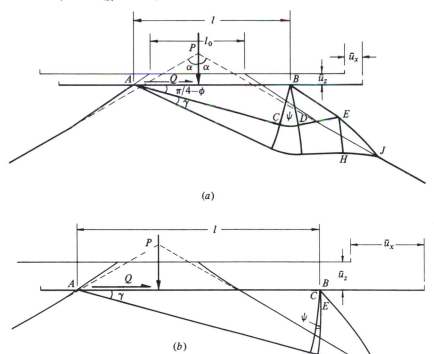

(a)

(b)

area increase is given by

$$\frac{A}{A_0} = \frac{l}{l_0} = \frac{p_0}{p} \tag{7.77}$$

where p is given by equations (7.71) and (7.74). The area growth for a wedge of angle $\alpha = 60°$ is shown in Fig. 7.16.

So far it has been assumed that adhesion between the wedge and the die is sufficient to prevent sliding at the interface. The shear stress at the interface is also plotted against Q/P in Fig. 7.16. If, due to contamination or lubrication, the shear strength of the interface is less than that of the solid k, then the process of plastic deformation and area growth will be interrupted by premature sliding at the interface. Fig. 7.16 is instructive in relation to metallic friction: it shows the importance of surface contamination upon the effective coefficient of limiting friction. For example, if the maximum shear stress which the interface can sustain is reduced by contamination to one half the shear strength of the metal (Q/P) is only 0.15. At the other extreme, if the strength of the interface approaches k, the analysis suggests that the coefficient of friction approaches 1.0, a value which is consistent with experiments on a chemically clean ductile material.

Collins (1980) has examined the situation where a wedge is deformed by a rigid flat die under the action of normal and tangential forces which increase

Fig. 7.16. Contact area growth of plastic wedge ($\alpha = 60°$) under the action of a constant normal load P and an increasing tangential load Q.

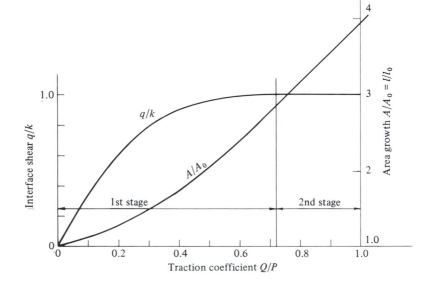

in a fixed ratio. At small or moderate ratios of Q/P the modes of deformation of the wedge are similar to those shown in Fig. 7.15. For a ratio $Q/P \to 1.0$ a different mode of deformation is possible in which a spiral chip is formed as shown in Fig. 7.17(b). Experiments suggest that this mode of deformation may also occur in the case of a constant normal load discussed previously, when $Q/P \to 1.0$, if the material is capable of strain hardening.

Fig. 7.17. Application of an oblique force to a plastic wedge. When $Q/P \to 1.0$ a 'curly chip' is formed (Collins, 1980).

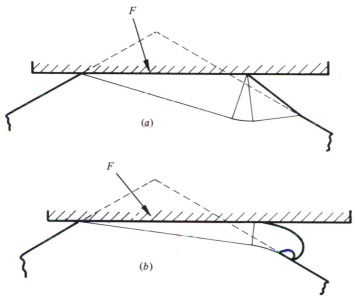

(b) Ploughing of a rigid-plastic surface by a rigid wedge

In this example we consider a rigid-perfectly-plastic half-space which is indented by a rigid wedge and is subsequently ploughed by the action of a tangential force applied to the wedge perpendicular to its axis. In the first instance we shall take the faces of the wedge to be frictionless. The deformation is determined by constructing slip-line fields and their associated hodographs progressively in the stages shown in Fig. 7.18. In stage (1) the application of the normal load P alone produces an indentation of depth c_0 as shown. The subsequent application of a steadily increasing tangential force Q, whilst keeping P constant, unloads the left-hand face of the wedge so that further deformation takes place by the wedge sliding down the left-hand shoulder of the initial indentation. The wedge continues to slide deeper until $Q = P \tan \alpha$, when the pressure on the left-hand face of the wedge has fallen to zero (stage 2). During subsequent deformation the force F on the active face of the wedge remains constant ($=P$ cosec α). As material from the trough is displaced into the prow, the wedge rides up on its own 'bow wave' such that the contact length h satisfies the relationship:

$$F = ph = 2k(1 + \psi)h = \text{constant} \tag{7.78}$$

As deformation proceeds (stages 3–6) the apex of the wedge approaches the free surface asymptotically. In the steady state a plastic wave is pushed along the surface, like a wrinkle in a carpet, giving a permanent shear displacement δ to the surface layer of the half-space. The steady-state behaviour ($\alpha < \pi/4$) is illustrated in Fig. 7.19(a). In this condition

$$\left. \begin{array}{l} p_s = 2k(1 + \psi_s) = 2k(1 - \pi/2 + 2\alpha) \\ P = p_s h_s \sin \alpha; \quad Q = p_s h_s \cos \alpha \end{array} \right\} \tag{7.79}$$

Through the progressive construction of the slip-line fields and hodographs the trajectory followed by the apex of the wedge can be found and is plotted in Fig. 7.20. In common with crushing a plastic wedge, the application of tangential force causes the surfaces firstly to sink together, thereby increasing the area of contact and decreasing the contact pressure. In the case of the indenting wedge, this first stage is followed by a period in which the wedge climbs back to the surface pushing a plastic 'wave' ahead of it along the surface.

Friction at the wedge face modifies the behaviour described above. Moderate friction causes the slip lines to meet the wedge face at $\pi/4 \pm \phi$, but severe friction results in adhesion between the wedge face and the solid. Initial penetration then follows the field shown in Fig. 6.5 or 6.6. As a tangential force is applied the wedge sinks in further, at 45° to begin with, and then at the angle of the wedge face. Proceeding as before progressive construction of the slip-line fields and hodographs enables the trajectory of the wedge and forces on the wedge face

Fig. 7.18. Ploughing of a rigid-plastic surface by a rigid frictionless wedge ($\alpha = 60°$), under the action of a steady normal load P and an increasing tangential force Q. Progressive stages of deformation.

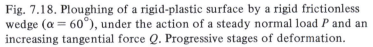
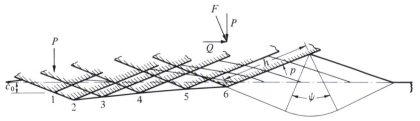

Fig. 7.19. Plastic bow wave produced by the steady-state sliding of a rigid wedge over a perfectly plastic surface: (*a*) frictionless; (*b*) with no slip at the wedge face.

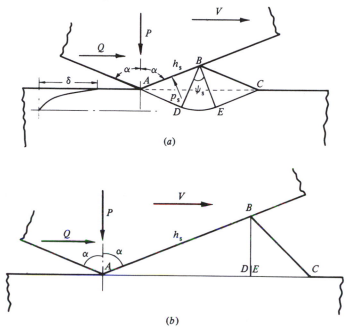

(*a*)

(*b*)

Fig. 7.20. Transient ploughing of a plastic surface by a rigid wedge: ratio of tangential to normal force Q/P; ratio of shear stress to normal pressure q/p on wedge face; depth of penetration δ/c_0.

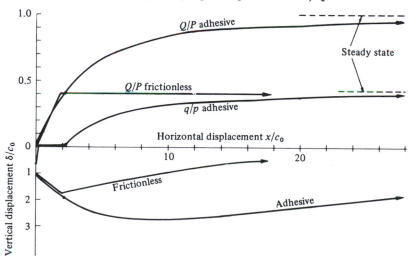

to be found (see Fig. 7.20). The adhesive wedge ploughs much deeper than the frictionless wedge and builds up a larger prow ahead. The situation in the limit when the wedge reaches the surface level is shown in Fig. 7.19(*b*). At this stage the ratio Q/P approaches unity and the prow is free to shear off along the slip line $ADEC$. The ratio q/p of shear to normal traction on the wedge face during the process is also plotted in Fig. 7.20. For adhesion to be maintained throughout, the coefficient of friction between the wedge face and the solid must exceed cot α.

The development of a prow of heavily deformed material in the manner just described has been demonstrated by Cocks (1962) and others in sliding experiments with clean ductile metals. Challen & Oxley (1979) refer to a plastic wave (Fig. 7.19(*a*)) as the 'rubbing mode' of deformation, since material is smeared along the surface without being detached. They refer to the deformation shown in Fig. 7.19(*b*), in which a prow builds up and may become detached, as the 'wear mode' and investigate the conditions of wedge angle and interfacial friction which lead to one mode or the other. They also show that a third mode – the 'cutting mode' – is possible; the wedge does not rise to the level of the free surface but produces a continuous chip in the manner of a cutting tool.

(c) Interaction of two rigid-plastic wedges

In the study of sliding friction between rough metal surfaces, the interaction of two wedges, which interfere and mutually deform as they pass, is of interest. The situation is different from the above examples in that the relative motion of the two wedges is taken to be purely tangential. In these circumstances both the tangential force Q and the normal force P vary throughout a contact cycle; indeed, with strong adhesion at the interface, P becomes *tensile* just before the wedges separate.

Slip-line fields for the initial deformation have been proposed by Green (1954). Complete cycles of contact have been analysed by Edwards & Halling (1968*a* & *b*), using the approximate 'upper bound' method, which are in reasonable agreement with experiments (see also Greenwood and Tabor, 1955).

(d) Ploughing by three-dimensional indenters

Most practical instances of ploughing, the action of a grit on a grinding wheel for example, involve three-dimensional deformation; a furrow is produced and material is displaced sideways. The simplest model of this process is that of a rigid cone or pyramid sliding steadily over the surface, but even these cases are difficult to analyse.

By assuming a simple mode of deformation in which a cap of material adheres to the front of the cone, Childs (1970) has used the plastic work method to obtain approximate values for the tangential force Q to plough a furrow of prescribed depth. Material from the furrow is ploughed into shoulders on either side. Good agreement was found with measurements of the ploughing force, but the height of the shoulders was observed to be appreciably less than in the theoretical deformation mode.

8

Rolling contact of elastic bodies

8.1 Micro-slip and creep

In Chapter 1 *rolling* was defined as a relative angular motion between two bodies in contact about an axis parallel to their common tangent plane (see Fig. 1.1). In a frame of reference which moves with the point of contact the surfaces 'flow' through the contact zone with tangential velocities V_1 and V_2. The bodies may also have angular velocities ω_{z1} and ω_{z2} about their common normal. If V_1 and V_2 are unequal the rolling motion is accompanied by *sliding* and if ω_{z1} and ω_{z2} are unequal it is accompanied by *spin*. When rolling occurs without sliding or spin the motion is often referred to as 'pure rolling'. This term is ambiguous, however, since absence of apparent sliding does not exclude the transmission of a tangential force Q, of magnitude less than limiting friction, as exemplified by the driving wheels of a vehicle. The terms *free rolling* and *tractive rolling* will be used therefore to describe motions in which the tangential force Q is zero and non-zero respectively.

We must now consider the influence of elastic deformation on rolling contact. First the normal load produces contact over a finite area determined by the Hertz theory. The specification of 'sliding' is now not so straightforward since some contacting points at the interface may 'slip' while others may 'stick'. From the discussion of incipient sliding in §7.2, we might expect this state of affairs to occur if the interface is called upon to transmit tangential tractions. A difference between the tangential strains in the two bodies in the 'stick' area then leads to a small apparent slip which is commonly called *creep*. The way in which creep arises may be appreciated by the example of a deformable wheel rolling on a relatively rigid plane surface. If, owing to elastic deformation under load, the tangential (i.e. circumferential) strain in the wheel is tensile, the surface of the wheel is stretched where it is in sticking contact with the plane. The wheel then behaves as though it had an enlarged circumference and, in one revolution, moves forward a distance greater than its undeformed perimeter by a fraction known as the *creep ratio*. If the tangential strain in the wheel is compressive the effect is reversed.

The phenomenon of creep was described by Reynolds (1875) in a remarkably perceptive paper. He recognised that the contact region would be divided into stick and micro-slip zones in a manner determined by the interplay of friction forces and elastic deformation, and supported his arguments by creep measurements using a rubber cylinder rolling on a metal plane and vice versa.†

The boundary conditions which must be satisfied in the stick and micro-slip regions of rolling contact will now be developed. In our coordinate system we shall take rolling to be about the y-axis so that, in the absence of deformation and sliding, material particles of each surface flow through the contact region parallel to the x-axis with a common velocity V known as the rolling speed. In addition the bodies may have angular velocities ω_{z1} and ω_{z2} due to spin. The application of tangential tractions and the resulting deformation introduce creep velocities δV_1 and δV_2, each having components in both the x and y directions, which are small compared with the rolling speed V. This is the Eulerian view in which the material moves while the field of deformation remains fixed in space. The velocity of a material element is also influenced by the state of strain in the deformed region. If the components of tangential elastic displacement at a surface point (x, y) are $\bar{u}_x(x, y, t)$ and $\bar{u}_y(x, y, t)$ the 'undeformed' velocity is modified by the components

$$\frac{\mathrm{d}\bar{u}_x}{\mathrm{d}t} = V\frac{\partial \bar{u}_x}{\partial x} + \frac{\partial \bar{u}_x}{\partial t}$$

and

$$\frac{\mathrm{d}\bar{u}_y}{\mathrm{d}t} = V\frac{\partial \bar{u}_y}{\partial x} + \frac{\partial \bar{u}_y}{\partial t}$$

Hence the resultant particle velocities at a general surface point, taking into account creep, spin and deformation, are given by the expressions:

$$v_x(x, y) = V + \delta V_x - \omega_z y + V\frac{\partial \bar{u}_x}{\partial x} + \frac{\partial \bar{u}_x}{\partial t} \tag{8.1a}$$

† Reynolds found that a rubber cylinder moved forward more than its undeformed circumference in one revolution and hence deduced that the circumferential strain was *tensile*. He explained this result by the influence of Poisson's ratio on the radial *compressive* strain produced by the normal load. However, it follows from the Hertz theory (1882a) that the tangential strain is, in general, *compressive* and for an incompressible material like rubber is zero. The explanation for the anomaly lies in Reynolds' use of a relatively thin rubber cover on a rigid hub for his deformable roller. In these circumstances the tangential strains in the cover are tensile and consistent with the experiments (see Bentall & Johnson, 1967).

and

$$v_y(x, y) = \delta V_y + \omega_z x + V \frac{\partial \bar{u}_y}{\partial x} + \frac{\partial \bar{u}_y}{\partial t} \tag{8.1b}$$

If the strain field does not change with time, which would be the case in *steady rolling* (i.e. uniform motion under constant forces), the final terms in these expressions vanish. The terms $(\partial \bar{u}_x/\partial x)$ and $(\partial \bar{u}_y/\partial x)$ arise from the state of strain in the surface, which can be found if the surface tractions are known. They are necessarily small compared with unity. The velocities of micro-slip between contacting points in steady rolling are then given by

$$\dot{s}_x(x, y) \equiv v_{x1} - v_{x2}$$

$$= (\delta V_{x1} - \delta V_{x2}) - (\omega_{z1} - \omega_{z2})y + V \left(\frac{\partial \bar{u}_{x1}}{\partial x} - \frac{\partial \bar{u}_{x2}}{\partial x} \right) \tag{8.2a}$$

and

$$\dot{s}_y(x, y) \equiv v_{y1} - v_{y2}$$

$$= (\delta V_{y1} - \delta V_{y2}) + (\omega_{z1} - \omega_{z2})x + V \left(\frac{\partial \bar{u}_{y1}}{\partial x} - \frac{\partial \bar{u}_{y2}}{\partial x} \right) \tag{8.2b}$$

For an elliptical contact area of semi-axes a and b, we rewrite (8.2) in non-dimensional form to give

$$\dot{s}_x/V = \xi_x - \psi y/c + \left(\frac{\partial \bar{u}_{x1}}{\partial x} - \frac{\partial \bar{u}_{x2}}{\partial x} \right) \tag{8.3a}$$

$$\dot{s}_y/V = \xi_y + \psi x/c + \left(\frac{\partial \bar{u}_{y1}}{\partial x} - \frac{\partial \bar{u}_{y2}}{\partial x} \right) \tag{8.3b}$$

where $\xi_x \equiv (\delta V_{x1} - \delta V_{x2})/V$ and $\xi_y \equiv (\delta V_{y1} - \delta V_{y2})/V$ are the *creep ratios* ψ is the non-dimensional *spin parameter* $(\omega_{z1} - \omega_{z2})c/V$ and $c = (ab)^{1/2}$.

In a stick region

$$\dot{s}_x = \dot{s}_y = 0 \tag{8.4}$$

In addition, the resultant tangential traction must not exceed its limiting value, viz.:

$$|q(x, y)| \leq \mu p(x, y) \tag{8.5}$$

where μ is the coefficient of limiting friction.

In a slip region, on the other hand,

$$|q(x, y)| = \mu p(x, y) \tag{8.6}$$

and the direction of q must oppose the slip velocity, viz.:

$$\frac{q(x, y)}{|q(x, y)|} = - \frac{\dot{s}(x, y)}{|\dot{s}(x, y)|} \tag{8.7}$$

Equations (8.4)–(8.7) specify the boundary conditions at the interface of two bodies in steady rolling contact; the first two apply in that part of the contact area where there is no slip and the second two apply in the zone of micro-slip. It will be appreciated from the examples which we shall discuss in the remainder of this chapter that one of the chief difficulties of such problems lies in finding the configurations of the stick and slip zones. In this respect the conditions at the boundaries of the contact area are significant. At the leading edge, where the material is flowing into the contact, the strain and hence the velocity must be continuous across the boundary. We saw in §2.5 that a discontinuity in tangential traction q at the boundary leads to a singularity in surface strain just outside the contact. It follows, therefore, that $q = 0$ at all points on the leading edge of the contact. The situation at the trailing edge is different. If we postulate a coefficient of friction sufficiently high to prevent any slip until the material emerges from the trailing edge, such that there is a discontinuity in traction, then the instantaneous change in strain and hence velocity at that point implies that, in reality, some of the stored elastic energy is irreversibly dissipated. For such a situation to arise in reverse at the leading edge would clearly contravene thermodynamic principles.

Creep of an elastic belt

The most elementary example of creep and micro-slip in rolling contact is provided by a flexible elastic belt which is transmitting a torque M between two equal pulleys (see Fig. 8.1). If $T_1(= T_0 + t)$ and $T_2(= T_0 - t)$ are the tensions on the tight side and slack side respectively,

$$T_1 - T_2 = 2t = M/R \qquad (8.8)$$

The belt slips on each pulley over an arc $R\phi$ given by the capstan formula:

$$e^{\mu\phi} = \frac{T_1}{T_2} = \frac{1 + M/2RT_0}{1 - M/2RT_0} \qquad (8.9)$$

where it is assumed that the torque M is sufficiently small for the arc of slip to be less than the arc of the belt (i.e. $\phi < \pi$). The question now arises: where is the slip arc located on each pulley? The tensile strain in an elastic belt of extensibility λ is given by

$$\epsilon = \lambda T$$

so that an element of belt of unstretched length dx, when under tension, has length $dl = (1 + \epsilon)\,dx$. The velocity of the element is thus given by

$$v = \frac{dl}{dt} = (1 + \epsilon)\frac{dx}{dt} = (1 + \lambda T)V \qquad (8.10)$$

where $V\,(= dx/dt)$ is the 'unstretched' velocity of the belt. This expression is

Fig. 8.1. Creep of a flexible belt transmitting a torque M between pulleys rotating at ω_1 and ω_2.

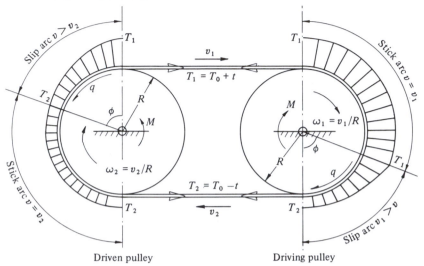

Driven pulley   Driving pulley

consistent with the general equation (8.1). Thus the velocity of the tight side v_1 is greater than that of the slack side v_2. We note that frictional traction q pulls the belt forward on the driving pulley, but drags it back on the driven pulley, as shown. Since friction must oppose the direction of slip (condition (8.7)) the pulley must be moving faster than the belt in the slip arc of the driver and slower than the belt in the slip arc of the driven pulley. The peripheral speed of the driving pulley must therefore be v_1, so that the stick arc is located where the belt runs onto the pulley. Similarly the peripheral speed of the driven pulley must coincide with v_2, whereupon the stick arc occurs at the run on to the driven pulley also. The creep ratio for the whole system may be defined by:

$$\xi \equiv \frac{(\omega_1 - \omega_2)R}{V} = \frac{v_1 - v_2}{V} = \lambda(T_1 - T_2) = \lambda M/R \qquad (8.11)$$

Thus the driven pulley runs slightly slower than the driving pulley in proportion to the transmitted torque. The loss of power is expended in frictional dissipation in the slip arcs. The features exhibited by this one-dimensional example arise in the more complex cases now to be considered.

8.2   Freely rolling bodies having dissimilar elastic properties

Two elastic bodies which are geometrically identical and have the same elastic properties are completely symmetrical about their interface. When they roll freely under the action of a purely normal force, no tangential traction or slip can occur, so that the contact stresses and deformation are given by the Hertz theory of static contact. In these circumstances the rolling process is completely reversible in the thermodynamic sense.

The problem of the stresses and micro-slip at the rolling contact of two bodies whose elastic constants differ, which was discussed qualitatively by Reynolds in 1875, has had to wait for nearly a century for a quantitative solution. The problem arises through a difference in the tangential strains in the two surfaces if the elastic constants of the bodies are different, which introduces tangential tractions and possible slip at the interface. This problem is the equivalent in rolling contact to the normal contact of dissimilar solids with friction discussed in §5.4.

(a) Freely rolling cylinders with parallel axes

The cylinders of radii R_1 and R_2 are pressed into contact by a force P per unit length which produces a contact area and contact pressure given by Hertz. In steady rolling it follows from condition (8.3) that the difference in tangential displacement gradients in a stick region is constant. Following the customary approach we shall assume first that friction is sufficient to prevent slip entirely, so that the strain difference is constant throughout the contact strip $(-a \leqslant x \leqslant a)$. The strain $\partial \bar{u}_x/\partial x$ at each contact surface due to distributions of tangential and normal tractions $q(x)$ and $p(x)$ is given by equation (2.25a). When this expression is substituted into equation (8.3a), with the slip velocity $\dot{s}_x$ taken to be zero,

$$\pi\beta p(x) + \int_{-a}^{a} \frac{q(s)}{x-s} \, ds = \tfrac{1}{2}\pi E^* \xi_x = \text{constant}, \quad -a \leqslant x \leqslant a \qquad (8.12)$$

This equation and (5.27) comprise coupled integral equations for $p(x)$ and $q(x)$. They have been solved by Bufler (1959) using the method of §2.7 with class III boundary conditions. We shall simplify the problem by neglecting the effect of tangential traction on the normal pressure which is then given by Hertz. Equation (8.12) now provides a single integral equation for $q(x)$. Following the treatment of the static problem in §5.4, $q(x)$ is divided into two components $q'(x)$ and $q''(x)$. The traction $q'(x)$ is that necessary to eliminate the difference in strains due to the normal pressure; it is given by equation (5.32). $q''(x)$ is the traction required to give rise to a constant strain difference ξ_x; it is determined by an integral equation of the form specified in (2.39) and (2.44) where $n = 0$ and $A = \pi E^* \xi_x/2$ with the solution

$$q''(x) = \frac{E^* \xi_x}{2} \frac{x}{(a^2 - x^2)^{1/2}} \qquad (8.13)$$

At the leading edge of a rolling contact the strain must be continuous as a material element proceeds from outside to inside. Hence q must be zero at the edge of contact. For this requirement to be satisfied the singular terms at $x = -a$ must vanish when $q'(x)$ and $q''(x)$ are added. This condition fixes the magnitude of the creep ratio to be

$$\xi_x = 4\beta p_0/\pi E^* = 2\beta a/\pi R \tag{8.14}$$

where $1/R = 1/R_1 + 1/R_2$, whereupon

$$q(x) = \frac{\beta}{\pi} p_0 (1 - x^2/a^2)^{1/2} \ln\left(\frac{a+x}{a-x}\right) \tag{8.15}$$

This traction is plotted in Fig. 8.2. For practical values of β (see Table 5.1) it is an order of magnitude smaller than the normal pressure and acts outwards on the more compliant surface and inwards on the more rigid one. Bufler's exact solution taking into account the influence of tangential traction on normal pressure gives results which differ only slightly from (8.14) and (8.15).

We now examine the possibility of slip. From equation (8.15) it is apparent that the ratio $q(x)/p(x)$ becomes infinite at $x = \pm a$, so that some slip is inevitable. To obtain an indication of the slip pattern it is instructive to

Fig. 8.2. Distributions of tangential traction in the rolling contact of dissimilar rollers ($\beta = 0.3$, $\mu = 0.1$). No slip: solid line – exact, Bufler (1959); chain line – approximate, eq. (8.15). Complete slip: broken line – $q(x) = \pm\mu p(x)$. Partial slip: line with spots – numerical solution, Bentall & Johnson (1967).

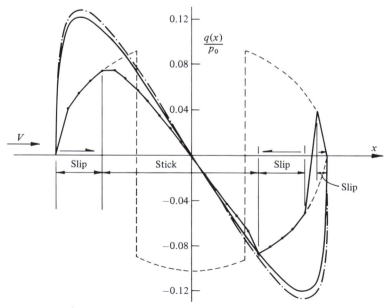

examine the other extreme case, in which the coefficient of friction is small, so that slip is unrestricted and the tangential traction is too small to affect the elastic deformation. The tangential strain in each surface is then given by the first term in equation (2.25a) where $p(x)$ is the Hertz pressure distribution. Substituting in (8.2) gives an expression for the slip velocity:

$$\frac{\dot{s}_x}{V} = \xi_x - (\beta a/4R)(1 - x^2/a^2)^{1/2} \tag{8.16}$$

If ξ_x has a suitable positive value the slip velocity, given by (8.16), is positive at the ends of the contact and negative in the centre. The frictional traction $\pm \mu p(x)$, though small, must change direction at two points symmetrically disposed about the origin. The location of these points, and hence the value of ξ_x, are determined by the condition that, in free rolling, the net tangential force is zero. This condition demands that the direction of slip reverses at $x = \pm 0.404a$ whereupon

$$\xi_x = 0.914\beta a/R \tag{8.17}$$

The true state of affairs must lie between the extremes of no-slip and complete slip. We might expect there to be two stick regions separating three regions in which there is slip in alternate directions. A numerical analysis by Bentall & Johnson (1967), using the method described in §5.9, has shown that this is the case. The solution is a function of the parameter (β/μ). The spread of the three micro-slip zones with increasing values of (β/μ) is shown in Fig. 8.3.

Fig. 8.3. Slip regions in the rolling contact of dissimilar rollers.

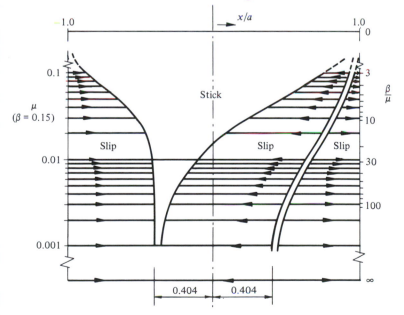

A typical distribution of tangential traction is shown in Fig. 8.2 where it is compared with the solutions for no-slip (equation (8.15)) and complete slip. It is interesting to note the rapid reversal of traction and slip direction experienced by contacting points as they pass towards the rear edge of the contact.

Rolling is no longer reversible; energy dissipation in the slip regions contributes to a moment M resisting rotation of the cylinders. This moment has been computed and is shown in Fig. 8.4.† As Reynolds predicted, the rolling resistance is low when μ is high, since micro-slip is prevented, and is again low when μ is small since the frictional forces are then small. Maximum resistance occurs at an intermediate value of $\mu \approx \beta/5$.

(b) Freely rolling spheres

Tractions at the interface of freely rolling dissimilar spheres can be approached along the same lines. We first assume no-slip and neglect the influence of tangential tractions on normal pressure. For no-slip, equations (8.3) require that the differences in displacement gradient $\{(\partial \bar{u}_{x1}/\partial x) - (\partial \bar{u}_{x2}/\partial x)\}$ and $\{(\partial \bar{u}_{y1}/\partial x) - (\partial \bar{u}_{y2}/\partial x)\}$ should be constant throughout the contact area and equal to $-\xi_x$ and $-\xi_y$ respectively. The tangential traction $q(x, y)$ which satisfies these conditions again may be divided into a component $q'(x, y)$ which eliminates the difference in tangential displacements due to normal pressure

† The tangential traction and slip, from which the dissipation has been calculated, have been found on the assumption that the pressure distribution is symmetrical and given by Hertz. In fact the asymmetry of the traction will give rise to a slight asymmetry in pressure which is responsible for the resisting moment.

Fig. 8.4. Rolling resistance of dissimilar rollers.

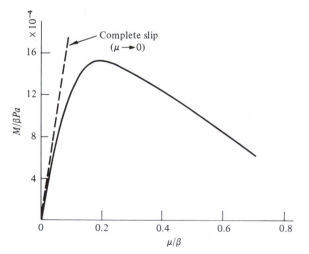

and a component $q''(x, y)$ which gives rise to the constant creep coefficients. The tangential displacements due to the Hertz pressure are radial and axi-symmetric: $q'(x, y)$ is therefore also radial and axi-symmetric $(= q'(r))$ and is given by equation (5.35). This traction contains a term $(a^2 - r^2)^{-1/2}$ which must be annulled at the leading edge by an equal and opposite term in the expression for $q''(x, y)$. In order to satisfy the conditions of no-slip (eq. (8.3) and (8.4)) the resulting displacements within the contact area must satisfy the condition $\partial \bar{u}_y / \partial x = 0$. These conditions are satisfied by a state of uniform bi-axial strain: $\partial \bar{u}_x / \partial x = \partial \bar{u}_y / \partial y = $ constant. The traction which gives rise to this strain can be found by the method of §3.7 with the result

$$q''(r) = \frac{2E^* \xi_x}{\pi r} \{(a^2 - r^2)^{1/2} - a^2 (a^2 - r^2)^{-1/2}\} \tag{8.18}$$

To remove the infinite traction at $r = a$ when $q''(r)$ is added to $q'(r)$ we take

$$\xi_x = \frac{\beta a}{\pi R} \tag{8.19}$$

whereupon

$$q(r) = \frac{\beta p_0}{\pi} \left\{ -\frac{a}{r} (a^2 - r^2)^{1/2} + \frac{1}{r} \int_r^a \frac{t^2}{(t^2 - r^2)^{1/2}} \ln \left(\frac{t+r}{t-r} \right) dt \right\} \tag{8.20}$$

An exact solution to this problem has been obtained by Spence (1968).

In common with the two-dimensional case there must be some slip at the edge of the contact circle but once slip has occurred the traction is no longer axi-symmetric and no solution is at present available.

Equation (8.19) has been checked by experiments in which the distance rolled by a ball in one revolution has been accurately measured and compared with the undeformed circumference of the ball. Measurements (a) with a Duralumin ball rolling between two parallel steel planes and (b) a steel ball between Duralumin planes ($\beta = \pm 0.12$) are shown in Fig. 8.5. The average of these two experiments agrees well with equation (8.19) for the creep calculated by neglecting slip. The small difference between the two experiments is due to various second-order effects: (i) points on the periphery of the ball lie at different radii from the axis of rotation. This effect, frequently referred to as 'Heathcote slip', is considered in §5. For a ball on a plane a creep ratio $\xi_x = -0.125(a/R)^2$ is predicted. (ii) the interface is curved such that the surface of the ball is compressed and that of the plane is stretched by an amount of order $\xi_x = -0.10(a/R)^2$. These estimates are consistent in sign and order of magnitude with the discrepancy between the creep measurements and equation (8.19) but they are hardly significant in comparison with the second-order errors introduced by the linear theory of elasticity.

Fig. 8.5. Free rolling creep of a ball on a plane of dissimilar material.
Cross – Duralumin ball on steel plane; circle – steel ball on Duralumin
plane; broken line – average experimental line; chain line – theory,
eq. (8.19).

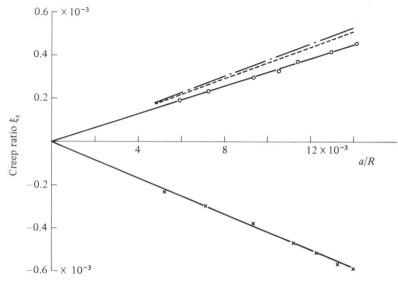

8.3   Tractive rolling of elastic cylinders

In this section we consider rolling cylinders which transmit a resultant
tangential (tractive) force through friction at the interface, such as the wheel of
a vehicle which is driving or braked. To eliminate the effects discussed in the last
section we shall consider elastically similar cylinders. The first solution to this
problem in its two-dimensional (plane strain) form was presented by Carter
(1926) and independently by Fromm (1927), Föeppl (1947), Chartet (1947) and
Poritsky (1950).

From our discussion of two stationary cylinders which transmit a tangential
force less than limiting friction (§7.2), we would expect the contact area to be
divided into zones of slip and stick. It is instructive to start by examining the
solution to the static problem in light of the conditions of stick and slip in
rolling contact established in §1 of this chapter. In the static case (Fig. 7.7)
there is a central stick region in which tangential displacement is constant, with
equal regions of micro-slip on either side. The strains in the stick region
$(\partial \bar{u}_x/\partial x = 0)$ satisfy condition (8.4) for no slip in rolling contact, and the
tractions shown in Fig. 7.7 satisfy conditions (8.5) and (8.6). However,
condition (8.7), which requires the direction of slip to oppose the traction
in a slip zone, is violated in the slip zone at the leading edge of the contact.

This result leads us to expect the stick region to be located adjacent to the leading edge of the contact area and for slip to be confined to a single zone at the trailing edge. The same conclusion was reached in the one-dimensional example of a belt being driven on a pulley.

The distribution of tangential traction in the static case comprises the superposition of two elliptical distributions $q'_x(x)$ and $q''_x(x)$ given by equations (7.23) and (7.25) respectively. The traction $q'_x(x)(=\mu p_0(1-x^2/a^2)^{1/2})$ produces a tangential strain within the contact strip, by equation (7.24),

$$\frac{\partial \bar{u}'_x}{\partial x} = -\frac{2(1-v^2)}{aE}\mu p_0 x \qquad (8.21)$$

We displace the centre of $q''(x)$ by a distance $d (=a-c)$ so that it is now adjacent to the leading edge, as shown in Fig. 8.6.

$$q''_x(x) = -\frac{c}{a}\mu p_0\{1-(x+d)^2/c^2\}^{1/2} \qquad (8.22)$$

Within the strip $(-a \leqslant x \leqslant c-d)$ it produces a tangential strain:

$$\frac{\partial \bar{u}''_x}{\partial x} = \frac{c}{a}\frac{2(1-v^2)}{cE}\mu p_0(x+d) \qquad (8.23)$$

Adding (8.23) to (8.21) gives the resultant tangential strain in the strip $(-a \leqslant x \leqslant c-d)$ to be

$$\frac{\partial \bar{u}_x}{\partial x} = \frac{2(1-v^2)}{aE}\mu p_0 d = \text{constant}$$

The tractions acting on each surface are equal and opposite so that the tangential strains in each surface are equal and of opposite sign. Thus substituting in equation (8.3), we see that the condition of no-slip (8.4) is satisfied in the strip $(-a \leqslant x \leqslant a-d)$ with the creep ratio given by:

$$\xi_x = -4(1-v^2)\mu p_0 d/aE \qquad (8.24)$$

The resultant traction $q_x(x) = q'_x(x) + q''_x(x)$ satisfies the conditions (8.5) and (8.6) in both slip and stick zones, and the direction of slip satisfies (8.7). As in the static case (eq. 7.29), the width of the stick region is determined by the magnitude of the tangential force, whence

$$\frac{d}{a} = 1 - \frac{c}{a} = 1 - (1-Q_x/\mu P)^{1/2} \qquad (8.25)$$

whereupon, by using the Hertz relationship for p_0, the creep ratio is given by

$$\xi_x = -\frac{\mu a}{R}\{1-(1-Q_x/\mu P)^{1/2}\} \qquad (8.26)$$

where $1/R = 1/R_1 + 1/R_2$. The relationship between ξ_x and Q_x, plotted in Fig. 8.7, is known as a 'creep curve'.

The action of a tractive force, however small, causes some micro-slip at the trailing edge of the interface. The slip region spreads forwards with increasing tangential force until, when $Q = \mu P$, the slip zone reaches the leading edge and complete sliding occurs.

Fig. 8.6. Tractive rolling contact of similar cylinders; (a) distribution of tangential tractions – chain line – no slip, eq. (8.27); (b) surface strains $\partial u_x/\partial x$.

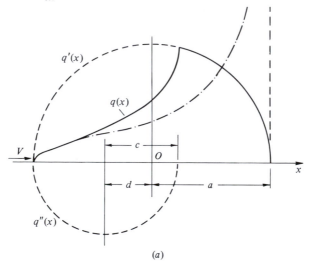

(a)

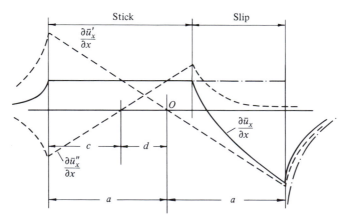

(b)

The state of stress in the cylinders due to a traction of the form $q'(x)$ has been discussed in §7.1. The stresses in the rolling contact case can be found by superposition of $q'_x(x)$ and $q''_x(x)$ (see §9.2).

Under conditions of high friction, such that $Q/\mu P$ is small, the slip region at the trailing edge becomes vanishingly small and the distribution of tangential traction approaches the limiting form:

$$q_x(x) = \frac{p_0}{2} \frac{a+x}{(a^2-x^2)^{1/2}} \frac{Q_x}{P} \tag{8.27}$$

The corresponding limit for the creep ratio is

$$\xi_x = aQ_x/2RP \tag{8.28}$$

This relationship corresponds to a linear creep curve (Fig. 8.7), whose gradient is referred to as the *creep coefficient*. The traction of (8.27) is zero at the leading edge; in consequence the strain is continuous from outside to inside the contact boundary, as can be seen from Fig. 8.6(b). At the trailing edge there is a singularity in traction just inside the contact and a singularity in strain just outside. In reality the singularities will be relieved by slip but we can think of it as a limit in which the elastic strain energy built up through the contact is dissipated instantaneously and irreversibly at the trailing edge. The rate of energy so dissipated is given by the product of the tractive force and the creep ratio, i.e.

$$\dot{W} = Q_x\xi_x V = aQ_x^2 V/2RP \tag{8.29}$$

Fig. 8.7. Creep curve for tractive rolling contact of cylinders. Solid line – Carter's (1926) creep curve, eq. (8.26); broken line – no slip, eq. (8.28); chain line – elastic foundation, eq. (8.69).

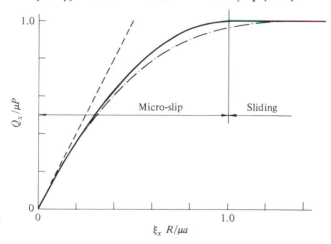

The results obtained in this section strictly apply only when the two cylinders are elastically similar, otherwise additional tangential tractions are present as discussed in the previous section. Under high friction conditions (small regions of micro-slip) the results of the two sections can be superposed. Generally the influence of the tractive force outweighs that of the difference in elastic constants whereupon the above results can be used with $E/(1-\nu^2)$ replaced by $2E^*$. Interaction between the two effects has been analysed numerically by Kalker (1971a).

When a tangential force Q_y acts parallel to the axis of the cylinders tangential tractions and micro-slip arise in the axial direction. The surface displacements and internal stresses produced by the distribution of tangential traction

$$q'_y(x) = \mu p_0 (1-x^2/a^2)^{1/2}$$

have been found in §2.9. The surface displacement gradients for $(-a \leqslant x \leqslant a)$ are

$$\frac{\partial \bar{u}_x}{\partial x} = 0, \quad \frac{\partial \bar{u}_y}{\partial x} = -\frac{\mu p_0}{G} \frac{x}{a} \tag{8.30}$$

This relationship is similar to that expressed by equation (8.21) except for the change in elastic constant from $(1-\nu^2)/E$ to $(1/2\pi G)$. The analysis of the present case is, therefore, completely analogous to that for a longitudinal tangential force. The contact region is divided into a stick region at the leading edge and slip region at the trailing edge. The axial traction is found by the superposition of q'_y and $q''_y(=-q'_x c/a)$ as before, where the extent of the slip region is given by (8.25). The axial creep ratio is given by substituting equation (8.30) for each surface in (8.3b) with the result:

$$\xi_y = -\left(\frac{1}{G_1} + \frac{1}{G_2}\right) \mu p_0 \{1 - (1-Q_y/\mu P)^{1/2}\} \tag{8.31}$$

Cases of combined axial and longitudinal traction have been examined by Heinrich & Desoyer (1967) and Kalker (1967a).

8.4   Rolling with traction and spin of three-dimensional bodies

Three-dimensional bodies in rolling contact may be called upon to transmit tangential forces, Q_x in the longitudinal and Q_y in the transverse directions, while being subjected to a relative angular velocity about the normal axis ($\Delta\omega_z \equiv \omega_{z1} - \omega_{z2}$) referred to as spin. The spin motion tends to twist the contact interface and hence also gives rise to tangential tractions and micro-slip. In the two-dimensional situations considered previously stick and slip zones comprised strips parallel to the y-axis. The three-dimensional case is much more complex. Contact is made on an elliptical area; the shape of the stick and slip zones are not known *a priori* and condition (8.7), which requires the tangential traction in a slip zone to oppose the relative slip, couples the effects of tangential forces and spin in a nonlinear manner.

In face of the difficulty caused by the unknown pattern of stick and slip, we will consider first the situation where the coefficient of friction is sufficiently great to prevent slip and take a lead from the two-dimensional case analysed in §3.

(a) Vanishing slip (μ → ∞): linear creep theory

If the coefficient of friction is sufficiently high, slip is limited to a vanishingly thin zone at the trailing edge of the contact. In this case we seek tangential tractions which satisfy the no-slip conditions (8.3), (8.4) and (8.5) throughout the contact area. For simplicity we shall begin with a circular contact subjected to a longitudinal tangential force Q_x which is the three-dimensional equivalent of the situation examined in the previous section. On the basis of the two-dimensional case (eq. (8.27)), we consider the traction

$$q_x(x,y) = \frac{Q_x}{2\pi a^2} \frac{a+x}{(a^2-r^2)^{1/2}} \qquad (8.32)$$

The tangential displacements within the contact circle $(r \leqslant a)$ produced by this traction may be found by substitution in equation (3.83a and b) and by performing the integrations, with the result:

$$\frac{\partial \bar{u}_x}{\partial x} = \frac{Q_x(4-3v)}{32Ga^2} \quad ; \quad \frac{\partial \bar{u}_y}{\partial x} = 0 \qquad (8.33)$$

These displacement gradients are constant throughout the contact region and hence satisfy the conditions of no-slip (8.3) and (8.4), with creep ratios:

$$\xi_x = -\frac{(4-3v)}{16Ga^2} Q_x ; \quad \xi_y = 0 \qquad (8.34)$$

Under the action of a transverse tangential force Q_y, the traction

$$q_y(x,y) = \frac{Q_y}{2\pi a^2} \frac{a+x}{(a^2-r^2)^{1/2}} \qquad (8.35)$$

satisfies the conditions of no-slip and results in a transverse creep:

$$\xi_y = -\frac{(4-v)Q_y}{16Ga^2} ; \quad \xi_x = 0 \qquad (8.36)$$

Due to the asymmetry of the traction distribution (8.35) it exerts a twisting moment about O given by

$$M_z = -\tfrac{1}{3}Q_y a = \{16/3(4-v)\}\xi_y$$

In the case of pure spin, distributions of traction can be found (Johnson, 1958*b*) which satisfy the conditions of no-slip, viz.:

$$q_x = \frac{8G(3-\nu)}{3\pi(3-2\nu)}\psi\,\frac{(a+x)y/a}{(a^2-r^2)^{1/2}} \tag{8.37a}$$

and

$$q_y = \frac{8G(1-\nu)}{3\pi(3-2\nu)}\psi\,\frac{(a^2-2x^2-ax-y^2)/a}{(a^2-r^2)^{1/2}} \tag{8.37b}$$

where ψ is the spin parameter defined on p. 244. These tractions correspond to zero resultant force ($Q_x = Q_y = 0$) but give rise to a resultant twisting moment M_z, which resists the spin motion, given by:

$$M_z = \frac{32(2-\nu)}{9(3-2\nu)}\,Ga^3\psi \tag{8.38}$$

When the displacement gradients due to the tractions of (8.37) are substituted into the conditions of no-slip (8.3) and (8.4), the creep ratios are found to be:

$$\xi_x = 0; \quad \xi_y = \frac{2(2-\nu)}{3(3-2\nu)}\,\psi \tag{8.39}$$

It is perhaps surprising at first sight that pure spin gives rise to transverse creep, but this phenomenon has been well supported by experiment. For details of the above analysis see Johnson (1958*a* & *b*).

The tentative distributions of traction for rolling with traction and spin proposed in equations (8.32), (8.35) and (8.37) do not satisfy the no-slip conditions completely. To do so they must vanish at all points on the leading edge of the contact. We see that, in each case, $q = 0$ at the 'leading point' $(-a, 0)$, but that elsewhere along the leading edge q is unbounded.

To overcome this difficulty Kalker (1964, 1967*a*) made use of the fact (see §3.7(*e*)) that a general traction:

$$q(x, y) = A_{mn}(x/a)^m(y/b)^n\{1 - (x/a)^2 - (y/b)^2\}^{-1/2} \tag{8.40}$$

gives rise to tangential displacements $\bar{u}_x$ and $\bar{u}_y$ each of which varies throughout the elliptical contact region as a polynomial in x and y of order $(m + n)$. By appropriate superposition of tractions of the form (8.40) the displacement gradients can be made to satisfy the condition of no-slip throughout the contact ellipse, while maintaining a zero value of q along the leading edge. In carrying out the necessary computations Kalker truncated the series at $m + n = 5$ and minimised the integrated traction round the leading edge, which amounted to making $q = 0$ at a finite number of points on the leading edge. From this point of view the results expressed above in equations (8.32)–(8.38) may be regarded as the first approximation to Kalker's solution, taking $m + n = 2$ and satisfying the condition $q = 0$ at one point $(-a, 0)$ only.

Since this is a linear theory there is no interaction between longitudinal and transverse forces and the effects can be superposed. The results may conveniently be summarised in three linear creep equations:

$$\frac{Q_x}{Gab} = C_{11}\xi_x \tag{8.41}$$

$$\frac{Q_y}{Gab} = C_{22}\xi_y + C_{23}\psi \tag{8.42}$$

$$\frac{M_z}{G(ab)^{3/2}} = C_{32}\xi_y + C_{33}\psi \tag{8.43}$$

where C_{11}, C_{22} etc. are non-dimensional creep coefficients found from the theory.

A shortened table of the creep coefficients found by Kalker (1967a) for elliptical contact areas of varying eccentricity is given in Appendix 4. The approximate values for a circular area obtained from equations (8.34), (8.36) and (8.38) are given for comparison.

(b) Complete slip

At the other extreme, when the creep and spin ratios become large and the coefficient of friction is small, the elastic deformation due to the tangential tractions becomes small. In the expressions for slip velocities (8.3) the elastic displacement terms can be neglected, so that slip vanishes at one point $P(x_p, y_p)$ only (see Fig. 8.8), where

$$x_p/a = -\xi_y/\psi \quad \text{and} \quad y_p/a = \xi_x/\psi \tag{8.44}$$

Fig. 8.8. Rolling with creep and spin: location of the 'spin pole', P.

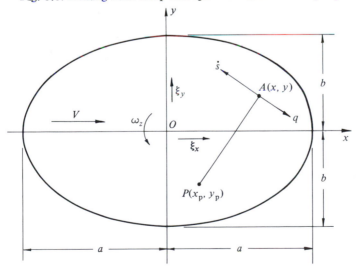

This point is known as the *spin pole*; it may lie inside or outside the contact area. Since elastic deformation in the x-y plane is neglected, the relative motion of the surfaces comprises a rigid rotation with angular velocity $(\omega_{z1} - \omega_{z2})$ about the spin pole. At any point $A(x, y)$ the resultant tangential traction $q(x, y)$ has the magnitude $\mu p(x, y)$ in the direction perpendicular to the line PA. The forces Q_x, Q_y and the moment M_z corresponding to any combination of creep and spin can be readily computed by numerical integration. Such computations have been carried out for circular contacts by Lutz (1955 *et seqq.*) and for elliptical contacts by Wernitz (1958) and Kalker (1967*a*). The influence of spin upon the creep curves under the action of tangential tractions, calculated on the basis of complete slip, is shown in Fig. 8.11. The interaction between spin and traction plays an important role in the operation of rolling contact friction drives.

(c) Partial slip: nonlinear creep theory

The distributions of traction calculated on the assumption of no slip are zero at the leading edge and rise progressively through the contact area to infinite values at the trailing edge. We would, therefore, expect slip to start at the trailing edge and to spread forward through the contact area with an increase in the tractive forces as in the two-dimensional contact analysed in §3. This conjecture is borne out by experiment. Observations by Ollerton & Haines (1963) of the contact ellipse between photo-elastic models show the stick region to be roughly 'lemon' shaped. One edge coincides with the leading edge of the contact ellipse and the boundary with the slip zone is a reflection of the leading edge. Experiments using a rubber ball rolling with spin on a transparent plane (Johnson, 1962*a*) revealed the stick zone to be 'pear' shaped with one boundary coincident with the leading edge of the contact circle. When stick and slip zones coexist in the contact area the distribution of traction and equations for creep have, so far, been found only approximately. Three methods have been tried.

In the first method, the stick zone under the action of a tangential force (without spin) is assumed to be an ellipse, similar to the contact ellipse, and touching it at the leading point $(-a, 0)$ shown in Fig. 8.9. This assumption has the merit of giving simple expressions in closed form for the traction and creep (Johnson, 1958*a*, *b*; Vermeulen & Johnson, 1964). For a circular contact area, transmitting Q_x alone, the creep is given by

$$\xi_x = -\frac{3\mu P(4 - 3\nu)}{16Ga^2} \left\{ 1 - \left(1 - \frac{Q_x}{\mu P}\right)^{1/3} \right\} \tag{8.45}$$

and when transmitting Q_y alone

$$\xi_y = -\frac{3\mu P(4-v)}{16Ga^2}\left\{1-\left(1-\frac{Q_x}{\mu P}\right)^{1/3}\right\} \tag{8.46}$$

When $Q \ll \mu P$ these equations reduce to the linear creep equations (8.34) and (8.36) which neglect slip. The assumption of an elliptical stick region is clearly in error since it does not include the leading edge of the contact circle.

This particular difficulty is avoided by a quite different approach to the problem suggested by Haines & Ollerton (1963) and developed by Kalker (1967b). In their method the contact area is divided into thin strips parallel

Fig. 8.9. Tractive rolling of an elliptical contact region under a longitudinal force Q_x. Broken line – elliptical stick zone, Johnson (1958a); chain line – strip theory, Haines & Ollerton (1963).

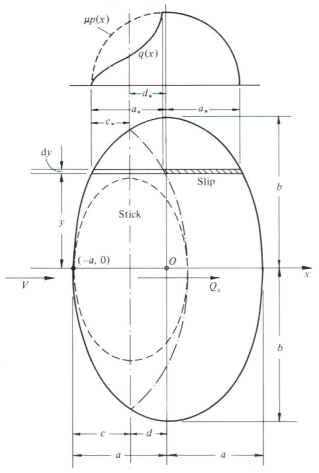

to the rolling direction (x-axis). The two-dimensional theory from §3 above is then applied to each strip, neglecting interaction between adjacent strips. We shall apply their method to an elliptical contact under the action of a longitudinal tangential force Q_x only (see Fig. 8.9).

A typical strip, distance y from the x-axis has a length $2a_*$ and supports a pressure

$$p(x) = p_0^*(a_*^2 - x^2)^{1/2}$$

where

$$\frac{p_0^*}{p_0} = \frac{a_*}{a} = (1 - y^2/b^2)^{1/2} \tag{8.47}$$

We now assume that Carter's theory for cylindrical contact can be applied to the strip. A stick region of length $2c_*$ is located adjacent to the leading edge. The creep ratio is given by equation (8.24), with p_0, d and a replaced by p_0^*, d_* and a_* respectively, whence

$$\xi_x = -\frac{2(1-\nu)}{Ga_*} \mu p_0^* d_* \tag{8.48}$$

Now the creep ratio ξ_x must be a constant for the contact as a whole, hence it follows from equation (8.48) that $d_*(= a_* - c_*)$ has the same value for all the strips in which a stick zone exists. Thus the mid-points of the stick zones lie on the straight line $x = -d_*$. The curve separating the stick and slip zones is therefore a reflexion of the leading edge in this line, giving rise to a lemon shaped stick zone as observed experimentally.

The distribution of traction on a strip follows Carter and is shown in Fig. 8.9. The tangential force dQ_x provided by a strip is determined by equation (8.25) i.e.

$$dQ_x = \mu P_*(1 - c_*^2/a_*^2)\, dy$$

$$= \frac{\pi}{2}\mu p_0 a(1 - y^2/a^2)\{1 - (1 - d_*/a_*)^2\}\, dy$$

The total force Q_x is found by integration of equation (8.49) over the contact area, noting that, when $y > a^2 - d_*^2$, the stick region vanishes so that the term $(1 - d_*/a_*)$ is put equal to zero, with the result:

$$Q_x/\mu P = \tfrac{3}{2}\xi_x \cos^{-1}(\xi_x) + \{1 - (1 + \tfrac{1}{2}\xi_x^2)(1 - \xi_x^2)^{1/2}\} \tag{8.49}$$

where $\xi_x = \xi_x G/\mu p_0$. For vanishingly small slip ($\mu \to \infty$) this expression becomes

$$Q_x = \frac{\pi^2}{4}\frac{Gab}{1-\nu}\xi_x \tag{8.50}$$

The value of the creep coefficient given by the strip theory (8.50) is independent of the shape of the contact ellipse. It is compared with Kalker's value for vanishingly small slip in Appendix 5. As might be expected the agreement is good when the ellipse is thin in the rolling direction ($b \gg a$), but the effect of neglecting interaction between the strips becomes serious for contacts in which $b < a$.

To apply the strip theory to contacts transmitting a transverse force Q_y or rolling with spin, use is made of the two-dimensional theory of transverse traction given in §2.9. For further details the reader is referred to Kalker (1967b). It is clear, however, that the strip theory is not satisfactory unless $b > a$; it breaks down completely when the spin motion is large. For these circumstances Kalker (1967a) devised a different approach based upon numerical techniques of optimisation.

The difficulty of problems involving micro-slip lies in the different boundary conditions which have to be satisfied for the stick and slip zones when the configuration of these zones is not known in advance. Kalker's approach to this difficulty is to combine the separate conditions of stick and slip into a single condition which is satisfied approximately throughout the contact area. If the tangential traction is denoted by the vector $\mathbf{q}$ and the slip velocity by the vector $\dot{\mathbf{s}}$, then we may combine the conditions of (8.4)–(8.7) in the statements

$$|\dot{\mathbf{s}}|\mathbf{q} + \mu p \dot{\mathbf{s}} = 0 \tag{8.51}$$

and

$$|\mathbf{q}| \leqslant \mu p \tag{8.52}$$

In a stick region $\dot{\mathbf{s}} = 0$ so that (8.51) is automatically satisfied; in a slip region $|\mathbf{q}| = \mu p$ in the opposite direction to $\dot{\mathbf{s}}$, so that (8.51) is again satisfied. Thus the correct distributions of slip and traction satisfy equations (8.51) and (8.52) throughout the whole contact area. A measure of the closeness with which any proposed distribution of traction satisfies the boundary conditions may be obtained by forming the integral over the contact area

$$I = \int_A (|\dot{\mathbf{s}}|\mathbf{q} + \mu p \dot{\mathbf{s}})^2 \, dA \tag{8.53}$$

Since the integral is positive everywhere and zero when the boundary conditions are satisfied, the value of I is always positive and approaches zero when the correct distribution of traction and corresponding slip are inserted. Thus, out of any class of traction distributions, the 'best fit' is that which minimises I. Well developed techniques of nonlinear programming are available to assist in performing this minimisation.

This approach, unlike those discussed previously, blurs the distinction between stick and slip zones, which are now identified *a posteriori*: where $|\dot{\mathbf{s}}| \cong 0$ is identified as a stick zone; where $|\mathbf{q}| \approx \mu p$ is identified as a slip zone.

Approximate distributions of traction may be found by the superposition of distributions of the form expressed in equation (8.40). Alternatively they may be made up of discrete traction elements in the manner described in §5.9. The tangential displacements $\bar{u}_x$ and $\bar{u}_y$ are calculated by the methods of §3.6 and substituted in equations (8.3) for the slip velocity $\dot{s}$. The optimum distribution of traction is then found from the minimisation of the integral I of (8.53).†
Values of Q_x, Q_y and M_z have been calculated for various combinations of creep and spin ξ_x, ξ_y and ψ with elliptical contacts of varying eccentricity. (See Kalker (1969) for a summary of results.)

Creep forces play an important role in the guidance and stable running of railway vehicles. They arise as shown in Fig. 8.10. The point of contact is taken to be at rest, so that the rail moves relative to it with the forward speed of the vehicle V_1. The wheel profiles are coned so that longitudinal creep ξ_x can arise when the two wheels of a pair are running on different radii. Longitudinal creep is also a consequence of driving or braking a wheel. Lateral creep ξ_y arises if, during forward motion of the wheelset, the plane of the wheel is

Fig. 8.10. Creep motion of a railway wheel. Longitudinal creep ratio: $\xi_x = (V_2 - V_1)/V_1$; lateral creep ratio: $\xi_y = \delta V_y/V_1 = \tan \phi$; spin parameter: $\psi = \omega(ab)^{1/2}/V_1 = \{(ab)^{1/2}/R\} \tan \lambda$.

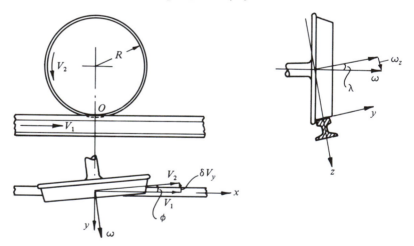

† More recently on grounds of versatility and dependability Kalker (1979) has abandoned the object function in the integral of equation (8.46) in favour of the complementary energy principle of Duvaut & Lions (1972) discussed in §5.9. In this approach the Eulerian formulation in §1 of this chapter is replaced by a Lagrangian system in which the moving contact area is followed and the traction is built up incrementally with time from some initial state until a steady state is approached. Such transient behaviour is discussed further in §6.

skewed through a small angle ϕ to the axis of the rail. Finally, since the common normal at the point of contact is tilted at the cone angle λ to the axis of rotation, the wheel has an angular velocity of spin $\omega_z = \omega \sin \lambda$ relative to the rail. For sufficiently small values of creep and spin the linear theory, embodied in equations (8.41)–(8.43), is adequate to determine the creep forces. At larger values the nonlinear theory, involving partial slip, must be used. For large creep and spin the creep forces are said to 'saturate' and their values are given by the 'complete slip' theory which does not depend upon elastic deformation tangential to the surface.

We shall conclude this section with an assessment of creep theory in relation to experimental observations. Surface tractions and associated internal stresses have been investigated by photo-elasticity using large epoxy-resin models in very slow rolling (Haines & Ollerton, 1963; Haines, 1964–5). The stick and slip zones were clearly visible. In the slip zone the traction closely follows Amonton's Law of friction as assumed in the theory. The measured traction in a circular contact transmitting a longitudinal force is compared with Carter's distribution (strip theory) and with Kalker's (1967a) continuous distribution in Fig. 8.11. The measured traction is very close in form to Carter's distribution, but the strip theory gives rise to an error in the size of the stick region. Kalker's method removes the sharp distinction between stick and slip, but in view of the small number of terms employed gives a remarkably good approximation to the measured traction.

Fig. 8.11. Tangential traction $q(x)$ on centre-line of circular contact transmitting a longitudinal force $Q_x = 0.72\mu P$. Solid line – numerical theory, Kalker (1967a); chain line – strip theory; circle – photo-elastic measurements.

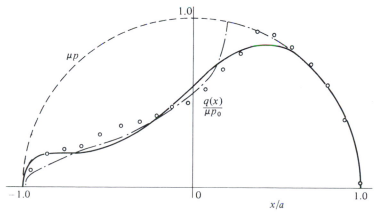

Creep experiments are usually performed by accurately measuring the distance traversed by a rolling element in precisely one revolution. Laboratory experiments by Johnson (1958*a* & *b*, 1959) in slow rolling using good quality surfaces are generally in good accord with present theory. The case of longitudinal creep is illustrated in Fig. 8.12 for a circular contact (a ball rolling on a plane). The influence of spin is governed by the non-dimensional parameter $\chi = \psi \bar{R}/\mu c$, where

$$1/\bar{R} = \tfrac{1}{4}\{(1/R_1') + (1/R_1'') + (1/R_2') + (1/R_2'')\} \quad \text{and} \quad c = (ab)^{1/2}$$

For a ball of radius R rolling on a plane, $\bar{R} = 2R$ and $c = a$. It is clear that increasing spin has the effect of reducing the gradient of the linear part of the creep curve, i.e. reducing the creep coefficient. The full lines denote Kalker's numerical nonlinear theory, which is well supported by the experiments. For no spin ($\chi = 0$), chain and dotted lines represent respectively the strip theory (eq. (8.49)) and Johnson's approximate theory (eq. (8.45)). The discrepancies, of opposite sign in each case, are not large, particularly in view of the practical uncertainty in the value of μ. Provided that the traction force Q_x is less than about 50% of its limiting value ($Q_x/\mu P < 0.5$), the linear theory, which assumes vanishingly small slip, provides a reasonable approximation. The predictions of Wernitz' complete slip theory, which neglect the tangential elastic compliance of the rolling bodies, have been added by the broken lines in Fig. 8.12. When there is no spin ($\chi = 0$) this theory is entirely inadequate since it

Fig. 8.12. Longitudinal creep combined with spin: theories and experiment (circular contact). Solid line – numerical theory, Kalker (1967*a*); large-dashed line – complete slip theory, Wernitz (1958); small-dashed line – approximate theory, eq. (8.45); chain line – strip , theory eq. (8.49).

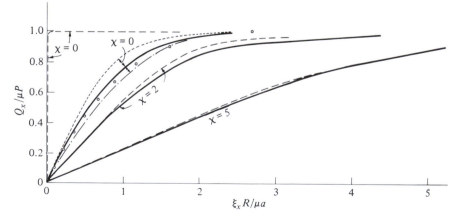

predicts zero creep when $Q_x < \mu P$. With increasing spin, however, it becomes more satisfactory and when $\chi = 5$ it is indistinguishable from Kalker's complete numerical theory.

The case of transverse creep is not so straightforward, since spin itself produces a transverse tangential force, known in the automobile industry as *camber thrust*. In consequence the creep curves are asymmetrical with respect to the origin (Fig. 8.13). This effect is not predicted at all by the theory of complete slip since it is entirely due to tangential elastic compliance of the surface. The complete slip theory, therefore, shows a very large error in this case, even when the spin is large. Once again Kalker's numerical results are well supported by experiment in the range where accurate measurements have been made.

The variation in the transverse force with spin when the transverse creep is zero, i.e. the camber thrust (given by the intersection of the creep curves in Fig. 8.13 with the axis $\xi_y = 0$), is plotted in Fig. 8.14. It rises to a maximum at $(\psi \bar{R}/\mu a) \approx 2$ and falls, with increasing spin, to zero as complete slip is approached. This force arises not only in cornering of a vehicle, but when the axis of rotation of a body is inclined, i.e. 'cambered', to the surface on which it rolls.

Fig. 8.13. Transverse creep combined with spin: theories and experiment (circular contact). Solid line – numerical theory; large-dashed line – complete slip theory; small-dashed line – approximate theory, eq. (8.46).

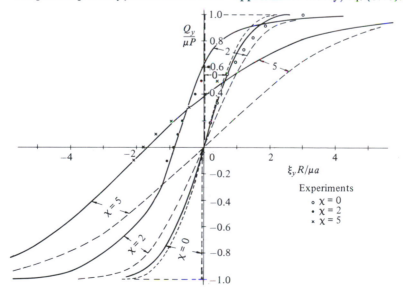

Fig. 8.14. Camber thrust: transverse tangential force due to spin (circular contact). Kalker's numerical theory compared with experiment. Broken line – linear theory, eq. (8.42).

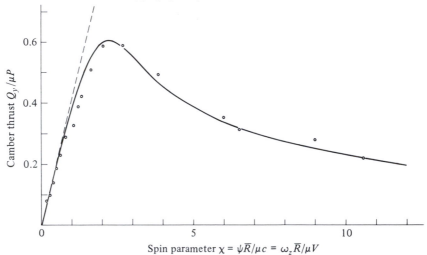

Under engineering conditions, such as are encountered on railway tracks for example, the creep coefficients are observed to be much less than their theoretical values (Hobbs, 1967). A serious cause of this discrepancy lies in the lubricating effect of contaminant films, particularly oil or grease, on the rolling surfaces (Halling & Al-Qishtaini, 1967). Surface roughness and vibration are also likely causes of reduced creep coefficients in practice.

8.5   A ball rolling in a conforming groove

A ball rolling in a groove whose cross-sectional radius of curvature ρ is fairly close to that of the ball itself presents a special case of importance in rolling bearing technology. Under normal load the contact area is an ellipse which is elongated in the transverse direction. With close conformity the contact area is no longer planar but shares the transverse curvature of the ball and groove; surface points in different transverse positions on the ball have different peripheral speeds which leads to micro-slip. To a reasonable approximation the peripheral velocity of the ball is given by

$$V_1 = \omega(R - y^2/2R) \tag{8.54}$$

where R is the radius of the ball, ω is its angular velocity. Thus the creep ratio can be expressed by

$$\xi(y) \equiv \frac{V_1 - V_2}{\omega R} = \left(1 - \frac{V_2}{\omega R}\right) - \frac{y^2}{2R^2} = \xi_0 - y^2/2R^2 \tag{8.55}$$

Under free rolling conditions there is no net tangential force exerted so that the contact area must be split in three zones: a central zone (y small) where the slip is positive and two outer zones (y large) where the slip is negative. Three parallel wear bands were observed in unlubricated ball-bearing races by Heathcote (1921), who developed a theory on the basis of complete slip, i.e. by neglecting the elastic compliance of the ball and race. He deduced that the rotation of the ball would be resisted by a frictional moment M given by

$$\frac{M}{RP} = 0.08\mu \frac{b^2}{R^2} \tag{8.56}$$

where $2b$ is the transverse width of the contact ellipse.

The influence of elastic compliance is not generally negligible however, and can be conveniently analysed by strip theory.

The dimensions of the contact ellipse a and b, and the maximum contact pressure p_0 are given by the Hertz theory. A typical strip is shown in Fig. 8.15 to which the Carter theory is applied. A stick region at the leading edge, whose centre is located at $x = -d$, is followed by a slip region at the trailing edge. Equation (8.24) for the creep ratio of the strip is now substituted into equation (8.55) to give

$$\frac{d}{a} = \Gamma(\gamma^2 - y^2/b^2) \tag{8.57}$$

Fig. 8.15. Strip theory applied to a ball rolling in a conforming groove. Pure rolling takes place on two lines at $y = \pm\gamma b$; elsewhere micro-slip occurs: backwards where $y < \gamma b$ and forwards where $y > \gamma b$.

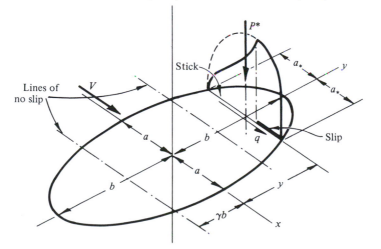

where

$$\Gamma = \frac{b^2 E^*}{4\mu p_0 R^2} = \frac{b^2 \rho E(e)}{2\mu a R(2\rho - R)}$$

and

$$\gamma^2 = 2R^2 \xi_0 / b^2$$

No micro-slip occurs at $y = \pm \gamma b$ and the extent of slip elsewhere is given by (8.57) provided $d/a_* < 1$. Strips at larger values of y slip completely whereupon $d/a_* = 1$. To determine γ we use the condition of no resultant tangential force. From equation (8.25) the tangential force exerted by the strip per unit width is

$$Q^* = \frac{\pi}{2} \mu p_0^* a_* \frac{d}{a_*} \left(2 - \frac{d}{a_*} \right) = \frac{\pi}{2} \mu p_0 a \frac{d}{a} \left(2 \frac{a_*}{a} - \frac{d}{a} \right) \tag{8.58}$$

For free rolling the total traction force

$$Q = \int_{-b}^{b} Q^* \, dy = 0 \tag{8.59}$$

This condition determines the value of γ, the position of the zero slip bands,

Fig. 8.16. Resisting moment on a ball rolling in a conforming groove, eq. (8.60). Broken line – limit for vanishingly small slip; chain line – complete slip, eq. (8.56), Heathcote (1921).

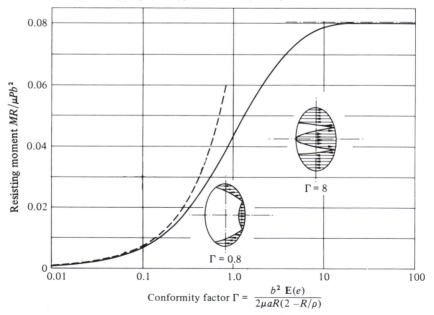

Conformity factor $\Gamma = \dfrac{b^2 E(e)}{2\mu a R(2 - R/\rho)}$

and hence the overall creep ratio ξ_0. The moment resisting rolling is then found by

$$M = \int_{-b}^{b} Q^*(R - y^2/2R)\,\mathrm{d}y = -\int_{-b}^{b} Q^*(y^2/2R)\,\mathrm{d}y \qquad (8.60)$$

where the integral is a function of the geometric parameter Γ. The result of computing this integral is shown in Fig. 8.16. When the conformity is close or the coefficient of friction is small, Γ is large, and the moment approaches the result for complete slip given by equation (8.56). Theoretical stick and slip regions are also shown in Fig. 8.16.

8.6 Transient behaviour in rolling

So far in this chapter we have considered the contact stresses which arise in steady rolling, that is when the forces and contact geometry do not change with time and when rolling has proceeded for a sufficient time for the stress field to be no longer influenced by the initial conditions. Various cases of unsteady rolling contact of cylinders in plane strain have been examined in a sequence of papers by Kalker (1969, 1970, 1971b).

Since the strain field is changing with time, the term $\partial \bar{u}_x/\partial t$ in equation (8.1a) for the particle velocity is no longer zero and appears in the expression for the velocity of slip. In plane strain there is no spin and no motion in the y direction so we can omit the suffix x, and we shall restrict the discussion to similar elastic bodies. Equation (8.3a) for the slip velocity then becomes

$$\dot{s}(x, t) = V\xi(t) + 2V\,\frac{\partial \bar{u}(x, t)}{\partial x} + 2\,\frac{\partial \bar{u}(x, t)}{\partial t}$$

$$= V\xi(t) + 2V\,\frac{\partial \bar{u}(x, t)}{\partial x} + 2\,\frac{\partial}{\partial t}\,(\bar{u}(x, t) - \bar{u}(0, t))$$

$$+ \frac{2\partial\bar{u}(0, t)}{\partial t} \qquad (8.61)$$

Difficulty arises in plane strain with the last term in equation (8.61), since the absolute value of $\bar{u}(0, t)$ depends upon the choice of datum for displacements. This choice must be governed by the bulk geometry of the bodies in any particular case. If the length of the contact perpendicular to the rolling direction is well defined and denoted by $2b$ ($b \gg a$), we can take as an approximation to $\bar{u}(0, t)$ the displacement at the centre of a rectangle $2b \times 2a$ due to a uniform traction $q = Q/4a$, i.e.

$$\bar{u}(0, t) = \frac{2(1 - v^2)Q(t)}{\pi E}\,\{1/(1 - v) + \ln(2b/a)\}$$

where Q is the tangential force per unit axial length. The value of $\bar{u}(0, t)$ will not be very sensitive to the precise distribution of traction. In those cases where the normal load, and hence a, remain constant

$$\frac{\partial \bar{u}(0, t)}{\partial t} = \frac{2(1 - \nu^2)}{\pi E} \left\{ \frac{1}{1 - \nu} + \ln\left(\frac{2b}{a}\right) \right\} \frac{dQ(t)}{dt} \tag{8.62}$$

This term enters into the calculation of the transient creep; but the difficulty does not arise if the traction only is required.

During rolling we expect the interface to have stick and slip zones governed by the conditions (8.4)–(8.7) in which the slip is now related to the elastic displacements by equation (8.61). The distributions of traction $q(x, t)$ and the value of the creep ratio $\xi(t)$ which satisfy these conditions must be found step by step, starting from given initial conditions, and following the prescribed loading history of the particular problem. The reader is referred to Kalker (1969, 1970, 1971b) for the technique of solution; the results of two cases only will be discussed here.

(a) Constant tractive force starting from rest

If a tractive force less than limiting friction is applied to cylinders at rest, micro-slip takes place at both edges of contact and the tangential traction is distributed according to equation (7.28), as shown in Fig. 8.17 ($l = 0$). Assuming no-slip ($\mu \to \infty$) the traction would rise to an infinite value at both edges, given by equation (7.22). The cylinders are now permitted to roll. In the steady-state slip is restricted to the trailing edge and the traction is distributed according to the sum of (7.23) and (8.22). The steady-state traction without slip is given by (8.27). Between the start of rolling and the steady state, the slip regions and tractions vary transiently with l, the distance rolled. Since inertia effects are ignored, $\partial/\partial t = V \partial/\partial l$. The traction distributions at various stages in the process are shown in Fig. 8.17 for the case of no slip ($\mu \to \infty$) and also for the case of partial slip in which $Q = 0.75\mu P$. With no slip the singularity in traction, which is initially located at the leading edge, moves through the contact with the rolling velocity and finally disappears at $l = 2a$ leaving a distribution of traction which is close to that in the steady state. When slip is permitted, the initial application of Q causes slip at both edges but, as rolling proceeds, no further slip occurs at the leading edge. The original traction distribution moves through the contact with the rolling velocity until it merges with the trailing slip zone at $l = 0.6a$. The steady state is virtually reached and established by $l = 1.6a$.

The behaviour might be described qualitatively by saying that the interfacial points, initially located at the leading edge, and their associated traction move

Fig. 8.17. Transient rolling from rest under the action of constant normal and tangential forces: distributions of tangential traction $q(x)$ with distance rolled l. Chain line – no slip ($\mu \to \infty$); solid line – with partial slip ($Q = 0.75\mu P$).

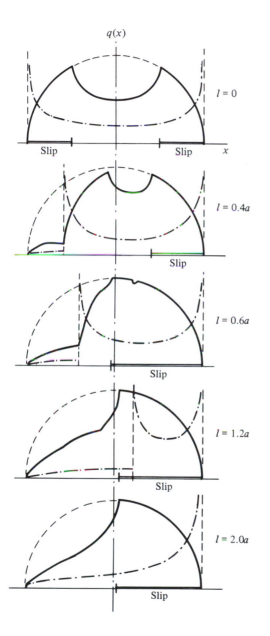

through the contact until they are swallowed up by the slip region at the trailing edge, at which instant a good approximation to the steady state has been reached.

(b) Oscillating tractive force: $Q(t) = Q^* \cos \omega t$

An oscillating tangential force whose period $2\pi/\omega$ is comparable with the time of passage of the surfaces through the contact zone $2a/V$ will induce

Fig. 8.18. Steady cyclic variations of traction due to an oscillating tangential force. $Q(t) = Q^* \cos \omega t$. *(a)* $\omega = V/a$; *(b)* $\omega = 2.405 \, V/a$.

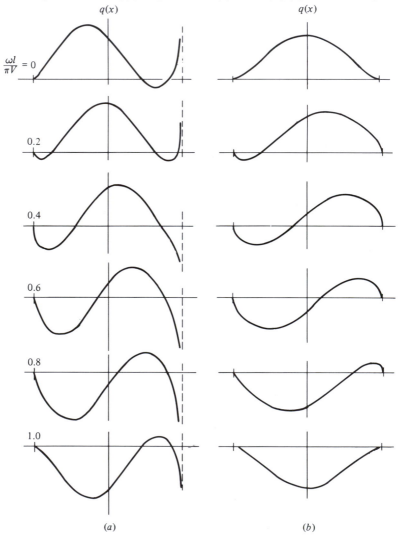

$$\frac{\omega l}{\pi V} = 0$$

0.2

0.4

0.6

0.8

1.0

(a)

(b)

tractions which are appreciably different from those in steady rolling. Transient effects occur at the start, which depend upon the initial state of traction, but after a few cycles the system settles down to a steady cyclic state. By way of example Fig. 8.18(a) shows the fifth cycle at a frequency $\omega = V/a$, from which it is evident that a steady cyclic state has been reached in which $q(\omega t + \pi) = -q(\omega t)$. No slip has been permitted in this example and, as expected, an infinite traction oscillating in sign occurs at the trailing edge, which must in fact be relieved by micro-slip. However Kalker has shown that there are a series of specific frequencies ω_n at which no infinite traction arises, given by

$$J_0(\omega_n a/V) = 0 \qquad (8.63)$$

where $J_0 \equiv$ Bessel function of the first kind. The traction distributions in the steady cyclic state corresponding to the first such frequency: $\omega_1 = 2.405 V/a$, are shown in Fig. 8.18(b). Under these conditions no slip occurs at any time provided $Q^*/\mu P$ is less than about 0.4.

The complexities of transient creep analysis have so far restricted exact solutions to two-dimensional cases. Three-dimensional problems involving lateral creep and spin in addition to longitudinal creep can be analysed approximately, however, using the elastic foundation model described in the next section.

8.7 Elastic foundation model of rolling contact

We saw in §4.3 how normal elastic contact could be greatly simplified by modelling the elastic bodies by a simple Winkler elastic foundation rather than by elastic half-spaces. The same expedient can be applied to the tangential tractions which arise in rolling contact. The two rolling bodies can be replaced by a rigid toroid having the same relative principal curvatures, rolling on an elastic foundation of depth h which in turn is supported on a flat rigid substrate. The elasticities of both bodies are represented by the moduli of the foundation: K_p in normal compression and K_q in tangential shear.

The shape and size of the contact ellipse and the contact pressure could be found by the Hertz theory, but it is more consistent to use the elastic foundation model in the manner described in §4.3. The tangential surface displacements $\bar{u}_x$ and $\bar{u}_y$ are related to the components of tangential traction by

$$q_x = (K_q/h)\bar{u}_x; \quad q_y = (K_q/h)\bar{u}_y \qquad (8.64)$$

Such a foundation is sometimes referred to as a 'wire brush' model since individual bristles might be expected to deform according to (8.64) independently of their neighbours. The conditions for slip and stick expressed in equations (8.1)–(8.7) still apply and, together with the condition that the traction is zero on the leading edge, are used to find the distribution of tangential traction throughout the contact area. Since, by (8.64), the traction at any point depends only upon the displacement at that point, the slip equations (8.3) can be integrated directly to find the traction. Under transient conditions the variation in the displacements with time must be followed step by step from the initial conditions to the steady state. The numerical procedure is very straightforward. Kalker (1973) has discussed the procedure in detail with a view to ensuring that the results from the foundation model correspond as closely as possible to exact solutions based on the elastic half-space.

As an example we will consider the tractive rolling contact of long cylinders analysed exactly in §3. The elastic displacements of both surfaces are combined in the displacement $\bar{u}_x$ of the foundation so that in a stick zone equations (8.3) and (8.4) reduce to

$$\xi_x + \frac{\partial \bar{u}_x}{\partial x} = \dot{s}_x/V = 0 \tag{8.65}$$

We can substitute for $\bar{u}_x$ from equation (8.64) to give a simple differential equation for $q(x)$ which, when integrated, with the condition that $q_x = 0$ at $x = -a$, gives

$$q_x = -(K_q \xi_x/h)(a + x) \tag{8.66}$$

Thus the traction increases linearly from the leading edge and, if slip is entirely prevented, it is released suddenly at the trailing edge. In contrast to the exact solution, this traction is still finite at the trailing edge. The total tangential force follows directly from the integration of (8.66) with the result:

$$Q_x = -2K_q \xi_x a^2/h \tag{8.67}$$

In practice slip will occur at the trailing edge where the pressure falls to zero but the traction given by (8.66) does not. For consistency we will take the parabolic pressure distribution given by elastic foundation theory in equation (4.58). A stick region of width $2c$ extends from the leading edge. At the point where slip begins ($x = 2c - a$)

$$q_x = (K_q/h)\xi_x 2c = \mu p = 4\mu(K_p/2Rh)(a - c)c$$

which gives

$$\lambda \equiv 2(1 - c/a) = K_q \xi_x R/K_p \mu a \tag{8.68}$$

The traction force, found from the sum of the traction in both stick and slip zones, is then given by

$$Q_x/\mu P = -\tfrac{3}{2}\lambda(1 - \tfrac{1}{2}\lambda + \tfrac{1}{12}\lambda^2) \tag{8.69}$$

This equation and equation (8.67) are the counterparts of the exact solutions

(8.26) and (8.28); they are compared in Fig. 8.7. In order for the creep coefficient, i.e. the linear gradient, of the foundation model to coincide with that of the half-space solution, the tangential modulus of the foundation K_q should be 2/3 the normal modulus K_p. Thus we should take $K_q a/h \approx 1.1E^*$.†

In a three-dimensional contact undergoing longitudinal creep only, each elemental strip parallel to the x-axis behaves like the two-dimensional contact analysed above. The traction rises linearly from zero at the leading edge until it reaches the value μp when slip starts. If all slip is prevented, integrating the traction over the contact ellipse gives

$$Q_x = \frac{8ab}{3}\left(\frac{K_q a}{h}\right)\xi_x \tag{8.70}$$

To agree with the exact result (8.41) the foundation modulus K_q must be given by

$$\frac{K_q a}{h} = \tfrac{3}{8}GC_{11} = \tfrac{3}{8}(1-\nu)E^*C_{11} \tag{8.71a}$$

where values of C_{11} for different elliptical shapes are quoted in Appendix 5. An identical expression, with C_{11} replaced by C_{22}, is obtained in pure lateral creep. To obtain agreement with exact theory in pure spin requires

$$\frac{K_q a}{h} = \frac{4}{\pi}\left(\frac{b}{a}\right)^{1/2}(1-\nu)E^*C_{23} \tag{8.71b}$$

It is clear that a single value of the foundation modulus will not secure agreement over the whole range of creep conditions or (b/a). However, if we take $(K_q a/h) = 1.1E^*$, as in the two-dimensional case, equation (8.71a) gives (for $\nu = 0.3$) $C_{11} = C_{22} = 4.2$ and equation (8.71b) gives $C_{23} = 1.2(a/b)^{1/2}$ which compare reasonably with the values in the table except where $b \ll a$.

If, in the case of pure longitudinal or lateral creep, slip is assumed to occur when the tangential traction reaches its limiting value μp (where p is given by eq. (4.54)) the boundary between the stick and slip regions is found to be a reflexion of the leading edge as observed by experiment.

8.8 Pneumatic tyres

A wheel having a pneumatic tyre is an elastic body which, in rolling contact with the ground, exhibits most of the phenomena of creep and micro-slip which have been discussed in this chapter. In fact, the tangential force and twisting moment due to lateral creep, usually referred to as the 'cornering

---

† Although we can use the combined modulus E^* to account for the elasticities of both bodies, the foundation model is incapable of handling the tractions which arise from a *difference* in their elastic properties as discussed in § 2.

force' and 'self-aligning torque' play a significant role in the steering characteristics of a road vehicle. Clearly the complex structure of a tyre does not lend
itself to the analytical treatment which is possible for solid isotropic bodies;
nevertheless simple one-dimensional models have been proposed which do
account for the main features of the observed behaviour.

A thin flexible membrane of toroidal shape with internal pressure, when
pressed into contact with a rigid plane surface, has a contact area whose approximately elliptical shape is related to the intersection of the plane with the undeformed surface of the toroid and whose area is sufficient to support the contact
force by the internal pressure. An aircraft tyre, which has very little tread,
approximates to a thin membrane. Referring to Fig. 8.19 the apparent dimensions
of the contact ellipse a' and b' are related to the vertical deflexion of the wheel by

$$a' = \{(2R - \delta)\delta\}^{1/2}, \quad b' = \{(w - \delta)\delta\}^{1/2}$$

The apparent area of contact is

$$a'b' = \pi\delta \{(w - \delta)(2R - \delta)\}^{1/2} 2\pi wR\delta \tag{8.72}$$

In reality the tyre is tangential to the flat surface at the periphery of the contact,
so that the real contact dimensions a and b are less than a' and b'. The true area
is found to be about 85% of the apparent value; the deflexion is found to be
approximately proportional to the load with 80–90% of the load taken by the
inflation pressure. The stiff tread and cross-section shape of a motor tyre, on
the other hand, result in a contact path which is roughly rectangular, having
a constant width equal to the width of the tread and having slightly rounded
ends. The load is transmitted from the ground to the rim through the walls as
shown in Fig. 8.20. When the ground reaction P is applied to the tyre the tension

Fig. 8.19. A pneumatic tyre as a thin inflated membrane.

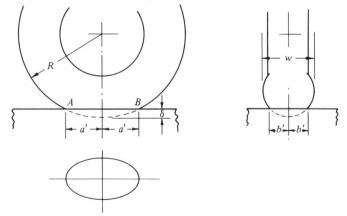

in the walls is decreased with a consequent increase in curvature, thereby exerting an effective upthrust on the hub.

On the membrane model the contact pressure distribution would be uniform and equal to the inflation pressure, whereas a solid tread would concentrate the pressure in the centre. The effect of bending stiffness of the tread is to introduce pressure peaks at the ends of the contact (see §5.8) and support from the walls gives high pressures at the edge. The relative importance of these different effects depends upon the design of tyre.

Creep in free rolling

Like a solid elastic wheel in contact with a rigid ground, a pneumatic tyre will exhibit longitudinal creep if the circumferential strain in the contact patch is different from that in the unloaded periphery. Following membrane theory, the centre-line of the running surface is shortened in the contact patch by the difference between the chord AB and the arc AB. By geometry this gives a strain, and hence a creep ratio:

$$\xi_x = \frac{\partial u_x}{\partial x} = -\delta/3R \tag{8.73}$$

This equation assumes that behaviour of the whole contact is governed by the centre-line strain and that there is no strain outside the contact, therefore the fact that it agrees well with observations must be regarded as somewhat fortuitous.

Transverse tangential forces from sideslip and spin

When the plane of a wheel is slightly skewed to the plane of rolling, described as 'sideslip' or 'yaw', transverse friction forces and moments are

Fig. 8.20. Contact of an automobile tyre with the road: (*a*) unloaded, (*b*) loaded.

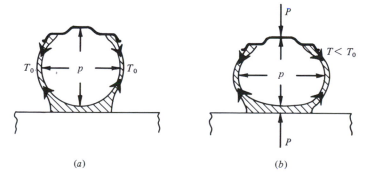

(*a*) (*b*)

brought into play; they also arise owing to spin when turning a corner or by cambering the axis of the wheel at an angle to the ground. The behaviour is qualitatively the same as that for solid bodies discussed in §4. The contact is divided into a stick region at the leading edge of the contact patch and a slip region at the trailing edge. The slip region spreads forward with increasing sideslip or spin.

A one-dimensional model of the resistance of a tyre to lateral displacement is shown in Fig. 8.21. The lateral deformation of the tyre is characterised by the lateral displacement u of its equatorial line, which is divided into the displacement of the carcass u_c and that of the tread u_t. Owing to the internal pressure the carcass is assumed to carry a uniform tension T. This tension resists lateral deflexion in the manner of a stretched string. Lateral deflexion is also restrained by the walls, which act as a spring foundation of stiffness K per unit length. The tread is also assumed to deflect in the manner of an elastic foundation ('wire brush') as discussed in §7. The tyre is deflected by a transverse surface traction $q(x)$ exerted in contact region $-a \leqslant x \leqslant a$. The equilibrium equation is:

$$K_c u_c - T \frac{\partial^2 u_c}{\partial x^2} = q(x) = K_t u_t \tag{8.74}$$

where K_t is the tread stiffness. The lateral slip velocity in the contact region is given by equation (8.3b). The ground is considered rigid ($u_2 = 0$) and the motion one-dimensional, so that we can drop the suffixes. Thus

$$\dot{s}/V = \xi + \psi(x/a) + \frac{du}{dx} \tag{8.75}$$

Fig. 8.21. The 'stretched string' model of the lateral deflexion of a tyre. The carcass and the tread resist lateral deflexion as elastic foundations of stiffness K_c and K_t.

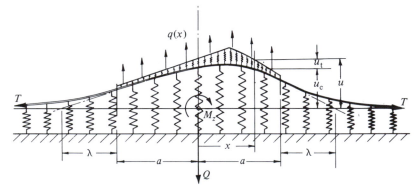

In a stick region $\dot{s} = 0$.

Various approaches to the solution of the problem have been proposed. Fromm (1943) took the carcass to be rigid ($u_c = 0$) so that all the deformation was in the tread ($u = u_t$). Equations (8.74) and (8.75) can then be solved directly throughout the contact region for any assumed pressure distribution. The carcass deflexions are clearly not negligible however and it is more realistic to follow von Schlippe (1941) and Temple (see Hadekel, 1952) who neglected the tread deflexion compared with the carcass deflexion ($u_t = 0, u = u_c$) as shown in Fig. 8.22. Equation (8.74) then becomes

$$u - \lambda^2 \frac{\mathrm{d}^2 u}{\mathrm{d}x^2} = q(x)/K_c \tag{8.76}$$

where the 'relaxation length' $\lambda = (T/K_c)^{1/2}$. To develop a linear theory we shall assume vanishingly small slip, so that in equation (8.75) $\dot{s} = 0$ throughout the contact region. Taking the case of sideslip first, from (8.75) the displacement within the contact region is given by

$$u = u_1 - \xi x$$

where u_1 is the displacement at the leading edge ($x = -a$). Outside the contact region $q(x) = 0$ so that the complementary solution to (8.76) gives

$$u = u_1 \exp\{(a+x)/\lambda\}$$

ahead of the contact and

$$u = u_2 \exp\{(a-x)/\lambda\}$$

at the back of the contact.

Fig. 8.22. Traction distribution for a tyre with yaw angle ξ and no slip in the contact patch: von Schlippe's theory.

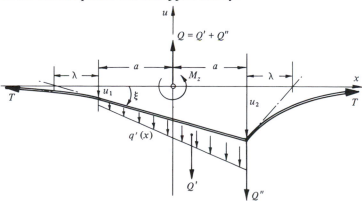

From our discussions in §4 of the location of slip and stick regions, it is clear that the displacement gradient must be continuous at the leading edge, for which $u_1 = -\lambda\xi$. The deflected shape of the equatorial line is shown in Fig. 8.22 together with the traction distribution. In the contact patch itself

$$q'(x) = K_c u = -K_c \xi(\lambda + a + x)$$

which corresponds to a force $Q' = -2K_c a(\lambda + a)$. At the trailing edge ($x = a$) the discontinuity in du/dx gives rise to an infinite traction $q''(a)$ which corresponds to a concentrated force Q'' of magnitude $-2K_c \xi\lambda(\lambda + a)$. The total cornering force is thus

$$Q = Q' + Q'' = -2K_c \xi a^2 \left(\frac{\lambda}{a} + 1\right)^2 \tag{8.77}$$

Taking moments about O gives the self-aligning torque to be

$$M_z = \int_{-a}^{a} q(x)x \, dx = -2K_c \xi a^3 \left(\frac{1}{3} + \frac{\lambda}{a} + \frac{\lambda^2}{a^2}\right) \tag{8.78}$$

As with solid bodies, the infinite traction at the trailing edge necessitates slip such that the deflected shape $u(x)$ has no discontinuity in gradient and satisfies the condition $q(x) = \mu p(x)$ within the slip region. Calculations of the cornering force Q and self-aligning torque M_z by Pacejka (1981) assuming a parabolic pressure distribution and taking $\lambda = 3a$ are shown in Fig. 8.23.

Fig. 8.23. Cornering force Q and self-aligning torque M_z by von Schlippe's theory ($\lambda = 3a$) from Pacejka (1981). Broken line – linear theory (no slip), eqs. (8.77), (8.78).

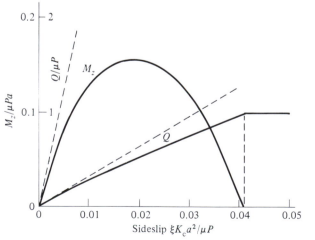

In the case of pure spin, equations (8.75) demands that the deflected shape in the stick region should be parabolic, i.e.

$$u = u_1 + \psi x^2/a$$

Proceeding as before to ensure continuity at $x = -a$ leads to a distribution of traction and to a resultant transverse force ('camber thrust')

$$Q = -2K_c \psi a^2 \left(\tfrac{1}{3} + \frac{\lambda}{a} + \frac{\lambda^2}{a^2} \right) \quad \text{and} \quad M_z = 0 \qquad (8.79)$$

Equations (8.77)–(8.79) are equivalent to the linear (small slip) equations (8.41)–(8.43) for solid bodies.

The analysis outlined above has been extended by Pacejka (1981) to include the elasticity of the tread. Some investigators have felt that the representation of the carcass by a 'string' in tension is inadequate and have included a term proportional to (d^4u/dx^4) in the equilibrium equation (8.74) to represent the flexural stiffness of the carcass (see Frank, 1965). However the influence on the overall conclusions of the theory is relatively minor since the values of the elasticity parameters of the tyre (K_c, K_t, λ etc.) have to be found by experiment rather than directly from the structure of the tyre.

Longitudinal creep due to driving or braking can be analysed in the same way. If the cover is regarded as an 'elastic belt' restrained circumferentially by the elastic walls, an equilibrium equation similar to (8.75) is obtained for the circumferential displacements of the belt. Recent reviews of tyre mechanics have been published by Clark (1971) and by Pacejka & Dorgham (1983).

9

Rolling contact of inelastic bodies

9.1   Elastic hysteresis

No solid is perfectly elastic. During a cycle of loading and unloading even within the so-called elastic limit some energy is dissipated. The energy loss is usually expressed as a fraction α of the maximum elastic strain energy stored in the solid during the cycle where α is referred to as the 'hysteresis loss factor'. For most metals stressed within the elastic limit the value of α is very small, usually less than 1%, but for polymers and rubber it may be much larger.

The material of bodies in freely rolling contact undergoes a cycle of loading and unloading as it flows through the region of contact deformation (see Fig. 9.1). The strain energy of material elements increases up to the centre-plane $(x = 0)$ due to the work of compression done by the contact pressure acting on the front half of the contact area. After the centre-plane the strain energy

Fig. 9.1. Deformation in rolling contact. An element of material experiences the cycle of reversed shear and compression A–B–C–D–E.

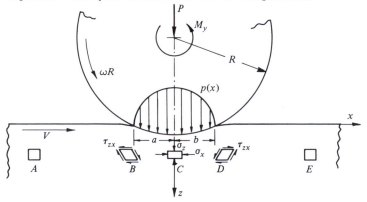

decreases and work is done against the contact pressure at the back of the contact. Neglecting any interfacial friction the strain energy of the material arriving at the centre plane in time dt can be found from the work done by the pressure on the leading half of the contact. For a cylindrical contact of unit width

$$dW = \omega\, dt \int_0^a p(x)x\; dx$$

where $\omega (= V/R)$ is the angular velocity of the roller. Taking $p(x)$ to be given by the Hertz theory

$$\dot{W} = \frac{2}{3\pi} Pa\omega \tag{9.1}$$

where P is the contact load. If a small fraction α of this strain energy is now assumed to be dissipated by hysteresis, the resultant moment required to maintain the motion is given by equating the net work done to the energy dissipated, then

$$M_y \omega = \alpha \dot{W} = \frac{2}{3\pi} \alpha Pa\omega$$

or

$$\mu_R \equiv \frac{M_y}{PR} = \alpha\, \frac{2a}{3\pi R} \tag{9.2}$$

where μ_R is defined as the coefficient of rolling resistance. Thus the resistance to rolling of bodies of imperfectly elastic material can be expressed in terms of their hysteresis loss factor. This simple theory of 'rolling friction' is due to Tabor (1955).† Performing the same calculation for an elliptical (or circular) contact area gives the result.

$$\mu_R \equiv \frac{M_y}{PR} = \alpha \tfrac{3}{16} \frac{a}{R} \tag{9.3}$$

where a is the half-width of the contact ellipse in the direction of rolling. For a sphere rolling on a plane, a is proportional to $(PR)^{1/3}$ so that the effective rolling resistance $F_R = M_y/R$ should be proportional to $P^{4/3}R^{-2/3}$. This relationship is reasonably well supported by experiments with rubber (Greenwood *et al.*, 1961) but less well with metals (Tabor, 1955).

The drawback to this simple theory is twofold. First, the hysteresis loss factor α is not generally a material constant. For metals it increases with strain (a/R), particularly as the elastic limit of the material is approached.

---

† The use of an elastic hysteresis loss factor in rolling is similar in principle to the use of a coefficient of restitution e in impact problems (see §11.5). The fractional energy loss in impact is given by $1 - e^2$.

Second, the hysteresis loss factor in rolling cannot be identified with the loss factor in a simple tension or compression cycle. The deformation cycle in rolling contact, illustrated in Fig. 9.1, involves rotation of the principal axes of strain between points *B*, *C* and *D*, with very little change in total strain energy. The hysteresis loss in such circumstances cannot be predicted from uniaxial stress data although a plausible hypothesis has been investigated for rubber by Greenwood *et al.* (1961) with reasonable success.

A rigid sphere rolling on an inelastic deformable plane surface would produce the same deformation cycle in the surface as a *frictionless* sphere *sliding* along the surface. In spite of the absence of interfacial friction the sliding sphere would be opposed by a resistance to motion due to hysteresis in the deformable body. This resistance has been termed the 'deformation component' of friction. Its value is the same as the rolling resistance F_R given by equation (9.3). Experiments by Greenwood & Tabor (1958) with a steel ball rolling and sliding on a well-lubricated rubber surface confirm this view. They suggest that the tread of motor tyres should be made from high hysteresis rubber to introduce a large deformation resistance when skidding on the rough surface of a road in wet and slippery conditions.

To formulate a more sophisticated theory of inelastic rolling contact it is necessary to define the inelastic stress–strain relations of the solids more precisely.

9.2   Elastic-plastic materials: shakedown

In this section we shall consider the behaviour in rolling contact of solids which are perfectly elastic up to a yield point: *Y* in simple tension or compression, *k* in simple shear. Beyond yield they deform in a perfectly plastic manner according to the stress–strain relations of Reuss.

(a) Onset of yield

In free rolling, within the elastic limit, the stresses in rolling contact are given by the Hertz theory, provided the two bodies are elastically similar. The effect of dissimilar elastic properties upon the elastic-plastic behaviour, however, is generally small and will be neglected. The onset of yield in free rolling is therefore the same as in frictionless normal contact discussed in §6.1. Yield first occurs at a point beneath the surface when the maximum contact pressure $p_0 = cY$, where c is a constant (≈ 1.6) whose exact value depends upon the geometry of the contact and the yield condition used according to equations (6.4), (6.5), (6.6), (6.8) or (6.9).

In tractive rolling the shear traction at the interface influences the point of first yield. The case of complete sliding, where $Q = \mu P$, has been examined in

§7.1 (see Figs. 7.3 & 7.4). With increasing friction the point of first yield approaches the surface. In tractive rolling when $Q < \mu P$, slip only takes place over the rear part of the contact area. The stress components $\bar{\sigma}_x$ and $\bar{\tau}_{xy}$ at the surface of rolling cylinders due to the tangential traction $q(x)$ have been found in §8.3 and are shown in Fig. 9.2. Varying the value of μ whilst keeping the traction coefficient Q/P constant causes the micro-slip zone to change in size and the distribution of traction to change as shown in Fig. 9.2(a). When the point of first yield lies beneath the surface it is not much influenced by changes in distribution of surface traction. When first yield occurs at the surface the critical point lies at the boundary between the stick and slip zones shown in Fig. 9.2(b). With increasing friction the contact pressure to reach first yield falls, as shown by the broken line in Fig. 9.4.

Fig. 9.2. Stresses at contact of cylinders rolling with tangential traction $Q_x = 0.2P$. (a) Tangential surface tractions for varying μ; (b) Surface stresses σ_x and τ_{zx} for $\mu = 0.3$.

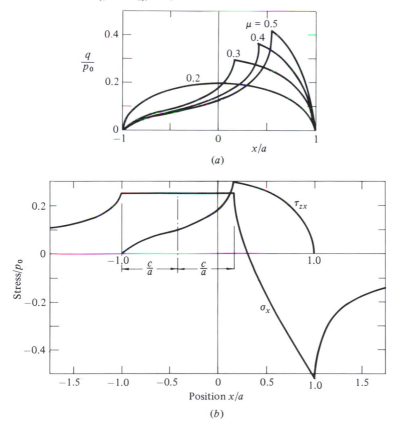

(b) Repeated rolling – shakedown

Most practical applications of rolling contact such as roller bearings or railway track have to withstand many repeated passes of the load. If, in the first pass, the elastic limit is exceeded some plastic deformation will take place and thereby introduce residual stresses. In the second passage of the load the material is subjected to the combined action of the contact stresses and the residual stresses introduced in the previous pass. Generally speaking such residual stresses are protective in the sense that they make yielding less likely on the second pass. It is possible that after a few passes the residual stresses build up to such values that subsequent passes of the load result in entirely elastic deformation. This is the process of *shakedown* under repeated loading, whereby initial plastic deformation introduces residual stresses which make the steady cyclic state purely elastic. To investigate whether shakedown occurs we can appeal to Melan's theorem (see Symonds, 1951) which states: if *any* time-independent distribution of residual stresses can be found which, together with the elastic stresses due to the load, constitutes a system of stresses within the elastic limit, then the system will shakedown. Conversely, if no such distribution of residual stresses can be found, then the system will not shakedown and plastic deformation will occur at every passage of the load.

We shall examine the case of an elastic cylinder rolling freely on an elastic-perfectly-plastic half-space (see Johnson, 1962b). If the elastic limit is not exceeded the contact area and the contact pressure are given by the Hertz theory. The stresses within the half-space are given by equations (4.49) and are shown by the full lines in Fig. 9.3 for a constant depth $z = 0.5a$. We now consider possible distributions of residual stress (denoted by suffix r) which can remain in the half-space after the load has passed. The assumption of plane deformation eliminates $(\tau_{xy})_r$ and $(\tau_{yz})_r$, and makes the remaining components independent of y. If the plastic deformation is assumed to be steady and continuous the surface of the half-space will remain flat and the residual stresses must be independent of x. Finally, for the residual stresses to be in equilibrium with a traction free surface $(\sigma_z)_r$ and $(\tau_{zx})_r$ cannot exist. The only possible system of residual stresses, therefore, reduces to

$$\left.\begin{array}{l} (\sigma_x)_r = f_1(z), \quad (\sigma_y)_r = f_2(z) \\ (\sigma_z)_r = (\tau_{xy})_r = (\tau_{yz})_r = (\tau_{zx})_r = 0 \end{array}\right\} \tag{9.4}$$

The principal stresses due to the combination of contact and residual stresses are given by

$$\sigma_1 = \tfrac{1}{2}\{\sigma_x + (\sigma_x)_r + \sigma_z\} + \tfrac{1}{2}[\{\sigma_x + (\sigma_x)_r - \sigma_z\}^2 + 4\tau_{zx}^2]^{1/2} \tag{9.5a}$$

$$\sigma_2 = \tfrac{1}{2}\{\sigma_x + (\sigma_x)_r + \sigma_z\} - \tfrac{1}{2}[\{\sigma_x + (\sigma_x)_r - \sigma_z\}^2 + 4\tau_{zx}^2]^{1/2} \tag{9.5b}$$

$$\sigma_3 = \nu\{\sigma_x + \sigma_z\} + (\sigma_y)_r \tag{9.5c}$$

Following Melan's theorem, we can choose the residual stresses to have any value at any depth in order to avoid yield. Thus $(\sigma_y)_r$ can be chosen to make σ_3 the intermediate principal stress. Then to avoid yield, by the Tresca criterion,

$$\tfrac{1}{4}(\sigma_1 - \sigma_2)^2 \leqslant k^2$$

where k is the yield stress in simple shear, i.e.

$$\tfrac{1}{4}\{\sigma_x + (\sigma_x)_r - \sigma_z\}^2 + \tau_{zx}^2 \leqslant k^2 \tag{9.6}$$

Examining this expression shows that it cannot be satisfied if τ_{zx} exceeds k, but it can just be satisfied with $\tau_{zx} = k$ if we choose $(\sigma_x)_r = \sigma_z - \sigma_x$. Thus the limiting condition for shakedown occurs when the maximum value of τ_{zx} anywhere in the solid just reaches k.

From equation (4.49) the maximum value of τ_{zx} is found to be $0.25p_0$ at points $(\pm 0.87a, 0.50a)$. Thus for shakedown to occur

$$p_0 \leqslant 4.00k \tag{9.7}$$

The same result is found if the von Mises criterion of yield is used. The residual stresses at a depth $0.50a$ necessary to ensure shakedown are

$$(\sigma_x)_r = -0.134p_0; \quad (\sigma_y)_r = -0.213p_0 \tag{9.8}$$

By von Mises criterion the value of p_0 for first yield is $3.1k$. Thus the ratio of the

Fig. 9.3. Rolling contact of elastic-plastic cylinders. Solid line – elastic stresses at depth $z = 0.5a$. Broken line – with addition of $(\sigma_x)_r$ and $(\sigma_y)_r$ for shakedown.

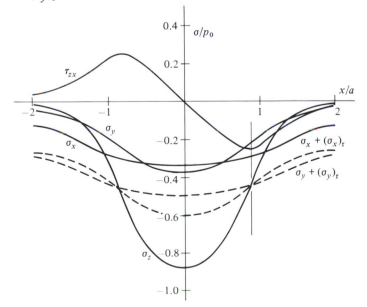

shakedown limit load to the elastic limit load is given by

$$\frac{P_S}{P_Y} = \frac{(p_0)_S^2}{(p_0)_Y^2} = 1.66 \tag{9.9}$$

from which it follows that the load must be increased by more than 66% above the first yield load to produce continuous deformation with repeated loading cycles. The modification to the stresses at $z = 0.5a$ by the introduction of the residual stresses is shown by the broken lines in Fig. 9.3.

A similar analysis can be made for tractive rolling if the elastic contact stress field is known (Johnson & Jefferis, 1963). Under conditions of complete slip or sliding the stresses in the half-space due to the frictional traction are given by equations (7.5)–(7.8). The residual stresses are still given by (9.4) and hence the shakedown limit is still determined by the maximum value of τ_{zx}. The point of maximum τ_{zx} lies below the surface provided $Q/P \leqslant 0.367$. At larger values of traction the critical stress lies in the surface layer. The influence of tangential traction on the shakedown limit is compared with its influence on initial yield in Fig. 7.4. The interval between the load for first yield and the shakedown limit load becomes narrower with increasing tangential force. The effect of partial slip on initial yield and shakedown is shown in Fig. 9.4. Shakedown of a wheel rolling on a rigid plane has been studied by Garg *et al.* (1974).

Fig. 9.4. Cylinders rolling with tangential traction $Q_x = 0.2P$. Broken line – first yield; solid line – shakedown. (Tresca yield criterion.)

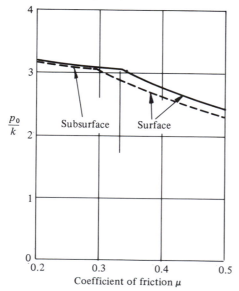

$$\frac{p_0}{k}$$

Coefficient of friction μ

The application of Melan's Theorem to three-dimensional rolling bodies is much more complicated since all components of residual stress can arise, and since a flat surface does not remain flat after deformation. If we consider the plane of symmetry ($y = 0$) of a ball rolling on an elastic-plastic half-space, by symmetry $(\tau_{xy})_r = (\tau_{yz})_r = 0$, but a residual shear stress $(\tau_{zx})_r$ can exist. However τ_{zx} due to a purely normal contact load has equal and opposite maxima on either side of $x = 0$. Thus yield cannot be inhibited by the addition of a uni-directional residual stress. It follows, therefore, that the shakedown limit is again governed by the maximum value of $(\tau_{zx})_r$.[†] If the reduction in contact pressure due to the formation of a shallow plastic groove is neglected then $(\tau_{zx})_{max} = 0.21p_0$, whereupon for shakedown

$$p_0 \leqslant 4.7k \tag{9.10}$$

By von Mises, the value of p_0 for first yield is $2.8k$, hence

$$\frac{P_S}{P_Y} = \left(\frac{4.7}{2.8}\right)^3 = 4.7 \tag{9.11}$$

which is a much larger factor than in the two-dimensional case.

The mechanism of shakedown may be appreciated qualitatively from Fig. 9.1. First yield occurs in the element at C on the centre-line by shear on planes at $45°$ to the axes: the element is compressed normal to the surface and attempts to expand parallel to it. Since all elements at that depth are plastically deformed in this way in turn, their lateral expansion must be annulled by the development of residual compressive stresses acting parallel to the surface. When these residual stresses are fully developed the elements no longer yield at C and normal compression of the surface ceases. The alternating 'orthogonal shear' τ_{zx} of the elements at B and D on the other hand cannot be reduced by the introduction of residual shear stress $(\tau_{zx})_r$. Hence it is the 'orthogonal shear' at B and D which governs the shakedown limit and the repeated plastic deformation which occurs when the shakedown limit is exceeded.

Two additional effects contribute to apparent shakedown of surfaces in rolling contact. The first, mentioned above, is the development of a groove with three-dimensional bodies which increases the contact area and reduces the contact pressure. Thus the true shakedown limit for a circular contact is somewhat greater than that given by (9.10). This process has been studied by Eldridge & Tabor (1955) for high loads when a deep groove is formed. With repeated rolling the depth and width of the groove reached steady values, from which it was concluded that subsequent deformation was entirely elastic. The stabili-

[†] It can be shown that a self-equilibrating system of residual stresses can be found such that all points in the half-space do not exceed the elastic limit.

sation of the groove dimensions does not guarantee a true shakedown state, however, since plastic shear parallel to the surface (orthogonal shear) may still be taking place.

The second effect which leads to apparent shakedown is a strain hardening. With repeated deformation the value of k may rise, such that a load which initially exceeds the shakedown limit subsequently lies within it. The theory is derived for ideally plastic solids and some difficulty arises in applying it to materials which strain-harden. Ponter (1976) has extended the theory to cover an idealised material which yields initially at $k = k'$, but is capable of 'kinematic hardening' up to an 'ultimate' value of $k = k''$ $(k'' > k')$.† For shakedown Melan's theorem must be satisfied with $k = k''$. At the same time yield must be avoided, with $k = k'$, by the loading stresses in combination with apparent residual stresses *which need not satisfy the equations of equilibrium.* In the two-dimensional case discussed above, the only possible system of residual stresses, given by equation (9.4), automatically satisfies equilibrium, so that the shakedown limit for a kinematically hardening material is still given by equation (9.7) with $k = k'$. Shakedown of a three-dimensional contact has been studied for a kinematically hardening material by Rydholm (1981) and for a perfectly plastic material by Ponter *et al.* (1985).

(c) Repeated rolling – cumulative deformation

When the load exceeds the shakedown limit, from the previous discussion we should expect orthogonal plastic shear to occur in the subsurface elements B and D (Fig. 9.1). The *elastic* stresses and strains at B and D are equal and opposite, but experiments by Crook (1957) and Hamilton (1963) have shown that in free rolling a net increment of permanent shear takes place in the sense of the outflowing element at D. In repeated rolling cycles the plastic deformation accumulates so that the surface layers are displaced 'forward', i.e. in the direction of flow, relative to the deeper layers (see Fig. 9.11(*a*)).

An approximate analysis of the elastic-plastic behaviour has been made (Merwin & Johnson, 1963) by using the Reuss stress–strain relations in conjunction with the *elastic distribution of strain.* In this way the stress components in an element at any depth may be computed as it flows through the strained region. By this technique the condition of compatibility of strains and the stress–strain relations are satisfied exactly but the equilibrium of stresses is only satisfied approximately. However, equilibrium of the residual stresses, expressed by equations (9.4), is maintained. The assumption that the strain field, including

---

† This idealisation of material behaviour implies that the initial yield locus of 'radius' k' is free to translate in stress space provided that it remains inscribed within a fixed yield locus of 'radius' k''.

plastic deformation, is the same as that without it, is likely to be a reasonable one while the plastic zone is fully contained beneath the surface and therefore is constrained by the surrounding elastic material. This will be the case provided that the load is not greatly in excess of the shakedown limit.

Starting with a stress-free body a number of cycles of load must be followed. The residual stresses build up very quickly and a steady state is virtually reached after four or five cycles.

The complete numerical analysis follows the variation in all the components of stress at each depth. However, the important components of stress and strain are τ_{zx} and γ_{zx}, and the mechanism of cumulative plastic deformation can be appreciated from a simple model which considers these components only. The steady-state stress–strain cycle is illustrated in Fig. 9.5. An element approach-

Fig. 9.5. Cumulative plastic shear in rolling contact: simplified orthogonal shear stress–strain cycle experienced by an element at depth $z = 0.5a$.

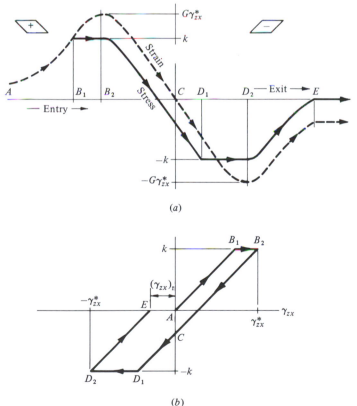

(a)

(b)

ing the loaded region deforms elastically from A to B_1 at which point the yield stress is reached ($\tau_{zx} = +k$). The element then deforms plastically at constant stress while the strain continues to increase to a maximum $\gamma_{zx} = \gamma_{zx}^*$ at B_2. From B_2, through C to D_1, the element unloads elastically and is deformed in the opposite ($-$ve) sense until the yield point ($\tau_{zx} = -k$) is reached at D_1. Reversed plastic deformation now takes place until the maximum negative strain, $\gamma_{zx} = -\gamma_{zx}^*$, is reached at D_2. The element then unloads until it is stress-free at E. It is not strain-free, however, since it has acquired an increment of negative residual strain $(\gamma_{zx})_r$. The forward displacement of the surface in one loading cycle is obtained by integrating $(\gamma_{zx})_r$ through the depth of the plastically deformed layer. Such calculations using the complete numerical analysis are compared with experimental measurements in Fig. 9.6.

The resistance to rolling can be calculated from computation of the total plastic work per unit distance rolled as elements flow through the plastic zone. On the first pass the plastic zone extends through a depth corresponding to that in which the elastic stresses, could they be realised, would exceed the plastic limit. In the steady state, continuous plastic deformation is restricted to a narrower layer (see Fig. 9.7). The resistance to rolling is then much less than

Fig. 9.6. Cumulative plastic shear in rolling contact: Merwin & Johnson theory (1963) compared with experiments: circle – Cu/Cu; triangle – Cu/Al; square – Cu/steel.

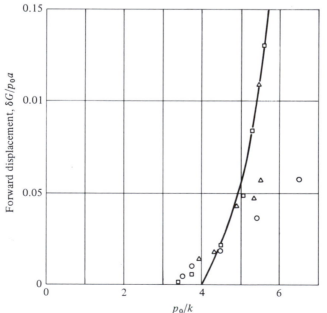

Fig. 9.7. Residual stresses induced in free rolling contact of Duralumin rollers. Circumferential stress $(\sigma_x)_r$: solid line – calculated; plus signs joined by dashes – measured. Axial stress $(\sigma_y)_r$: broken line – calculated; chain line – measured.

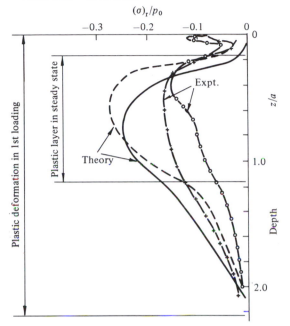

in the first pass, although it is still much larger than would normally be expected from elastic hysteresis (see Fig. 9.13).

Finally the analysis predicts the residual stresses $(\sigma_x)_r$ and $(\sigma_y)_r$ which are introduced by plastic deformation. The calculated variation with depth for $p_0 = 4.8k$ is shown in Fig. 9.7. Measurements by Pomeroy & Johnson (1969) of the circumferential and axial components of residual stress in an aluminium alloy disc due to free rolling contact are also plotted in Fig. 9.7. In their main features the agreement between theory and experiment is satisfactory; both components of stress are compressive; they arise in the layer in which the elastically calculated stresses exceed the elastic limit; the maximum values coincide roughly with the depth at which τ_{zx} is a maximum. The values of the measured stresses are appreciably lower than calculated. This discrepancy is likely to be due largely to the lack of plane-strain conditions in the experiment.

9.3 Rolling of a rigid cylinder on a perfectly plastic half-space

At high loads the plastic zone beneath a roller spreads to the free surface so that large plastic strains are possible. In these circumstances the stresses may be analysed by neglecting the elastic strains and making use of the theory of rigid-perfectly-plastic solids. In general the roller supports a normal load P, a tangential force Q and a moment M_G applied to the centre of the roller, shown in the positive sense in Fig. 9.8. A driving wheel has M_G positive and Q negative whereas a braked wheel has M_G negative and Q positive; in free rolling $Q = 0$. The mode of plastic deformation depends upon the magnitude and sign of the applied forces.

We shall consider first the case of Q positive, and assume that friction is sufficient to prevent slip at the interface. Incompressibility of the material and plane deformation require that, in the steady state, the level of the surface is unchanged by the roller. An approximate slip-line field and hodograph proposed by Mandel (1967) for this case are shown in Fig. 9.9(a). The cap of material ODC adheres to the roller and rotates with it about an instantaneous centre I fixed to the half-space. A velocity discontinuity follows the slip line $ABCO$. It is clear from the hodograph that the velocity of the free surface at D is some-what greater than at A so that the actual surface AD must be slightly concave and in consequence the slip lines DB and DC cannot be perfectly straight as shown. However the errors will not be large provided that the ratio of contact size to roller radius a/R is not too large.

The geometry of the field is specified by the two independent parameters: (a/R) and the angle α. The pressure on DC is then given by

$$p_{DC} = k(1 + \pi/2 + 2\alpha - 2\psi) \tag{9.12}$$

and that on OC by

$$p_{OC} = k(1 + \pi/2 + 2\theta - 2\psi) \tag{9.13}$$

By integrating the stresses along DC and CO the forces P and Q and the moment

Fig. 9.8. Rolling contact of a rigid cylinder on a rigid perfectly-plastic half-space.

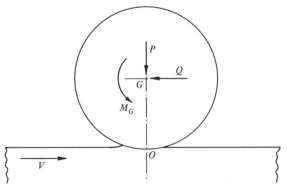

Fig. 9.9. Slip-line fields for a rigid cylinder rolling on a rigid-perfectly-plastic half-space (Mandel, 1967). (*a*) Mode I, (*b*) Mode II, Mode III.

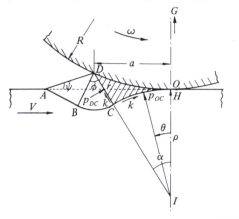

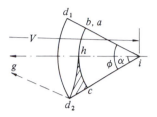

(*a*)

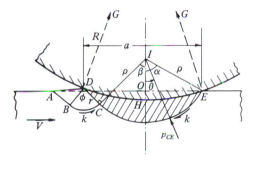

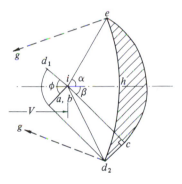

(*b*)

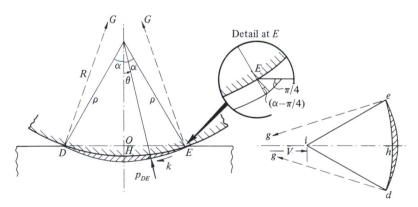

(*c*)

M_G can be found. Mandel simplifies his analysis still further for small values of a/R by assuming that the deformed surface ADO remains flat whereupon the angle $\psi = 0$. The solution is then a function of the single independent variable α only. This approximation is reasonable when I is located close to O and α is large, which is the case in the vicinity of free rolling. Under conditions of severe braking, on the other hand, the angle α becomes small and ψ is no longer small in comparison. In the limiting case of a locked wheel which slides without rotating, I moves to an infinite distance beneath the surface, $\alpha \to 0$, and $\psi \to \pi/4$. The limiting slip-line field comprises a small wedge of material which shears along the line AO. This is the same situation as that of a sliding wedge discussed in §7.6(*b*) and shown in Fig. 7.19(*b*). In this limit

$$P = Q = k(a + a^2/2R) \tag{9.14}$$

Taking moments about O

$$M_O = M_G + QR = \tfrac{1}{2}ka^2(1 + a/R + a^2/4R^2) = \tfrac{1}{2}P^2/k \tag{9.15}$$

The values of P, Q and M_O have been calculated for a range of values of α using the geometry of the field shown in Fig. 9.9(*a*) together with the equations (9.12) and (9.13). The results are plotted in non-dimensional form as $M_O k/P^2$ against the ratio Q/P and give the curve on the right-hand side of Fig. 9.10. The relationship has been computed for a value of $a/R = 0.2$, but this curve is almost independent of the value a/R, provided that a/R is not too large.

The pattern of plastic deformation is determined by the hodograph. To a first approximation surface points in the deforming region have a horizontal velocity from right to left relative to the body of the solid given by $ih - \omega\rho$. The forward velocity of the roller centre $V = \omega(R + \rho)$, so that the time of passage of a surface point through the deforming region from A to O is given by

$$T = AH/V \approx (\sqrt{(2)}\,r + a)/\omega(R + \rho)$$

where $r = DC$ and $\rho = IO$ as shown in Fig. 9.9(*a*). The plastic displacement of this surface Δ is thus given by $\Delta = -\omega\rho T$, i.e.

$$\frac{\Delta}{R} = -\frac{\rho(a + \sqrt{(2)}\,r)}{R(R + \rho)} \tag{9.16}$$

The surface of the half-space is displaced permanently *backwards* by the rolling action. This is in contrast with the behaviour at lighter loads, described in the previous section, where the action of the surrounding elastic material causes the surface to be displaced *fowards*. (See Fig. 9.11).

Two special cases call for comment.

(i) The condition of free rolling is specified by $Q/P = 0$ ($\alpha = 73.5°$), whereupon

$$M_O = M_G = 0.104P^2/k \tag{9.17}$$

(ii) The situation of a wheel driven forward by a horizontal force through a frictionless bearing at G (i.e. $M_G = 0$) is specified by $M_O = QR$ and is located therefore in Fig. 9.10 by the intersection of the curve by a straight line of gradient Rk/P. The resistance to rolling in this case is slightly greater than in free rolling.

An exact analysis of the deformation in Mode I shown in Fig. 9.9(a) has been made by Collins (1978), in which the free surface AD takes in correct concave profile and the slip lines DC and DB are appropriately curved. The exact results are shown by the broken lines in Fig. 9.10. They depend upon M_G as well as upon Q, but the difference from Mandel's approximate theory is not large. It has been pointed out by Petryk (1983) that even Collins' exact solutions are not unique, but the possible range of variation is not large except at high loads when the limit of complete shear ($Q/P = 1$) is approached.

As I approaches O and the angle α increases, the material to the right of CO becomes overstressed and deforms plastically. The new mode of deformation has not been found with certainty. Mandel proposes the mode shown in Fig. 9.9(b) with a velocity discontinuity along the line DCE and a stress discontinuity at E. The cap DCE adheres to the roller and rotates with it about an instantaneous centre I which now lies above O. The pressure along CE is given by:

$$p_{CE} = k\{1 + 2\alpha - 2\theta - \sin(\alpha - \pi/4)\} \qquad (9.18)$$

and the angle β is determined by equating the values of the pressure at C given

Fig. 9.10. Rolling moments and forces given by the slip-line fields of Fig. 9.9.

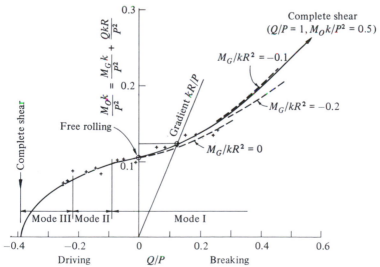

Fig. 9.11. Plastic deformation in rolling contact: (*a*) moderate loads –
cumulative forward displacement (see §9.2(*c*)); (*b*) heavy loads –
backward displacement in one pass (see eq. (9.16)).

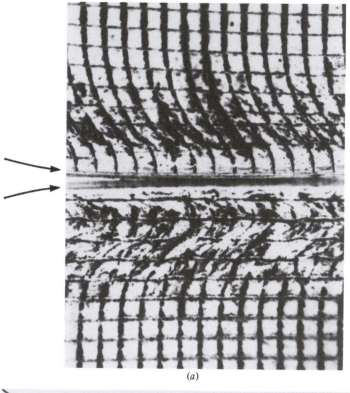

(*a*)

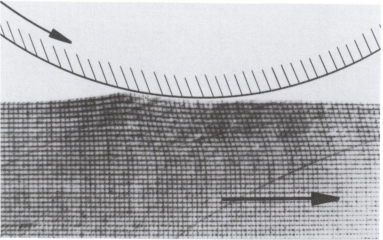

(*b*)

by equations (9.12) and (9.18). Mandel estimates that the change from the first to the second mode of deformation takes place at $Q/P = -0.09$. The change is accompanied by a large increase in contact area and a change in sign of the permanent displacement of the surface which is now given by

$$\frac{\Delta}{R} = \frac{a(\rho \cos \alpha - r^2/a)}{R(R - \rho \cos \alpha)} \tag{9.19}$$

However the moment M_O decreases in a continuous manner between the two modes as shown in Fig. 9.10.

With an increase in driving moment on the cylinder the mode of deformation changes again, when $\alpha \leqslant \pi/4$, to a single velocity discontinuity along the arc DE as shown in Fig. 9.9(c). A vanishingly small fan of angle $(\pi/4 - \alpha)$ is located with its apex at E, whereupon the pressure on the slip line DE is

$$p_{DE} = k(1 + \pi/2 - 2\theta) \tag{9.20}$$

Integrating the stresses along this slip line completes the variation in M_O with Q/P shown in Fig. 9.10. Complete shear, which corresponds to the spinning of a driving wheel, occurs when $\alpha \to 0$, i.e. when $Q/P \to -1/(1 + \pi/2)$.

The analysis outlined above assumes complete sticking along the arc of contact. In each mode of deformation the contact pressure is least at the trailing edge of contact, so slip would be expected to initiate there. A modification to the first mode of deformation, which includes micro-slip at the rear of the arc of contact, has been given by Mandel (1967), and also by Segal (1971). The effect is to reduce the values of M_O in Fig. 9.10 slightly and to restrict the limiting value of Q/P. A complete analysis for the case of zero interfacial friction has been presented by Marshall (1968). In this case M_G must be zero and Marshall finds that

$$M_O = QR = \frac{P^2 k}{2(2 + \pi)} \left\{ 1 + \frac{P}{6(2 + \pi)kR} \right\} \tag{9.21}$$

Taking a typical value of $P/kR = 1.0$, equation (9.21) gives $M_O/P^2 k = 0.100$. Drawing a line of slope 1.0 on Fig. 9.10 gives a value of $M_O/P^2 k = 0.125$, so that complete adhesion increases the resistance to rolling by 25% above the frictionless value.

Rolling resistance measurements by Mandel with a steel cylinder rolling on a lead surface are included in Fig. 9.10. The value of k for the lead has been chosen to fit the experiments to the theory at the free rolling point ($Q = 0$), but the theoretical variation in M_O with Q/P at constant load is well supported by the observations. They confirm that there is no discontinuity in M_O in changing from driving to braking. Absolute measurements of rolling resistance and surface displacements by Johnson & White (1974) with steel rolling on copper for the special cases of $Q = 0$ and $M_G = 0$ gave the results shown in

Table 9.1

		$Q = 0$		$M_G = 0$	
P/kR		M_G/RP	Δ/R	Q/P	Δ/R
0.73	Theory	0.076	−0.013	0.085	−0.030
	Experiment	0.047	−0.007	0.053	−0.012
1.05	Theory	0.110	−0.017	0.130	−0.042
	Experiment	0.078	−0.016	0.085	−0.032

Table 9.1. The general trend of the measurements follows the results of Mandel's theory but the observed magnitude of both the rolling resistance and the surface displacements is less than predicted. The discrepancies are almost certainly due to the influence of elastic deformation, and they emphasise that a rigid-perfectly-plastic analysis sets an upper limit on the rolling resistance and the amount of permanent deformation.

9.4 Rolling contact of viscoelastic bodies

When the stress in the material of a body in rolling contact is influenced by the *rate* of strain, the contact stresses and deformation will depend upon the speed of rolling. The simplest time-dependent constitutive relations for a material are those described as linear viscoelastic. They have been discussed in §6.5 in relation to normal contact. Even so, the application of the linear theory of visco-elasticity to rolling is not simple, since the situation is not one in which the viscoelastic solution can be obtained directly from the elastic solution. The reason for the difficulty is easy to appreciate. During rolling the material lying in the front half of the contact is being compressed, whilst that at the rear is being relaxed. With a perfectly elastic material the deformation is reversible so that both the contact area and the stresses are symmetrical about the centre-line. A viscoelastic material, on the other hand, relaxes more slowly than it is com-pressed so that the two bodies separate at a point closer to the centre-line than the point where they first make contact. Thus, in Fig. 9.12, $b < a$ and recovery of the surface continues after contact has ceased. The geometry of the rolling contact problem in viscoelasticity is different, therefore, from that in the perfectly elastic case so that the viscoelastic solution cannot be obtained directly from the elastic solution. Furthermore the point of separation ($x = b$) cannot be pre-scribed; it has to be located subsequently as the point where the contact pressure falls to zero.

In view of these difficulties we shall present a one-dimensional treatment in which a viscoelastic solid is modelled by a simple viscoelastic foundation of parallel compressive elements which do not interact with each other (May *et al.*, 1959). We shall consider the contact of a rigid frictionless cylinder of radius R rolling on such a foundation, as shown in Fig. 9.12.

The rolling velocity is V and the cylinder makes first contact with the solid at $x = -a$. Since there is no shear interaction between the elements of the foundation, the surface does not depress ahead of the roller. To the usual approximation, for $a \ll R$, the strain (compressive) in an element of the foundation at x is given by

$$\epsilon = -(\delta - x^2/2R)/h, \quad -a \leqslant x \leqslant b \tag{9.22}$$

where δ is the maximum depth of penetration of the roller.

If the foundation were perfectly elastic with modulus K, the contact would be symmetrical $(b = a)$ and the stress in each element σ would be $K\epsilon$. The pressure distribution under the roller and the total load would then be given by equations (4.58) and (4.59). For a viscoelastic material the elastic modulus K is replaced by a relaxation function $\Psi(t)$ as explained in §6.5. Thus by equation (6.51) the stress in the viscoelastic element at x is given by

$$p(x, t) = -\sigma = -\int_0^t \Psi(t - t') \frac{\partial \epsilon(t')}{\partial t'} \, dt' \tag{9.23}$$

In steady rolling $\partial/\partial t = V\partial/\partial x$, thus, from equation (9.22) for the strain,

$$\partial \epsilon/\partial t = Vx/Rh$$

Substituting in (9.23) and changing the variable from t to x, we get

$$p(x) = -\frac{1}{Rh} \int_{-x}^x x' \Psi(x - x') \, dx' \tag{9.24}$$

Fig. 9.12. Rolling of a rigid cylinder on a viscoelastic foundation.

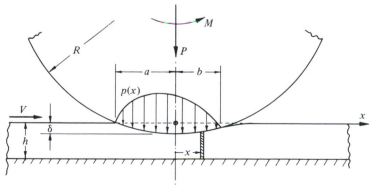

To proceed further we must specify the relaxation function for the material. Two simple examples incorporating delayed elasticity and steady creep were discussed in §6.5 (see Fig. 6.20). We shall make use of the first of these simple models, whose relaxation function is given by equation (6.54), and shall write

$$\Psi(t) = K(1 + \beta e^{-t/T}) \tag{9.25}$$

This material has an initial dynamic elastic response with modulus $K(1 + \beta)$, but under static (relaxed) conditions the modulus is K and T the relaxation time. Substituting this relaxation function into equation (9.24), changing the variable from t to x and performing the integration give an expression for the pressure distribution:

$$p(x) = \frac{Ka^2}{Rh} \left[\tfrac{1}{2}(1 - x^2/a^2) - \beta\varsigma(1 + x/a) \right.$$
$$\left. + \beta\varsigma(1 + \varsigma)\{1 - e^{-(1 + x/a)/\varsigma}\} \right] \tag{9.26}$$

where $\varsigma = VT/a$, which represents the ratio of the relaxation time of the material to the time taken for an element to travel through the semi-contact-width a. It is sometimes referred to as the 'Deborah Number'.

The pressure is zero at $x = -a$ and falls to zero again where $x = b$; this latter condition determines the value of b/a as a function of β and ς.

The normal load is found from

$$P = \int_{-a}^{b} p(x)\,dx = \frac{Ka^3}{Rh} F_P(\beta, \varsigma) \tag{9.27}$$

and, since the pressure is now asymmetrical, rolling is resisted by a moment given by

$$M = -\int_{-a}^{b} xp(x)\,dx = \frac{Ka^4}{Rh} F_M(\beta, \varsigma) \tag{9.28}$$

Thus the coefficient of rolling resistance may be expressed as:

$$\mu_R = M/PR = \frac{a}{R} F_R(\beta, \varsigma) \tag{9.29}$$

Computations have been carried out for a material in which $\beta = 1.0$; values of b/a and μ_R are plotted as a function of ς_0 ($=VT/a_0$) in Fig. 9.13, where a_0 is the static semi-contact-width. This diagram displays the significant features of visco-elastic rolling contact.

At slow rolling speeds, when the contact time is long compared with the relaxation time of the material ($\varsigma_0 \ll 1$), the pressure distribution (9.26) and load (9.27) approximate to the results for a perfectly elastic material of modulus K given in equations (4.58) and (4.59). The moment M approaches zero.

At very high speeds ($\zeta_0 \gg 1$), the pressure distribution and load again approximate to the elastic results but this time with a 'dynamic' foundation modulus $K(1 + \beta)$. Relaxation effects are important only when the contact time is roughly equal to the relaxation time of the material ($\zeta_0 \sim 1$). It is under these conditions that the contact becomes appreciably asymmetric and the maximum resisting moment arises. A similar analysis for a rigid sphere rolling on a viscoelastic foundation has been made by Flom & Bueche (1959).

The above analysis has been presented for a rigid cylinder rolling on a viscoelastic plane. As in elastic theory, it is equally valid to the same degree of approximation if the cylinder is viscoelastic and the plane rigid. The analysis again holds for two viscoelastic bodies if an equivalent relaxation function is used for a series combination of material elements of each body. An appropriate value for the ratio K/h of the foundation can be obtained by comparing the static deformation with Hertz as discussed in §4.3.

Fig. 9.13. Rolling of a rigid cylinder on a viscoelastic half-space ($\beta = 1$). Solid line – full solution, Hunter (1961). Large-dashed line – viscoelastic foundation model, eqs (9.26)–(9.29). Small-dashed line – loss tangent, eq. (9.36). (*a*) Contact dimensions; (*b*) rolling resistance.

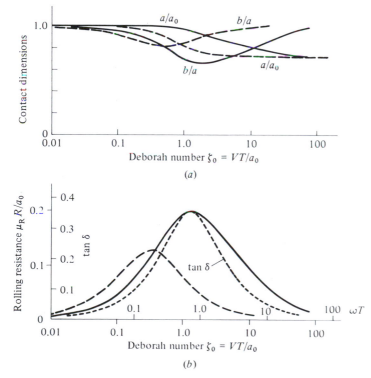

(*a*)

(*b*)

A full solution for the rolling contact of a rigid cylinder on a viscoelastic half-space has been presented by Hunter (1961) for materials having a single relaxation time. A different method, which is capable of handling two visco-elastic bodies having general relaxation functions, has been developed by Morland (1967a & b) but considerable computing effort is necessary to obtain numerical results.

For a rigid cylinder rolling on a simple material whose relaxation function is given by (9.25) Hunter's and Morland's solutions are identical. Results for a material whose Poisson's ratio v is constant and $\beta = 1$ are plotted in Fig. 9.13, where they are compared with the one-dimensional foundation model. The qualitative behaviour of the simple model is similar to that given by the full analysis; the contact is most asymmetrical and the rolling resistance is a maximum at a Deborah Number of about unity. The rolling moment peak is lower for the model since energy dissipated in shear between the elements is ignored and the peak occurs at a somewhat lower value of VT/a_0 since the length of the deforming zone in the model is less than that in the half-space.

Materials having more than one relaxation time would exhibit a rolling resistance peak whenever the time of passage through the contact coincided with one of the relaxation times.

9.5 Rolling friction

Ideally rolling contact should offer no resistance to motion, but in reality energy is dissipated in various ways which give rise to 'rolling friction'. Much of the analysis discussed in this chapter and the previous one has been the outcome of attempts to elucidate the precise mechanism of rolling resistance, so that it would seem appropriate to conclude this chapter with a summary of our present understanding. The various sources of energy dissipation in rolling may be classified into (a) those which arise through micro-slip and friction at the contact interface, (b) those which are due to inelastic properties of the material and (c) those due to roughness of the rolling surfaces. We shall consider each in turn.

'Free rolling' has been defined as motion in the absence of a resultant tangential force. Resistance to rolling is then manifest by a couple M_y which is demanded by asymmetry of the pressure distribution: higher pressures on the front half of the contact than the rear. The trailing wheels of a vehicle, however, rotate in bearings assumed frictionless and rolling resistance is overcome by a tangential force Q_x applied at the bearing and resisted at the contact interface. Provided that the rolling resistance is small ($Q_x \ll P$) these two situations are the same within the usual approximations of small strain contact stress theory, i.e. to first order in (a/R). It is then convenient to write the rolling resistance as a non-dimensional coefficient μ_R expressed in terms of the rate of energy dissipation $\dot{W}$, thus

$$\mu_R \equiv \frac{M_y}{PR} = \frac{Q_x}{P} = \frac{\dot{W}}{PV} \tag{9.30}$$

The quantity $\dot{W}/V$ is the energy dissipated per unit distance rolled.

(a) Micro-slip at the interface

Micro-slip has been shown to occur at the interface when the rolling bodies have dissimilar elastic constants (§8.2). The resistance from this cause depends upon the difference of the elastic constants expressed by the parameter β (defined by eq. (5.3)) and the coefficient of slipping friction μ. The resistance to rolling reaches a maximum value of

$$\mu_R = \frac{M_y}{PR} \approx 15 \times 10^{-4}\beta(a/R) \tag{9.31}$$

when $\beta/\mu \approx 5$. Since, for typical combinations of materials, β rarely exceeds 0.2, the rolling resistance from this cause is extremely small.

It has frequently been suggested that micro-slip will also arise if the *curvatures* of two bodies are different. It is easy to see that the difference in strain between two such surfaces will be second-order in (a/R) and hence negligible in any small strain analysis. A special case arises, however, when a ball rolls in a closely conforming groove. The maximum rolling resistance is given by (see §8.5)

$$\mu_R = \frac{M_y}{PR} = 0.08\mu(a/R)^2(b/a)^2 \tag{9.32}$$

The shape of the contact ellipse (b/a) is a function of the conformity of the ball and the groove; where the conformity is close, as in a deep groove ball-bearing, $b \gg a$ and the rolling resistance from this cause becomes significant.

In tractive rolling, when large forces and moments are transmitted between the bodies, it is not meaningful to express rolling resistance as Q_x or M_y/R. Nevertheless energy is still dissipated in micro-slip and, for comparison with free rolling, it is useful to define the effective rolling resistance coefficient $\mu_R = \dot{W}/VP$. This gives a measure of the loss of efficiency of a tractive drive such as a belt, driving wheel or variable speed gear. In the case of the belt drive (§8.1) for example, a moment M is transmitted between the two shafts but the driven shaft runs slightly slower than the driver, the difference between input and output power being accounted for by the energy dissipated in micro-slip given by

$$\dot{W}/V = M\left(\frac{V_1}{R_1} - \frac{V_2}{R_2}\right)/V = M\xi/R \tag{9.33}$$

where ξ is the creep ratio. This approach applies also to the tractive rolling of other elastic bodies. For rolling cylinders transmitting a tangential force Q_x the creep is given by equation (8.26) whence

$$\mu_R \equiv \dot{W}/VP = \frac{\mu Q_x}{P} \{1 - (1 - Q_x/\mu P)^{1/2}\} \frac{a}{R}$$

$$\begin{cases} \approx \frac{1}{2}(Q_x/P)^2(a/R) & \text{for} \quad (Q_x/P) \ll \mu & (9.34a) \\ = \mu^2(a/R) & \text{for} \quad (Q_x/P) = \mu) & (9.34b) \end{cases}$$

on the point of gross sliding. Similar expressions for the effective rolling resistance of spherical bodies transmitting tangential forces Q_x or Q_y can be obtained from the creep equations (8.45) and (8.46). If the tractive force is small compared with the limiting friction force it is clear from (9.34a) that the energy loss is small, but as sliding is approached the loss may become important if the coefficient of friction is high.

Finally micro-slip is introduced by spin. The angular velocity of spin ω_z is resisted by a spin moment M_z, given by equation (8.43) provided ω_z is small. At large spins the moment rises to a maximum value $3\pi\mu Pa/16$ for a circular contact area, whereupon

$$(\mu_R)_{\max} = M_z\omega_z/VP = \frac{3\pi\mu}{16}\left(\frac{\omega_z R}{V}\right)\frac{a}{R} \qquad (9.35)$$

which accounts for a serious loss of efficiency of rolling contact variable speed drives.

(b) Inelastic deformation of the material

Except in the special cases mentioned above resistance to free rolling is dominated by inelastic deformation of one or both bodies. In this case the energy is dissipated within the solids, at a depth corresponding to the maximum shear component of the contact stresses, rather than at the interface. With materials having poor thermal conductivity the release of energy beneath the surface can lead to high internal temperatures and failure by thermal stress (Wannop & Archard, 1973).

The behaviour of metals is generally different from that of non-metals. The inelastic properties of metals (and hard crystalline non-metallic solids) are governed by the movement of dislocations which, at normal temperatures, is not appreciably influenced either by temperature or by rate of deformation. Lower density solids such as rubber or polymers tend to deform in a visco-elastic manner which is very sensitive to temperature and rate of deformation.

The rolling friction characteristics of a material which has an elastic range of stress followed by rate-independent plastic flow above a sharply defined yield stress, typical of hard metals, are shown in Fig. 9.14. Non-dimensional rolling resistance $(G\mu_R/k)$ is plotted against non-dimensional load (GP/k^2R) for the contact of a rigid cylinder on a deformable solid, where G is the shear modulus and k the yield stress in shear of the solid. At low loads the deformation is predominantly elastic and the rolling resistance is given by the elastic hysteresis equation (9.2). The hysteresis loss factor α, found by experiment, is generally of the order of a few per cent.

The elastic limit is reached in the first traversal at a load given by equations (6.5) and (6.7) but, after repeated traversals, continuous plastic deformation takes place only if the load exceeds the shakedown limit given by equation (9.9). The resistance due to contained plastic deformation has been calculated by Merwin & Johnson (1963) for loads which do not greatly exceed the shakedown limit. At high loads, when the plastic zone is no longer contained, i.e. the condition of full plasticity is reached, the rolling resistance may be estimated by the rigid-plastic theory of Mandel. The onset of full plasticity cannot be precisely defined but, from our knowledge of the static indentation behaviour where full plasticity is reached when $P/2a \approx 2.6$ and $Ea/YR \approx 100$, it follows that $GP/k^2R \approx 300$. Extrapolating the elastic-plastic results completes the picture.

Fig. 9.14. Rolling resistance of a rigid cylinder on an elastic-plastic solid.

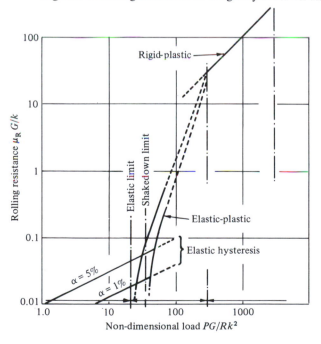

Experiments by Hamilton (1963) suggest that the elastic-plastic theory, which neglects losses in the elastically deformed material, under-estimates the rolling resistance, whilst experiments by Johnson & White (1974) suggest that the rigid-plastic theory, which neglects elastic deformation, overestimates the resistance. Nevertheless the general agreement is satisfactory and the figure shows that a steep rise in rolling resistance is to be expected when a continuously deforming plastic zone develops beneath the surface.

The characteristics of viscoelastic materials are somewhat different. In the previous section we saw how the rolling resistance of a simple linear viscoelastic solid could be analysed. Unfortunately most real viscoelastic materials are nonlinear and, further, their relaxation behaviour cannot usually be expressed in terms of a single relaxation time such as in the models shown in Fig. 6.20. However a useful empirical approach is possible using expressions (9.2) and (9.3) for rolling resistance in terms of an elastic hysteresis factor α. The most common method of measuring the hysteresis properties of viscoelastic materials is to measure the dissipation in cyclic strain as a function of frequency. The results of such measurements are usually expressed as a loss tangent $\tan \delta$, where δ is the phase angle between stress and strain. To correlate values of $\tan \delta$ with rolling resistance we compare the hysteresis theory with the full analysis given in §4 for the simple material whose relaxation function is given by (9.25). For such a material the loss tangent is given by

$$\tan \delta(\omega) = \beta \omega T / \{1 + (1 + \beta)\omega^2 T^2\} \tag{9.36}$$

where T is the relaxation time of the material. This relationship for $\beta = 1$ is compared with the variation of rolling resistance with Deborah Number $\zeta_0 (= VT/a_0)$ in Fig. 9.13(b). The curves are similar in shape. Their peaks can be made to coincide if ωT is put equal to 1.83ζ, i.e. the period of cyclic strain is put approximately equal to the time of passage of a material element through the contact zone. Their peak values agree if the value of α in equation (9.2) is taken to be about $2.6 \tan \delta$. In this way the rolling resistance for viscoelastic materials may be estimated from measurements of the loss tangent in cyclic strain, provided that the shear stress level at which the measurements are made is roughly comparable with that in rolling contact.

The variation of hysteresis loss with temperature in viscoelastic materials has been found to be related to the variation with frequency such that a unique curve is obtained when $\tan \delta$ is plotted against $a_T \omega$, where a_T is the Williams, Landen & Ferry shift factor defined by (see Ward, 1971)

$$\ln a_T = \frac{C_1(\theta - \theta_s)}{C_2 + (\theta - \theta_s)} \tag{9.37}$$

$\theta_s = \theta_g + 50$, where θ_g is the glass transition temperature and C_1 and C_2 are constants for the polymer. By the use of (9.37) the variation of hysteresis loss with temperature can be deduced from the variation with frequency. Measurements by Ludema & Tabor (1966) of the variation of rolling friction with temperature at a constant speed were found to follow the variation of hysteresis loss with temperature at a constant frequency.

(c) Surface roughness

It is an everyday experience that resistance to rolling of a wheel is greater on a rough surface than on a smooth one, but this aspect of the subject has received little analytical attention. The surface irregularities influence the rolling friction in two ways. Firstly they intensify the real contact pressure so that some local plastic deformation will occur even if the bulk stress level is within the elastic limit. If the mating surface is hard and smooth the asperities will be deformed plastically on the first traversal but their deformation will become progressively more elastic with repeated traversals. A decreasing rolling resistance with repeated rolling contact has been observed experimentally by Halling (1959). The second way in which roughness influences resistance is through the energy expended in surmounting the irregularities. It is significant with hard rough surfaces at light loads. The centre of mass of the roller moves up and down in its forward motion which is therefore unsteady. Measurements of the resistance force by Drutowski (1959) showed very large, high frequency fluctuations: energy is dissipated in the rapid succession of small impacts between the surface irregularities. It is the equivalent on a small scale of a wagon wheel rolling on a cobbled street. Because the dissipation is by impact the resistance from this cause increases with the rolling speed.

10

Calendering and lubrication

10.1 An elastic strip between rollers

Many processes involve the passage of a strip or sheet of material through the nip between rollers. In this section we consider the strip to be perfectly elastic and investigate the stresses in the strip, the length of the arc of contact with the roller, the maximum indentation of the strip and the precise speed at which it feeds through the nip in relation to the surface speed of the rollers. If the strip is wide and the rollers are long in the axial direction it is reasonable to assume plane deformation.

The static indentation of a strip by rigid frictionless cylinders was considered briefly in §5.8. The stresses in an elastic strip due to symmetrical bands of pressure acting on opposite faces have been expressed by Sneddon (1951) in terms of Fourier integral transforms. The form of these integrals is particularly awkward (e.g. eq. (5.65)) and most problems require elaborate numerical computations for their solution. However, when the thickness of the strip $2b$ is much less than the arc of contact $2a$ an elementary treatment is sometimes possible. The situation is complicated further by friction between the strip and the rollers. We can analyse the problem assuming (a) no friction ($\mu = 0$) and (b) complete adhesion ($\mu \to \infty$), but our experience of rolling contact conditions leads us to expect that the arc of contact will, in fact, comprise zones of both 'stick' and 'slip'.

We will look first at a strip whose elastic modulus is of a similar magnitude to that of the rollers, and write

$$C \equiv \frac{(1 - \nu_1^2)/E_1}{(1 - \nu_2^2)/E_2} = \frac{1 + \alpha}{1 - \alpha} \tag{10.1}$$

where α is defined by equation (5.3a) and 1, 2 refers to the strip and the rollers respectively. If the strip is thick ($b \gg a$) it will deform like an elastic half-space and the contact stresses will approach those discussed in §8.2. With equal elastic constants the deformation will be Hertzian, for unequal elasticity friction will introduce tangential tractions which, in the absence of slip, will be given by equation (8.15).

At the other extreme, when $b \ll a$, the deformation is shown in Fig. 10.1. The compression of the roller is now much greater than that of the strip so that the pressure distribution again approximates to that of Hertz, viz.

$$p(x) = \frac{2P}{\pi a}(1 - x^2/a^2)^{1/2} \tag{10.2}$$

The strip is assumed to deform with plane sections remaining plane so that the compression at the centre of the strip is given by

$$d = \frac{b(1 - \nu_1^2)p(0)}{E_1} = \frac{2b(1 - \nu_1^2)P}{\pi a E_1}$$

If the deformed surfaces of the strip are now approximated by circular arcs of radius R', then

$$\frac{1}{R'} = \frac{2d}{a^2} = \frac{4b(1 - \nu_1^2)P}{\pi a^3 E_1} \tag{10.3}$$

The rollers are flattened from a radius R to R', so that by equation (4.43)

$$a^2 = \frac{4P(1 - \nu_2^2)}{\pi E_2}\bigg/\left(\frac{1}{R} - \frac{1}{R'}\right) \tag{10.4}$$

Eliminating R' from (10.3) and (10.4) gives

$$\left(\frac{a}{a_0}\right)^2 = 1 + C\frac{b}{a} \tag{10.5}$$

Fig. 10.1. A thin elastic strip nipped between elastic rollers.

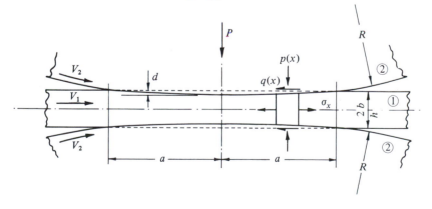

where $a_0 = \{4PR(1-\nu_2^2)/E_2\}^{1/2}$ is the semi-contact-width for a vanishingly thin strip.

With frictionless rollers the longitudinal stress in the strip σ_x is either zero or equal to any external tension in the strip. Due to the reduction in thickness, the strip extends longitudinally, whilst the roller surface compresses according to the Hertz theory, so that in fact frictional tractions $q(x)$ arise (acting inwards on the strip) whether or not the materials of the strip and rollers are the same. For equilibrium of an element of the strip

$$\frac{d\sigma_x}{dx} = \frac{1}{b}q(x) \tag{10.6}$$

Slip between the rollers and the strip is governed by equation (8.3). If there is no slip equation (8.3) reduces to

$$\frac{\partial \bar{u}_{x1}}{\partial x} - \frac{\partial \bar{u}_{x2}}{\partial x} = -\xi \tag{10.7}$$

where ξ is the creep ratio $(V_1 - V_2)/V_2$ of the strip relative to the periphery of the rollers. The longitudinal strain in the strip is given by

$$\frac{\partial \bar{u}_{x1}}{\partial x} = \frac{1-\nu_1^2}{E_1}\left\{\sigma_x + \frac{\nu_1}{1-\nu_1}p(x)\right\} \tag{10.8}$$

and the surface strain in a roller within the contact arc is given by equation (2.25a). For a thin strip we can take the pressure distribution $p(x)$ to be Hertzian given by (10.2). The expressions (10.8) and (2.25a) for the strains in the strip and rollers respectively can then be substituted into equation (10.7) which, together with (10.6), provides an integral equation for the tangential traction $q(x)$. This integral equation is satisfied by the traction:

$$q(x) = \left(1 - \frac{4\beta}{1+\alpha}\right)\frac{b}{2a}p_0\frac{x}{(a^2-x^2)^{1/2}} \tag{10.9}$$

where β is defined in equation (5.3b). This expression for $q(x)$ is satisfactory away from the edges of the contact, but the infinite values at $x = \pm a$ are a consequence of assuming plane sections remain plane. The traction falls to zero, in fact, at the edges.

A complete numerical analysis of this problem has been made by Bentall & Johnson (1968) for a range of values of b/a; the results are shown in Fig. 10.2. The contact pressure is close to a Hertzian distribution for all values of b/a. The frictional traction is zero at the extremes of both thick and thin strips; it reaches a maximum when $b/a \approx 0.25$. Although $q(x)$ falls to zero at $x = \pm a$, in the absence of slip the ratio $q(x)/p(x)$ reaches high values. This implies that some micro-slip is likely at the edges of the contact.

The pattern of micro-slip depends upon the relative elastic constants of the strip and rollers. If the materials are the same, or if the rollers are more flexible ($\beta \leqslant 0$) the tangential traction always acts inwards on the strip and a pattern of three slip regions similar to that with two dissimilar rollers in contact (Fig. 8.2) is obtained whatever the thickness of the strip. The distribution of traction and also the stress difference $|\sigma_x - \sigma_z|$ on the centre-plane of the strip are shown in Fig. 10.3 for $\beta = 0$, $b/a = 0.10$ and $\mu = 0.1$. Slip occurs in the same direction at entry and exit and a reversed slip region is located towards the exit. Slip affects neither the contact width a nor the indentation depth d, but the creep ratio is quite sensitive to the coefficient of friction (see Fig. 10.2).

When the strip is more flexible than the rollers ($\beta > 0$) the frictional traction acts outwards on thick strips and inwards on thin ones, so that the pattern of slip depends upon the strip thickness, but is similar to that shown in Fig. 10.3 if the strip is thin.

In the above discussion we have taken for granted that the strips and rollers have elastic constants which have a comparable magnitude, for example metal strips nipped by metal rollers. A somewhat different picture emerges when the rollers are relatively rigid compared with the strip, particularly if the strip is incompressible, for example rubber sheet nipped between metal rollers. With a sufficiently thick strip the contact stresses approach those for a rigid cylinder indenting an elastic half-space. A thin strip nipped between *frictionless* rollers is similar to an elastic layer supported on a frictionless base indented by a friction-less cylinder, which has been discussed in §5.8. But when friction between the roller and the strip is taken into account simple solutions based on homogeneous deformation are unsatisfactory.

Fig. 10.2. An elastic strip between elastic rollers: semi-contact-width a, penetration d and creep ratio ξ. ($C = 1$, $\beta = 0$, $\nu_1 = \nu_2 = 0.3$).

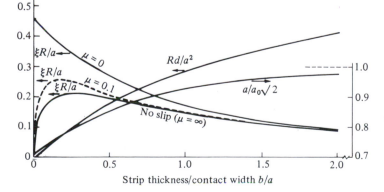

Strip thickness/contact width b/a

A complete numerical solution is presented in Fig. 10.4 for $C = 1000$ and $v_1 = 0.5$ assuming no slip. For an incompressible material in contact with a relatively rigid one $\beta = 0$, so that the half-space solution, which is the limit for $b \gg a$, does not involve any frictional traction and the stresses and deformation are given by Hertz. Thinner strips tend to be squeezed out longitudinally and inward acting tangential tractions arise. The ratio of $q(x)/p(x)$, shown in Fig. 10.4(b), indicates the magnitude of the coefficient of friction which is necessary to prevent slip at the edges of the contact. For $b \approx 0.25a$ the pressure distribution shown in Fig. 10.4(a) is approximately parabolic, as suggested by equation (5.71). For $b \approx 0.1a$ the pressure distribution for an incompressible material becomes bell-shaped, roughly as given by equation (5.75). Further

Fig. 10.3. An elastic strip between rollers ($C = 1$, $\beta = 0$, $v_1 = v_2 = 0.3$). (a) Distributions of pressure $p(x)$ and tangential traction $q(x)$. (b) Stress difference $|\sigma_x - \sigma_z|$.

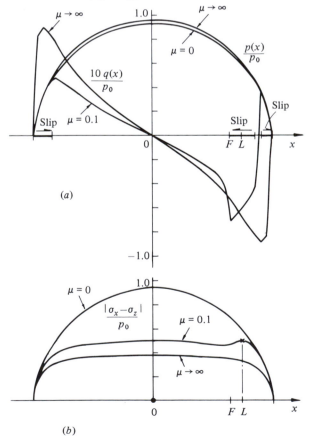

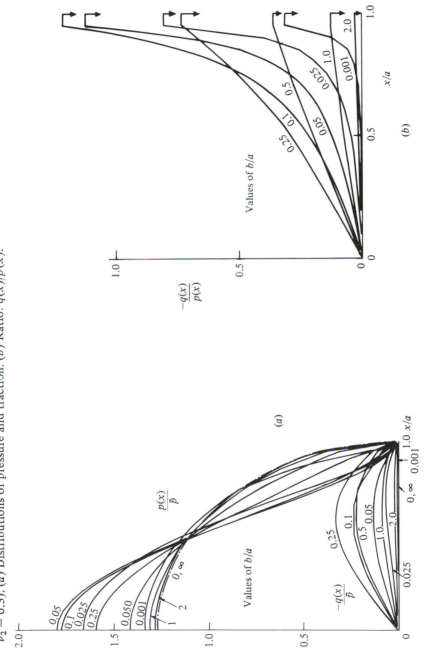

Fig. 10.4. An incompressible elastic strip nipped between relatively rigid rollers with no slip ($C = 1000$, $\beta = 0$, $\nu_1 = 0.5$, $\nu_2 = 0.3$). (a) Distributions of pressure and traction. (b) Ratio: $q(x)/p(x)$.

Fig. 10.5. An incompressible elastic strip between relatively rigid rollers ($C = 1000, \beta = 0, \nu_1 = 0.5, \nu_2 = 0.3$). Semi-contact-width a, penetration d and creep ratio ξ. Solid line – no slip ($\mu = \infty$); broken line – frictionless ($\mu = 0$).

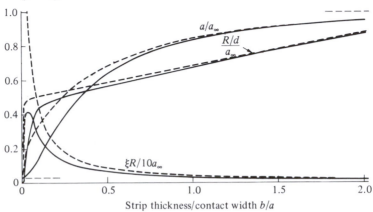

Strip thickness/contact width b/a

reduction in thickness of the strip results in deformation of the rollers becoming significant. In the limit when b is vanishingly small, the deformation is confined to the rollers, the stresses are again given by Hertz for the contact of two equal cylinders, so that the frictional traction also vanishes in this limit. The variations of contact width, penetration and creep ratio with strip thicknesses are plotted in Fig. 10.5.

10.2   Onset of plastic flow in a thin strip

In the metal industries thin sheet is produced from thick billets by plastic deformation in a rolling mill. We shall consider this process further in §3 but first we must investigate the conditions necessary to initiate plastic flow in a strip nipped between rollers. A thick billet is similar to a half-space so that the initial yield occurs (by the Tresca criterion) when the maximum elastic contact pressure p_0 reaches $1.67Y$ (eq. (6.4)), where Y is the yield stress of the billet in compression. A thin strip between rollers, as shown in Fig. 10.1, will yield when

$$|\sigma_x - \sigma_z|_{max} = Y\dagger \tag{10.10}$$

With frictionless rollers σ_x is approximately zero and $\sigma_z = -p$, so that yield in this case occurs when $p_0 \approx Y$, which is lower than for a thick billet. However it

---

† This criterion assumes that σ_y is the intermediate principal stress. With very thin strips that is no longer the case so that yield, in fact, initiates by lateral spread. However plane strain conditions restrict such plastic deformation to a negligible amount.

is a fact of experience that very high contact pressures are necessary to cause plastic flow in a thin strip. The frictional traction acts inwards towards the mid-point of the contact (see eq. (10.9) for a strip which sticks to the rollers) and results in a compressive longitudinal stress σ_x which inhibits yield.

Detailed calculations of the stresses on the mid-plane of the strip have been made by Johnson & Bentall (1969) for $\mu = 0, 0.1$ and for no slip ($\mu \to \infty$). A typical variation of $|\sigma_x - \sigma_z|$ through the nip is shown in Fig. 10.3(b). The effect of friction on $|\sigma_x - \sigma_z|_{max}$ is very marked. By using the yield criterion (10.10) the load to cause first yield P_Y is found for varying thicknesses of strip h and the results plotted non-dimensionally in Fig. 10.6. The influence of friction in producing a rise in the load to cause yield in thin strips is most striking.

The initiation of yield does not necessarily lead to measurable plastic defor-mation. If the plastic zone is fully contained by elastic material the plastic strains are restricted to an elastic order of magnitude. The point of initial yield (point of $|\sigma_x - \sigma_z|_{max}$) in the strip lies towards the rear of the nip in the middle slip zone marked L in Fig. 10.3. In this slip zone the strip is moving faster than the rollers. If there is to be any appreciable permanent reduction in thickness of the strip it must also emerge from the nip moving faster than the rollers. For this to happen the second stick zone and the final reversed slip zone of the elastic solution (Fig. 10.3(a)) must be swept away. The middle slip zone,

Fig. 10.6. Load to cause first yield P_Y and the load to cause uncontained plastic reduction P_F in the rolling of strip of thickness h.

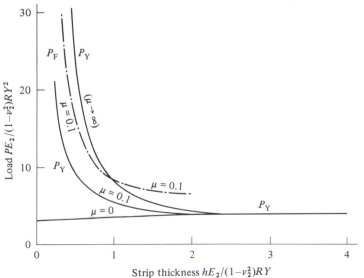

in which plastic reduction is taking place, will then extend to the exit of the nip. The distribution of traction and the corresponding variation in $|\sigma_x - \sigma_z|$ compatible with a single slip zone at exit have been calculated. $|\sigma_x - \sigma_z|_{max}$ has equal maximum values at the beginning and end of the no-slip zone, so that plastic flow begins at F in Fig. 10.3. Putting $|\sigma_x - \sigma_y|_{max} = Y$ in this case leads to a value of the load P_F at which uncontained plastic flow commences. The variation of P_F with strip thickness (taking $\mu = 0.1$) is included in Fig. 10.6. It shows that the load to initiate measurable plastic reduction is almost double that to cause first yield. The effect of friction in preventing the plastic flow of very thin strips is again clearly demonstrated. The superimposition of an external tension in the strip reduces the longitudinal compression introduced by friction and makes yielding easier.

10.3 Plastic rolling of strip

When a metal strip is passed through a rolling mill to produce an appreciable reduction in thickness, the plastic deformation is generally large compared with the elastic deformation so that the material can be regarded as being rigid-plastic. In the first instance the elastic deformation of the rolls may also be neglected. For continuity of flow, the rolled strip emerges from the nip at a velocity greater than it enters, which is in inverse proportion to its thickness if no lateral spread occurs. Clearly the question of sticking and slipping between the rolls and the strip, which has been prominent in previous chapters, arises in the metal rolling process. In hot rolling the absence of lubricant and the lower flow stress of the metal generally mean that the limiting frictional traction at the interface exceeds the yield stress of the strip in shear so that there is no slip in the conventional sense at the surface.

It is for the condition of no slip encountered in hot rolling that the most complete analyses of the process have so far been made. We saw in the previous section that interfacial friction inhibits plastic reduction, so that in cold rolling the strip is deliberately lubricated during its passage through the rolls in order to facilitate slip. At entry the strip is moving slower than the roll surfaces so that it slips backwards; at exit the strip is moving faster so that it slips fowards. At some point in the nip, referred to as the 'neutral point' the strip is moving with the same velocity as the rolls. At this point the slip and the frictional traction change direction. In reality, however, we should not expect this change to occur at a point. In the last section, when a thin *elastic* strip between *elastic* rollers was being examined, we saw that plastic deformation and slip would initiate at entry and exit; in between there is a region of no slip and no plastic deformation. It seems likely therefore that a small zone of no slip will continue to exist even when appreciable plastic reduction is taking place in the nip as a whole. Current theories of cold rolling, which are restricted to the idea of a 'neutral point', must be regarded as 'complete slip' solutions in the sense discussed in Chapters 8 and 9.

The complete solution of a problem involving the plane deformation of a rigid-perfectly-plastic material calls for the construction of a slip-line field. So far this has been achieved only for the condition of no slip, which applies to hot rolling. Before looking at these solutions we shall examine the elementary theories, with and without slip, which derive from von Kármán (1925).

The geometry of the roll bite, neglecting elastic deformation, is shown in Fig. 10.7. The mean longitudinal (compressive) stress in the strip is denoted by $\bar{\sigma}_x$ and the transverse stress at the surface by $\bar{\sigma}_z$. Equilibrium of the element gives

$$\bar{\sigma}_z \, dx = (p \cos \phi + q \sin \phi)R \, d\phi \tag{10.11}$$

and

$$d(h\bar{\sigma}_x) = (p \sin \phi - q \cos \phi)2R \, d\phi \tag{10.12}$$

In this simple treatment it is assumed that in the plastic zone $\bar{\sigma}_x$ and $\bar{\sigma}_z$ are

Fig. 10.7

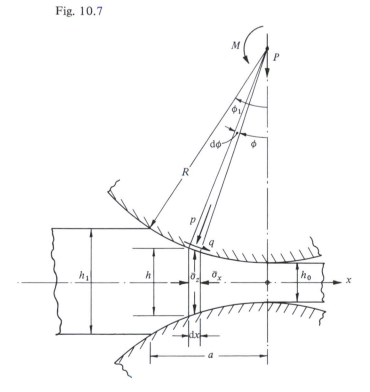

related by the yield criterion

$$\bar{\sigma}_z - \bar{\sigma}_x = 2k \tag{10.13}$$

This simplification implies a homogeneous state of stress in the element which is clearly not true at the surface of the strip where the frictional traction acts. Nevertheless by combining equations (10.11), (10.12) and (10.13) we obtain

$$\frac{d}{d\phi}\{h(p + q \tan \phi - 2k)\} = 2R(p \sin \phi - q \cos \phi) \tag{10.14}$$

which is von Kármán's equation. It is perfectly straightforward to integrate this equation numerically (see Alexander, 1972) to find the variation in contact pressure $p(\phi)$ once the frictional conditions at the interface are specified. Before electronic computers were available, however, various simplifications of von Kármán's equation were proposed to facilitate integration. For relatively large rolls it is reasonable to put $\sin \phi \approx \phi$, $\cos \phi \approx 1$ etc. and to retain only first order terms in ϕ. The roll profile is then approximated by

$$h \approx h_0 + R\phi^2 \approx h_0 + x^2/R \tag{10.15}$$

Making these approximations in (10.14), neglecting the term $q \tan \phi$ compared with p, and changing the position variable from ϕ to x give

$$h \frac{dp}{dx} = 4k \frac{x}{R} + 2q \tag{10.16}$$

In addition, it is consistent with neglecting second order terms in ϕ to replace h by the mean thickness $\bar{h}(= \frac{1}{2}(h_0 + h_i))$. To proceed, the frictional traction q must be specified.

(a) Hot rolling – no slip

For hot rolling, it is assumed that q reaches the yield stress k of the material in shear throughout the contact arc. Equation (10.16) then becomes

$$\bar{h} \frac{dp}{dx} = 2k \left(2\frac{x}{R} \pm 1 \right) \tag{10.17}$$

The positive sign applies to the entry region where the strip is moving slower than the rolls and the negative sign applies to the exit. Integration of (10.17), taking $\bar{\sigma}_x = 0$ at entry and exit, gives the pressure distribution:

At entry

$$\frac{\bar{h}}{a}\left(\frac{p}{2k} - 1 \right) = (1 + x/a) - \frac{a}{R}(1 - x^2/a^2) \tag{10.18a}$$

and at exit

$$\frac{\bar{h}}{a}\left(\frac{p}{2k} - 1 \right) = -x/a + \frac{a}{R}\frac{x^2}{a^2} \tag{10.18b}$$

The pressure at the neutral point is common to both these equations, which locates that point at

$$\frac{x_n}{a} = -\tfrac{1}{2} + \frac{a}{2R} \tag{10.19}$$

The total load per unit width is then found to be

$$\frac{P}{ka} = \frac{1}{ka} \int_{-a}^{0} p(x)\,dx \approx 2 + \frac{a}{h}\left(\tfrac{1}{2} - \tfrac{1}{3}\frac{a}{R}\right) \tag{10.20}$$

and the moment applied to the rolls is found to be

$$\frac{M}{ka^2} = \frac{1}{ka^2} \int_{-a}^{0} xp(x)\,dx \approx 1 + \tfrac{1}{4}\frac{a}{h}\left(1 - \frac{a}{R}\right) \tag{10.21}$$

This analysis is similar to the theory of hot rolling due to Sims (1954), except for the factor $\pi/4$ which Sims introduces on the right-hand side of equation (10.13) to allow for the non-homogeneity of stress. It is clear from the above expressions for force and torque that the 'aspect ratio' $\bar{h}/a$ is the primary independent variable: the parameter a/R, which is itself small in the range of validity of this analysis (ϕ small), exerts only a minor influence. Equations (10.20) and (10.21) for force and torque are plotted as dotted lines in Fig. 10.10.

The approach outlined above, in which the yield condition (10.13) is applied to the *average* stresses acting on the section of strip, makes equation (10.14) for the contact forces statically determinate, but the actual distributions of stress and deformation within the strip remain unknown. In reality the stresses within the strip should follow a statically admissible slip-line field and the deformation should follow a hodograph which is compatible with that field. To ensure such compatibility is far from easy. It was first achieved by Alexander (1955) by using a graphical trial-and-error method for a single configuration ($\bar{h}/a = 0.19$, $a/R = 0.075$) and by assuming that

$$-k \leqslant q \leqslant +k$$

everywhere throughout the arc of contact. The slip-line field and hodograph are shown in Fig. 10.8(*b*) and (*c*). In the centre of the roll bite there is a cap of undeforming material attached to the roller over the arc *CD*. The tangential traction $|q| < k$ in this arc. There is also a thin sliver of undeforming material on the arc *AB* at entry. The material is deforming plastically in the zones *ABCNF* and *DEGN*. A velocity discontinuity follows the slip lines *AB*, *CN* and *ND*. There is 'quasi-slip' between the rolls and the strip on the areas *BC* and *DE* which takes the form of a 'boundary layer' of intense shear at the yield stress k. The state of stress in the strip is obtained by following the slip lines from the entry at *A* or the exit at *E*. At the neutral point *N* the stress is the same from whichever end it is approached. The pressure distribution over the contact arc is shown by the full lines in Fig. 10.8 where it is compared with the simple theory given by equations (10.18) (shown dotted).

Slip-line fields for other geometric configurations have since been constructed by Crane & Alexander (1968) for thin strips and by Dewhurst *et al.* (1973) for thicker strips. It transpires that the form of the solution depends almost entirely upon the aspect ratio $\bar{h}/a$ and hardly at all upon the roll radius parameter a/R. A sample of the different fields is shown in Fig. 10.9. Alexander's original solution (Fig. 10.8) applies to thin strips. At a value of $\bar{h}/a \approx 0.29$ the quasi-

Fig. 10.8. Hot rolling of strip ($\bar{h}/a = 0.19$, $a/R = 0.075$). (*a*) Pressure distribution: solid line – from the slip-line field, Alexander (1955); broken line – from eq. (10.18); (*b*) Slip-line field (Mode I: $\bar{h}/a < 0.29$); (*c*) Hodograph.

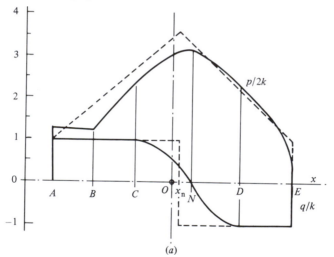

(*a*)

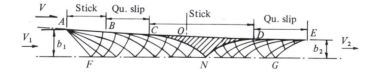

(*b*)

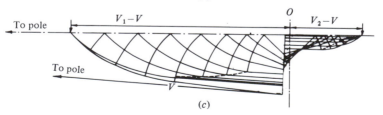

(*c*)

slip region *BC* just vanishes and the velocity discontinuity which follows the slip line *AN* now lies entirely within the material (Fig. 10.9(*a*)). A second critical aspect ratio occurs at $\bar{h}/a \approx 0.43$, when the quasi-slip region *DE* just vanishes and the rigid cap covers the whole arc of contact. The fields for greater values of $\bar{h}/a$ have been found by Dewhurst *et al.* (1973). The rigid cap *AGE* no longer penetrates to the centre-line of the strip, but a velocity discontinuity follows the slip lines *AN* and *NE* (Fig. 10.9(*b*)). At $\bar{h}/a \approx 0.72$, the arc *GE* contracts to a single point, whereupon the rigid zone takes the form shown in Fig. 10.9(*c*), with a velocity discontinuity along its boundary *AE*.

In view of the insensitivity of the fields to roll radius (i.e. to a/R), the roll force coefficient P/ka and torque coefficient M/ka^2 can be plotted as unique curves against the aspect ratio $\bar{h}/a$, as shown in Fig. 10.10. A minimum in both force and torque is obtained when the aspect ratio is about unity. With thin strips friction at the roll surfaces inhibits yield through high hydrostatic pressure in the centre of the roll bite; with thick strips, higher contact pressures are required to cause plastic flow through the thickness of the strip. When a further critical strip thickness is reached the roll pressure required to cause yield through the strip is greater than that to cause plastic flow only in the surface layers in the manner discussed in §9.3. By extending the slip-line field shown in Fig. 9.9(*a*) into the solid this critical thickness is found to be ~8.8*a*.

Fig. 10.9. Slip-line fields for hot rolling. (*a*) Mode II: $0.29 < \bar{h}/a < 0.43$; (*b*) Mode III: $0.43 < \bar{h}/a < 0.72$; (*c*) Mode IV: $0.72 < \bar{h}/a < 8.8$.

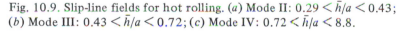

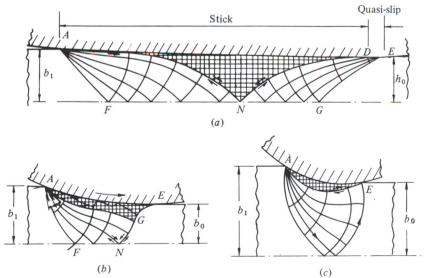

We must now examine the assumption that friction at the roll face is sufficient to prevent slip. In all cases the contact pressure is least at exit. In Crane & Alexander's solutions for thinner strips a coefficient of friction up to 0.7 is required to develop a traction of magnitude k in the zone DE (Fig. 10.8(b)).† In Dewhurst's solutions for thicker strips there is a small range of conditions, where G approaches E (Fig. 10.9(b)), which results in the roll pressure at E becoming negative. In practice some slip in the vicinity of E will remove this anomaly.

(b) Cold rolling – with slip

We turn now to the case of cold rolling, where the strip is taken to slip relative to the rolls at all points in the arc of contact, so that in equation (10.16) we write $q = \pm \mu p$. Replacing h by the mean thickness $\bar{h}$ gives a linear differential equation for the contact pressure:

$$\frac{\mathrm{d}(p/2k)}{\mathrm{d}X} \pm \frac{2\mu a}{\bar{h}}\left(\frac{p}{2k}\right) = \frac{2a^2 X}{R\bar{h}} \tag{10.22}$$

where $X = x/a$. The negative sign applies at entry and the positive sign at exit.

† Denton & Crane (1972) have proposed a modified theory to allow for slip at exit.

Fig. 10.10. Variation of roll force P and roll torque M with aspect ratio $\bar{h}/a$. Solid line – from slip-line fields; broken line – eqs. (10.20) and (10.21); cross in circle – Alexander's solution (Fig. 10.8).

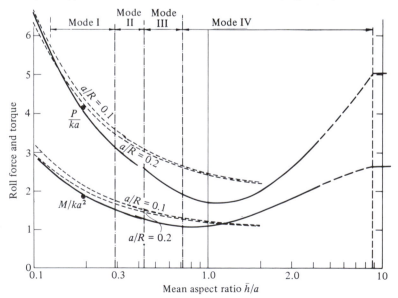

Integrating this equation, with $p/2k = 1$ at $X = 0$ and $X = -1$, gives the pressure in the entry zone to be

$$p/2k = \{1 + (\gamma/\lambda) - \gamma\} \, e^{\lambda(1+X)} - \gamma X - \gamma/\lambda \qquad (10.23a)$$

and in the exit zone

$$p/2k = (1 + \gamma/\lambda) \, e^{-\lambda X} + \gamma X - \gamma/\lambda \qquad (10.23b)$$

where $\lambda = (2\mu a/\bar{h})$ and $\gamma = (a/\mu R)$.

Equating the pressures given by ($10.23a$ and b) yields the position of the neutral point:

$$x_n/a \equiv X_n \approx -\tfrac{1}{2} + \frac{\gamma}{2\lambda}(1 - e^{-\lambda/2}) \qquad (10.24)$$

Integrating the pressure over the arc of the contact gives an expression for the roll force:

$$\frac{P}{ka} = \frac{4(\lambda + \gamma)}{\lambda^2} \{\exp(-\lambda X_n) - 1\} + \gamma + (4\gamma X_n/\lambda) - 2\gamma X_n^2 \qquad (10.25)$$

Similarly for the roll moment:

$$\frac{M}{ka^2} = -\frac{4(\lambda + \gamma)}{\lambda^2} X_n \exp(-\lambda X_n) + \frac{\gamma}{\lambda}(1 - 4X_n^2 + 4X_n/\lambda)$$

$$+ \tfrac{2}{3}\gamma(1 + 2X_n^3) - 2/\lambda \qquad (10.26)$$

Provided that γ is not too large the values of roll force and roll moment given by the closed form expressions (10.24), (10.25) and (10.26) are very close to the numerical integration of von Kármán's equation (10.14) by Bland & Ford (1948). Owing to the dominance of the exponential terms in (10.25) and (10.26) it follows that the force and moment are governed predominantly by the aspect ratio parameter $\lambda(= 2\mu a/\bar{h})$ while the influence of $\gamma(= a/\mu R)$ is relatively small. The reduction in thickness of the strip through the rolls is related to the parameters λ and γ by

$$r \equiv \frac{h_i - h_o}{h_i} = \frac{\lambda\gamma}{2 + \lambda\gamma/2}$$

This brief résumé of the mechanics of plastic deformation of a strip passing between rolls has omitted many aspects of the problem which are important in the technological process. In cold rolling, strain hardening of the strip during deformation is usually significant. It can be included in the theory in an approximate way by permitting the yield stress k to be a prescribed function of deformation and hence of x in equation (10.17). Internal heating due to plastic work also influences the value of k. The rolls flatten appreciably by elastic deformation. It is usual to calculate this by assuming that the contact pressure is distributed according to the Hertz theory, whereupon the rolls deform to a circular arc of modified radius R' which is related to R by equation (10.4). However the insensitivity of the deformation of the strip to roll radius suggests that this is not a serious effect except in the case of very thin hard strips where the elastic deformation of both strip and rolls is important. Finally thick billets will 'spread' laterally during rolling so that the deformation will not be plane, particularly towards the edges of the strip.

10.4 Lubrication of rollers

For engineering surfaces to operate satisfactorily in sliding contact it is generally necessary to use a lubricant. Even surfaces in nominal rolling contact, such as in ball bearings, normally experience some micro-slip, which necessitates lubrication if surface damage and wear are to be avoided. A lubricating fluid acts in two ways. Firstly it provides a thin protective coating to the solid surfaces, preventing the adhesion which would otherwise take place and reducing friction through an interfacial layer of low shear strength. This is the action known as 'boundary lubrication'; the film is generally very thin (it may be only a few molecules thick) and the behaviour is very dependent upon the physical and chemical properties of both the lubricant and the solid surfaces. The lubricant may act in a quite different way. A relatively thick coherent film is drawn in between the surfaces and sufficient pressure is developed in the film to support the normal load without solid contact. This action is known as 'hydrodynamic lubrication'; it depends only upon the geometry of the contact and the viscous flow properties of the fluid. The way in which a load-bearing film is generated between two cylinders in rolling and sliding contact will be described in this section. The theory can be applied to the lubrication of gear teeth, for example, which experience a relative motion which, as shown in §1.5, is instantaneously equivalent to combined rolling and sliding contact of two cylinders.

A thin film of incompressible lubricating fluid, viscosity η, between two solid surfaces moving with velocities V_1 and V_2 is shown in Fig. 10.11. With thin, nearly parallel films velocity components perpendicular to the film are negligible so that the pressure is uniform across the thickness. At low Reynolds' Number (thin film and viscous fluid) inertia forces are negligible. Then, for two-dimensional steady flow, equilibrium of the shaded fluid element gives

$$\frac{\partial p}{\partial x} = \frac{\partial \tau}{\partial z} = \frac{\partial}{\partial z}\left(\eta \frac{\partial v}{\partial z}\right) = \eta \frac{\partial^2 v}{\partial z^2} \tag{10.27}$$

where v is the stream velocity. Since $\partial p / \partial x$ is independent of z, equation (10.27) can be integrated with respect to z. Putting $v = V_2$ and V_1 at $z = 0$ and h gives a parabolic velocity profile as shown, expressed by

$$v(z) = \frac{1}{2\eta} \frac{dp}{dx} (z^2 - hz) + (V_1 - V_2)(z/h) + V_2 \tag{10.28}$$

The volume flow rate F across any section of the film is

$$F = \int_0^h v(z)\,dz = -\frac{h^3}{12\eta}\left(\frac{dp}{dx}\right) + (V_1 + V_2)\frac{h}{2}$$

For continuity of flow F is the same for all cross-sections, i.e.

$$F = (V_1 + V_2)\frac{h^*}{2}$$

where h^* is the film thickness at which the pressure gradient dp/dx is zero. Eliminating F gives

$$\frac{dp}{dx} = 6\eta(V_1 + V_2)\left(\frac{h - h^*}{h^3}\right) \tag{10.29}$$

This is Reynolds' equation for steady two-dimensional flow in a thin lubricating film. Given the variation in thickness of the film $h(x)$, it can be integrated to give the pressure $p(x)$ developed by hydrodynamic action. For a more complete discussion of Reynolds' equation the reader is referred to books on lubrication (e.g. Cameron, 1966). We shall now apply equation (10.29) to find the pressure developed in a film between two rotating cylinders.

(a) Rigid cylinders

The narrow gap between two rotating rigid cylinders is shown in Fig. 10.12. An ample supply of lubricant is provided on the entry side. Within the

Fig. 10.11

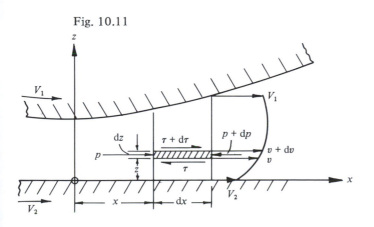

region of interest the thickness of the film can be expressed by

$$h \approx h_0 + x^2/2R \qquad (10.30)$$

where $1/R = 1/R_1 + 1/R_2$ and h_0 is the thickness at $x = 0$. Substituting (10.30) into (10.29) gives

$$\frac{dp}{dx} = \frac{6\eta(V_1 + V_2)}{h_0^2}\left\{\frac{1 - (h^*/h_0) + (x^2/2Rh_0)}{(1 + x^2/2Rh_0)^3}\right\} \qquad (10.31)$$

By making the substitution $\gamma = \tan^{-1}\{x/(2Rh_0)^{1/2}\}$ equation (10.31) can be integrated to give an expression for the pressure distribution:

$$\frac{h_0^2}{(2Rh_0)^{1/2}}\frac{p}{6\eta(V_1 + V_2)} = \frac{\gamma}{2} + \frac{\sin 2\gamma}{4}$$

$$-\sec^2\gamma^*\left(\frac{3\gamma}{8} + \frac{\sin 2\gamma}{4} + \frac{\sin 4\gamma}{32}\right) + A$$

$$(10.32)$$

where $\gamma^* = \tan^{-1}\{x^*/(2Rh_0)^{1/2}\}$ and x^* is the value of x where $h = h^*$ and $dp/dx = 0$. The values of γ^* and A are found from end conditions.

We start by taking zero pressure at distant points at entry and exit, i.e. $p = 0$ at $x = \pm\infty$. The resulting pressure distribution is shown by curve A in

Fig. 10.12. Lubrication of rigid rollers. Broken line – pressure distribution with a complete film. Solid line – pressure distribution without negative pressure.

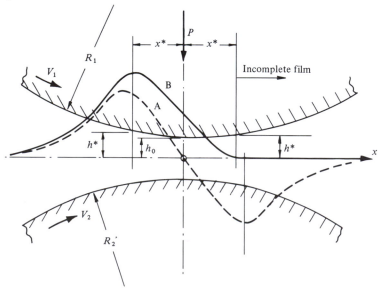

Fig. 10.12. It is positive in the converging zone at entry and equally negative in the diverging zone at exit. The total force P supported by the film is clearly zero in this case. However this solution is unrealistic since a region of large negative pressure cannot exist in normal ambient conditions. In practice the flow at exit breaks down into streamers separated by fingers of air penetrating from the rear. The pressure is approximately ambient (i.e. zero) in this region. The precise point of film breakdown is determined by consideration of the three-dimensional flow in the streamers and is influenced by surface tension forces. However it has been found that it can be located with reasonable success by imposing the condition

$$\frac{dp}{dx} = p = 0$$

at that point. When this condition, together with $p = 0$ at $x = -\infty$ is imposed on equation (10.32) it is found that $\gamma^* = 0.443$, whence $x^* = 0.475(2Rh_0)^{1/2}$. The pressure distribution is shown by curve B in Fig. 10.12. In this case the total load supported by the film is given by

$$P = \int_{-\infty}^{x^*} p(x)\,dx = 2.45(V_1 + V_2)R\eta/h_0 \tag{10.33}$$

In most practical situations it is the load which is specified; equation (10.33) then enables the minimum film thickness h_0 to be calculated. For effective hydrodynamic lubrication h_0 must not be less than the height of the inevitable surface irregularities. We see from equation (10.33) that the load-bearing capacity of the film is generated by *rolling* action characterised by the combined velocities $(V_1 + V_2)$. If the cylinders rotate at the same peripheral speed in *opposite* directions $(V_1 + V_2)$ is zero, no pressure is developed and the film collapses.

(b) Elastic cylinders

At all but the lightest loads the cylinders deform elastically in the pressure zone so that the expression for film profile becomes

$$h(x) = h_0 + x^2/2R + \{\bar{u}_{z1}(x) - \bar{u}_{z1}(0)\} + \{\bar{u}_{z2}(x) - \bar{u}_{z2}(0)\}$$

where the normal elastic displacements of the two surfaces $\bar{u}_{z1}$ and $\bar{u}_{z2}$ are given by equation (2.24b). Thus

$$h(x) = h_0 + (x^2/2R) - \frac{2}{\pi E^*}\int_{-\infty}^{\infty} p(s)\ln\left|\frac{x-s}{s}\right|\,ds \tag{10.34}$$

This equation and Reynolds' equation (10.29) provide a pair of simultaneous equations for the film shape $h(x)$ and the pressure $p(x)$. They can be combined into a single integral equation for $h(x)$ which has been solved numerically by Herrebrugh (1968). The deformed shape is shown in Fig. 10.13(b). This film shape is then substituted into Reynolds' equation to find the pressure distribution $p(x)$, as shown in Fig. 10.13(a). The solution depends upon a single non-dimensional parameter $J \equiv \{P^2/\eta R(V_1 + V_2)\pi E^*\}^{1/2}$, which is a measure of the ratio of the pressure generated hydrodynamically in the film to the pressure to produce the elastic distortion. The parameter is zero for rigid cylinders and increases with increasing elasticity. It is evident from Fig. 10.13 that the effect of elastic deformation is to produce a film which is convergent over most of its effective length like a tilting pad thrust bearing. On the other hand, as the

Fig. 10.13. Lubrication of elastic rollers with an isoviscous lubricant. A: $J = 0$ (rigid), B: $J = 0.536$, C: $J = 7.42$, D: $J = 143$. (a) Pressure distribution $p(x)$; (b) film shape $h(x)$.

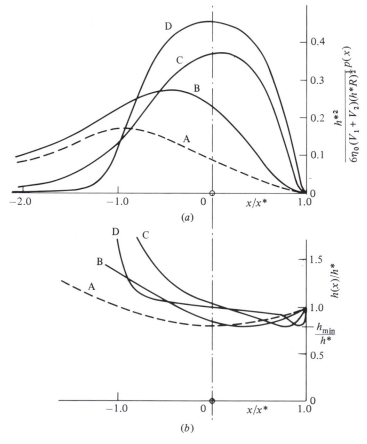

elastic flattening becomes large compared with the film thickness, the pressure distribution approaches that of Hertz for unlubricated contact.

From the point of view of effective lubrication it is the minimum film thickness h_{min} which is important. In all cases $h_{min} \approx 0.8h^*$. The variation in the non-dimensional film thickness $H \equiv Ph_{min}/\eta R(V_1 + V_2)$ with the parameter J is given in Table 10.1. For rigid cylinders ($J = 0$) the minimum thickness is h_0, given by equation (10.33). We see from the table that, by permitting a more favourable shape for generating hydrodynamic pressure, elastic rollers give thicker films under the same conditions of speed and loading.

This mode of lubrication behaviour in which the elastic deformation of the solid surface plays a significant role in the process is known as *elastohydrodynamic lubrication*. (See Dowson & Higginson, 1977).

(c) Variable viscosity

So far we have considered the viscosity η to be a constant property of the lubricating fluid, but in fact the viscosity of most practical lubricants is very sensitive to changes in pressure and temperature. In a non-conforming contact the pressures tend to be high so that it is not surprising that the increase in vicosity with pressure is also a significant factor in elastohydrodynamic lubrication. Particularly during sliding, frictional heating causes a rise in temperature in the film which reduces the viscosity of the film. However, for reasons which will become apparent later, it is possible to separate the effects of pressure and temperature. To begin with, therefore, we shall consider an isothermal film in which the variation in viscosity with pressure is given by the equation

$$\eta = \eta_0 \, e^{\alpha p} \tag{10.35}$$

where η_0 is the viscosity at ambient pressure and temperature and α is a constant pressure coefficient of viscosity. This is a reasonable description of the observed variation in viscosity of most lubricants over the relevant pressure range. Substituting this relationship into Reynolds' equation (10.29) gives

$$e^{-\alpha p} \frac{dp}{dx} = 6\eta_0 (V_1 + V_2) \left(\frac{h - h^*}{h^3} \right) \tag{10.36}$$

Table 10.1

$J \equiv \left\{ \dfrac{P^2}{\eta R(V_1 + V_2)\pi E^*} \right\}^{1/2}$	0 (rigid)	0.536	2.34	7.42	26.9	143
$H \equiv \dfrac{Ph_{min}}{\eta R(V_1 + V_2)}$	2.45	2.91	4.11	6.05	9.51	17.6

This modified Reynolds' equation for the hydrodynamic pressure in the field must be solved simultaneously with equation (10.34) for the effect of elastic deformation on the film shape. Numerical solutions to this problem have been obtained by Dowson *et al.* (1962). Typical film shapes and pressure distributions are shown in Fig. 10.14. The inclusion of the pressure viscosity coefficient α introduces a second non-dimensional parameter

$$K \equiv \{\alpha^2 P^3 / \eta_0 R^2 (V_1 + V_2)\}^{1/2}$$

Comparing Figs. 10.13 and 10.14 shows that the pressure–viscosity effect has a marked influence on the behaviour. Over an appreciable fraction of the contact area the film is approximately parallel. This follows from equation (10.36). When the exponent αp exceeds unity the left-hand side becomes small, hence $h - h^*$ becomes small, i.e. $h \approx h^* = $ constant. The corresponding pressure distribution is basically that of Hertz for dry contact, but a sharp pressure peak occurs on the exit side, followed by a rapid drop in pressure and thinning of the film where the viscosity falls back to its ambient value η_0. Dowson *et al.* (1962) have shown that allowing for the compressibility of the lubricant attenuates the peak to some extent. A more realistic equation for the variation of viscosity

Fig. 10.14. Numerical solution of the elastohydrodynamic eqs. (10.34) and (10.36) from Dowson *et al.* (1962). Values of J and K: A 0.54, 18; B 1.7, 58; C 5.4, 180; D 17, 580; E 54, 1800; F ∞, ∞ (dry). (*a*) Pressure distribution, (*b*) Film shape.

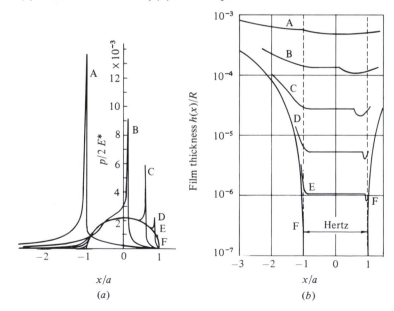

with pressure together with some shear thinning of the lubricant is likely to reduce the intensity of the theoretical peak still further.

Nevertheless these characteristic features of highly loaded elastohydro-dynamic contacts – a roughly parallel film with a constriction at the exit and a pressure distribution which approximates to Hertz but has a sharp peak near the exit – are now well established by experiment (see Crook, 1961; and Hamilton & Moore, 1971). The minimum film thickness is about 75% of the thickness in the parallel section.

In the range of their computations Dowson & Higginson's results can be expressed approximately by

$$H = 1.4K^{0.54}J^{0.06} \tag{10.37}$$

Under these conditions the film thickness is weakly dependent upon the elasticity parameter J. The question now arises: under what conditions is it appropriate to neglect elastic deformation and/or variable viscosity? This question has been considered by Johnson (1970b) which has led to the following guidelines. If the parameter $J < 0.3$ and $K < 0.7$ deformation and variable viscosity are negligible and the analysis in section (a) is adequate. The relative importance of variable viscosity and elasticity depends upon a parameter $(K^2/J^3)^{1/4}$. For values of this parameter less than about 0.4, changes in viscosity are negligible compared with elastic effects and Herrebrugh's analysis given in section (b) above is appropriate. This condition is likely to be met only with rubber or other highly elastic polymers (using typical lubricants) where large elastic deformations are obtained at relatively low pressures. Dowson & Higginson's results on the other hand require that $(K^2/J^3)^{1/4}$ should have a value in excess of about 1.5. This is the engineering regime of metal surfaces lubricated with typical mineral oils. Values of non-dimensional film thickness H with appropriate values of K, taking $(K^2/J^3)^{1/4} = 4$, are given in Table 10.2.

Table 10.2

$\left(\dfrac{K^2}{J^3}\right)^{1/4} = 4$							
	K	0 (isoviscous)	50	100	500	1000	5000
	J	0 (rigid)	2.1	3.4	9.9	15.8	46.2
Dowson & Higginson eq. (10.37)	H	2.45	12	18	46	68	175
Grubin eq. (10.43)	H	–	14	21	53	80	203

It is evident from this table that the combined effect of elastic deformation and variable viscosity is to increase the minimum film thickness by one to two orders of magnitude compared with the equivalent thickness for an isoviscous fluid and rigid rollers.

Although solutions of the elastohydrodynamic equations on the computer provide valuable data, they do not give the same insight into the mechanics of the process as an early approximate treatment by Grubin (1949), extended later by Greenwood (1972). Grubin realised that in a high pressure contact with a pressure-sensitive lubricant, such that αp_0 appreciably exceeds unity ($\alpha p_0 > 5$, say), Reynolds' equation (10.36) demands that the film must be nearly parallel in the high pressure region with a thickness $h \approx h^*$. He assumed therefore that the elastic deformation of the rollers is the same as in dry contact and that a parallel film of thickness h^* exists over the length $2a$ of the Hertz flat. The elastic displacements outside the parallel zone are found by substituting the Hertz pressure distribution into equation (2.24b), whereupon the film shape in the entry zone is given by

$$h(x) = h^* + \frac{a}{2R}\left[\frac{x}{a}\left(\frac{x^2}{a^2} - 1\right)^{1/2} - \ln\left\{\frac{x}{a} + \left(\frac{x^2}{a^2} - 1\right)^{1/2}\right\}\right]$$

A good approximation to this cumbersome expression in the relevant region is

$$h(x) \approx h^* + \frac{a^2}{3R}\left(\frac{2\chi}{a}\right)^{3/2} \tag{10.38}$$

where $\chi\,(= -(x + a))$ is measured from the edge of the parallel zone $x = -a$ (see Fig. 10.15). We now define a reduced pressure p' by

$$p' = (1 - e^{-\alpha p})/\alpha$$

or

$$p = -\frac{1}{\alpha}\ln(1 - \alpha p') \tag{10.39}$$

Substituting (10.38) and (10.39) into the hydrodynamic equation (10.36) gives

$$\frac{dp'}{d\chi} = -6\eta_0(V_1 + V_2)\left[\frac{(a^2/3R)(2\chi/a)^{3/2}}{\{(a^2/3R)(2\chi/a)^{3/2} + h^*\}^3}\right]$$

or, by writing $(2\sqrt{2}\,a^2/3Rh^*)^{2/3}(\chi/a) = \xi$,

$$\frac{dp'}{d\xi} = -6\eta_0(V_1 + V_2)\left(\frac{3Rh^*}{2\sqrt{2}\,a^2}\right)^{2/3}\frac{a}{h^{*2}}\frac{\xi^{3/2}}{(\xi^{3/2} + 1)^3} \tag{10.40}$$

This equation can be integrated directly for the build-up in pressure in the entry region, taking $p' = 0$ at $\xi = \infty$. At the start of the parallel section ($\xi = 0$)

$$p_0' = 6\eta_0(V_1 + V_2)\left(\frac{2Rh^*}{2\sqrt{2}\,a^2}\right)^{2/3}\frac{a}{h^{*2}}\int_0^\infty \frac{\xi^{3/2}}{(\xi^{3/2} + 1)^3}\,d\xi$$

Now

$$\int_0^\infty \frac{\xi^{3/2}\,d\xi}{(\xi^{3/2} + 1)^3} = 0.255$$

so that

$$h^* = 1.417 \left\{ \frac{\eta_0(V_1 + V_2)}{p_0'} \right\}^{3/4} \frac{R^{1/2}}{a^{1/4}} \tag{10.41}$$

It is clear from equation (10.39) that the reduced pressure p' cannot exceed $1/\alpha$ otherwise the actual pressure becomes infinite. This condition sets a lower limit on the value of the film thickness h^*, hence

$$h^* > 1.417 \{\alpha\eta_0(V_1 + V_2)\}^{3/4} R^{1/2} a^{-1/4} \tag{10.42}$$

Furthermore, since Grubin's treatment is restricted to high pressures in which αp_0 appreciably exceeds unity, $\alpha p_0'$ will approach unity very closely and hence the right-hand side of (10.42) will be a good approximation to the actual thickness of the film in the parallel section. Remembering that, by Hertz, $a^2 = 4PR/\pi E^*$, equation (10.42) can be rewritten in the form

$$H = 0.89 K^{0.75} J^{-0.25} \tag{10.43}$$

Fig. 10.15. Elastohydrodynamic lubrication of rollers: the Grubin–Greenwood idealisation.

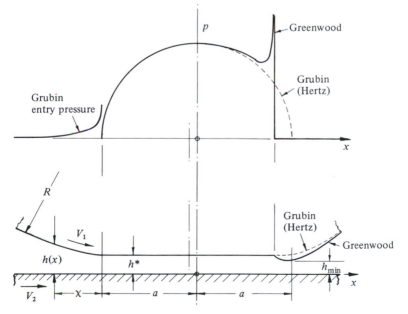

Values of H deduced from this approximate expression are compared with those from the computer solutions in Table 10.2. The agreement is reasonably good.

The analysis so far has been concerned entirely with the converging region at entry and has demonstrated that the film thickness in the parallel zone is determined to good approximation by the flow in this region. It also shows that the film thickness is relatively insensitive to load. Increasing the load increases the length $2a$ of the parallel zone, but this has only a marginal effect on the shape of the entry zone and hence upon the film thickness.

The assumption in Grubin's theory of a parallel zone of thickness approximately equal to h^* cannot be correct at exit. Here the pressure gradient must become negative, for which Reynolds' equation demands that h must fall below h^*. This is the reason for the constriction in the film at exit which is a feature of all elastohydrodynamic film profiles. Greenwood (1972) has extended Grubin's analysis to cover the exit zone by postulating a slightly shortened parallel zone. The pressure distribution within the parallel region required to produce this form of elastic deformation is found from equation (2.45). It is illustrated in Fig. 10.15. The elastic pressure is zero at the entry to the parallel zone, but rises to a sharp singularity at the end of the flat followed by a constriction in the film. Both the pressure spike and the constriction in the film reflect the characteristic features of the computer solutions for values of $\alpha p_0 > 5$. At practical speeds the entry conditions are independent of the exit conditions so that the value of h^* given by equation (10.42) is unchanged by Greenwood's modification. The minimum thickness which occurs in the exit constriction is found to be 75–80% of h^*.

The mechanism of elastohydrodynamic lubrication with a pressure-dependent lubricant is now clear. Pressure develops by hydrodynamic action in the entry region accompanied by a very large increase in viscosity. The film thickness at the end of the converging zone is limited by the necessity of maintaining a finite pressure. This condition virtually determines the film thickness in terms of the speed, roller radii and the viscous properties of the lubricant. Increasing the load increases the elastic flattening of the rollers with only a minor influence on the film thickness. The highly viscous fluid passes through the parallel zone until the pressure and viscosity collapse at the exit, which requires a thinning of the film. The inlet and exit regions are effectively independent; they meet at the end of the parallel zone with a discontinuity in slope of the surface which is associated with a sharp peak in pressure.

We can now return to the effect of temperature on viscosity. Viscous dissipation occurs in the entry region even without sliding, i.e. when $V_1 = V_2$. The dissipation gives rise to a resistance to rolling (Crook, 1963) and to a rise in temperature, which both increase with viscosity and rolling speed. Studies of viscous heating at entry by Murch & Wilson (1975) have shown that it will not affect the film thickness appreciably until the parameter $(V_1 + V_2)^2 (d\eta_0/d\theta)/K$ exceeds unity, where $(d\eta_0/d\theta)$ is the rate of change of viscosity with temperature and K is the thermal conductivity of the lubricant. Experiments have demonstrated that the appropriate values of η_0 and α to use in the theory are those at the temperature of the rolling surfaces.

When sliding accompanies rolling ($V_1 \neq V_2$) the whole film is sheared, giving rise to a resultant tractive force and much more severe viscous heating than in pure rolling. The value of the traction and the consequent temperature rise depend upon the shear properties of the lubricant in the high-pressure zone. There is clear evidence of non-Newtonian behaviour in this region and appropriate constitutive equations for the fluid at high pressure, such as those suggested by Johnson & Tevaarwerk (1977), are necessary in order to predict the tractive forces. Such calculations are beyond the scope of this book. Fortunately the film is established in the entry zone and shear heating in the parallel zone occurs too late to affect its thickness appreciably. Measurements of the film thickness (*a*) by Dyson *et al.* (1956) using electrical capacitance and (*b*) by Wymer & Cameron (1974) using optical interferometry give good support for the isothermal theory both with and without sliding.

The elastohydrodynamic lubrication of point contacts has been studied by Archard & Cowking (1956), Cheng (1970) and Hamrock & Dowson (1977) leading to formulae for the film thickness.

11

Dynamic effects and impact

11.1 Stress waves in solids

So far in this book we have discussed contact problems in which the rate of loading is sufficiently slow for the stresses to be in statical equilibrium with the external loads at all times during the loading cycle. Under impact conditions, on the other hand, the rate of loading is very high and dynamic effects may be important: in rolling and sliding contact at high speed the inertia of the material elements as they 'flow' through the deforming region may influence the stress field. In this chapter we shall examine the influence of inertia forces on a number of contact problems.

Inertia forces are incorporated in the mechanics of deformable solids by the addition in the stress equilibrium equations (2.1) of terms equal to the product of the density of the material ρ and the acceleration of the material element $\partial^2 u/\partial t^2$ at the point in question. When these modified equilibrium equations are combined with the equations of compatibility and the elastic stress–strain relations, solutions for stresses and displacements are obtained which may be interpreted as pulses or waves which travel through the solid with characteristic speeds (see Timoshenko & Goodier, 1951; Kolsky, 1953; or Graff, 1975).

We shall introduce the concept of a stress wave by considering the one-dimensional example of compression waves in a thin elastic rod (see Fig. 11.1). In this simple treatment we shall consider a stress pulse of intensity $-\sigma$ travelling along the rod from left to right with a velocity c_0. In time dt the wave front moves a distance dx $(= c_0 \, dt)$ and the element, of mass $\rho A \, dx$, acquires a velocity v under the action of the pressure pulse, where ρ is the density of the material and A is the cross-sectional area of the rod. The momentum equation for the element is thus

$$-\sigma A \, dt = (\rho A \, dx)v = \rho A c_0 v \, dt$$

i.e.

$$\sigma = -\rho c_0 v \qquad (11.1)$$

The element will have become compressed by du ($= v \, dt$), so that the strain in the element is:

$$-\frac{du}{dx} = -\frac{v}{c_0} = -\frac{\sigma}{E} \qquad (11.2)$$

Eliminating σ and v from equations (11.1) and (11.2) gives an expression for the velocity of the pulse:

$$c_0 = (E/\rho)^{1/2} \qquad (11.3)$$

which is a characteristic of the material. Since elastic strains are generally small, it is clear from equation (11.2) that the velocity v of particles of the rod is much less than the velocity of the pulse c_0. We note that in a compressive wave, such as we have been considering, the particles move in the same direction as the wave; whereas in a tension wave they move in the opposite direction.

To illuminate our discussion in the next section of the impact of solid bodies, it is instructive now to consider the wave motion set up in a thin elastic rod (Fig. 11.1) fixed at one end and struck on the other by a rigid block of mass M, moving with velocity V. Any tendency of the rod to buckle will be ignored. Immediately after impact the left-hand end of the rod acquires the velocity of the block V and a compression wave propagates along the rod with velocity c_0 given by (11.3). The initial compressive stress in the rod, given by equation (11.1) is $-\rho c_0 V$. The block decelerates under the action of the compressive force in the rod at the interface with the block. The sequence of events then depends upon the mass of the striker M compared with the mass of the rod ρAL.

A light striker is rapidly brought to rest by the compression in the rod; the pressure of the rod on the block decreases as the velocity of the block decreases. There is then a large variation in stress along the rod from $-\rho c_0 V$ at the wave front to a small value at the interface with the block. Meanwhile the pressure

Fig. 11.1. Impact of a rigid mass M on the end of an elastic rod. A compressive wave of intensity $-\sigma$ propagates along the rod with velocity c_0.

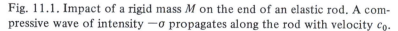
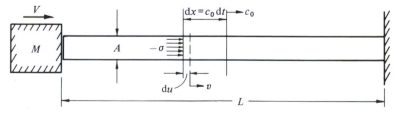

wave is reflected at the fixed end of the rod. When the reflected wave returns to the free end it accelerates the block and is itself partially reflected. Thus the block rebounds from the end of the rod with a velocity less than V and the rod is left in a state of vibration. The maximum stress in the rod as a result of the impact is $\rho c_0 V$, which is independent of M. It occurs first at the instant of impact and again when the reflected wave impinges on the block.

At the other extreme, if the mass of the striker is much larger than that of the rod, the pressure wave is reflected up and down the rod many times before the block is brought to rest. The state of stress in the bar at any instant is approximately uniform throughout its length and the sudden changes in stress associated with the passage of the stress wave in the rod are small compared with the general stress level. The stress in the rod can then be found to a good approximation by ignoring dynamic effects in the rod and treating it as a 'light spring'. The maximum stress in the rod, which occurs at the instant the striker comes to rest, can be found by equating the maximum strain energy stored in the bar to the loss of kinetic energy of the striker, with the result:

$$\sigma_{max} = V(ME/AL)^{1/2}$$

which does depend upon M and is very different from the previous result. An analysis of a dynamic problem on these lines in which inertia forces in the deforming material are neglected is usually referred to as *quasi-static* since the external dynamic loads are taken to be in 'equilibrium' with a statically deter-mined stress field.

The two extreme cases described above arise when the mass of the striker is either much smaller or much larger than the total mass of the rod; put another way these conditions correspond to a time for the mass to come to rest which is either short or long compared with the time for a stress wave to travel to the end of the rod and back. When these times are similar the behaviour is much more complex. This problem and others involving the longitudinal impact of rods are discussed by Goldsmith (1960) and Johnson (1972).

So far we have assumed perfectly elastic behaviour, but impact stresses are generally high and inelastic deformation plays an important part in practical impacts. For elastic behaviour the stress given by equation (11.1) must be less than the yield stress Y, for which the impact velocity

$$v \leqslant Y/\rho c_0 \tag{11.4}$$

The longitudinal wave speed in steel, given by (11.3), is about 5200 m/s. Taking $Y = 300$ N/mm^2, the maximum impact velocity for elastic deformation is 7.5 m/s. At speeds below this value, elastic hysteresis in the steel causes the elastic waves to attenuate slowly with distance travelled. Above this speed the end of the bar becomes plastically deformed and the elastic wave travelling at c_0 is followed by a slower moving plastic wave.

In extended three-dimensional elastic bodies two types of wave motion are possible: (i) *dilatational* (or pressure) waves in which the material elements fluctuate in volume without shear deformation and (ii) *distortional* (or shear) waves in which the elements distort without change in volume. The speeds of propagation of these waves in isotropic materials are given by:

$$\text{dilatation: } c_1 = \left\{ \frac{2(1-\nu)G}{(1-2\nu)\rho} \right\}^{1/2} \tag{11.5}$$

$$\text{distortion: } c_2 = (G/\rho)^{1/2} \tag{11.6}$$

where G is the elastic shear modulus.

For $\nu = 0.25$, $c_1 = \sqrt{3}\, c_2$. If the wave front is planar, which would be approximately so at a large distance from a point source, the motion of the material particles in a dilatational wave is in the direction of propagation of the wave front and the waves are sometimes described as *longitudinal.* In a distortional wave, on the other hand, the particles move at right angles to the direction of propagation of the wave so that the waves are then referred to as *transverse.*

Where the solid body is bounded by a plane or near-planar surface, such as we are concerned with in elastic contact problems, waves known as Rayleigh waves may be propagated along the surface with a velocity

$$c_3 = \alpha c_2 = \alpha(G/\rho)^{1/2} \tag{11.7a}$$

where α is the root of the equation

$$(2-\alpha^2)^4 = 16(1-\alpha^2)(1-\alpha^2 c_2^2/c_1^2) \tag{11.7b}$$

The value of α depends upon Poisson's ratio; for $\nu = 0.25$, $\alpha = 0.919$, and for $\nu = 0.5$, $\alpha = 0.955$, so that the speed of surface waves is just slightly less than that of distortion waves. Values of the wave velocities in a few common solids are given in Table 11.1

11.2 Dynamic loading of an elastic half-space

The starting point for our consideration of the static loading of an elastic half-space in Chapters 2 and 3 was the action of a concentrated force

Table 11.1. *Elastic wave velocities* (m/s)

		Steel	Copper	Aluminium	Glass	Rubber
1-dimensional tens./compr.	c_0	5200	3700	5100	5300	46
Dilatational	c_1	5900	4600	6300	5800	1100
Distortional	c_2	3200	2300	3100	3400	27
Rayleigh	c_3	3000	2100	2900	3100	26

applied normal to the surface. The stresses and deformations due to distributed loads could then be found by superposition. The equivalent problems in dynamic loading are those of a concentrated line or point force P which is (a) applied suddenly and then maintained constant - a step - or (b) an impulse

$$\hat{P} = \left[\int_0^{\Delta t} P \, dt \right]_{\Delta t \to 0}$$

or (c) an harmonically varying force $P = P^* \cos \omega t$. Each of these fundamental solutions can be used to build up a distributed load by superposition with respect to position on the surface and a time-varying pulse $P(t)$ by superposition with respect to time.

The wave motion initiated in a half-space by a step point force has been analysed by Pekeris (1955) and is depicted in Fig. 11.2. Following the application of the load at time $t = 0$, spherical wave fronts of pressure (P) waves and shear (S) waves propagate from the point of application of the load with

Fig. 11.2. Wave motion in an elastic half-space caused by a step load P_0. (a) Wave fronts, (b) Normal displacement at the surface $z = 0$. ($\nu = \frac{1}{4}$.)

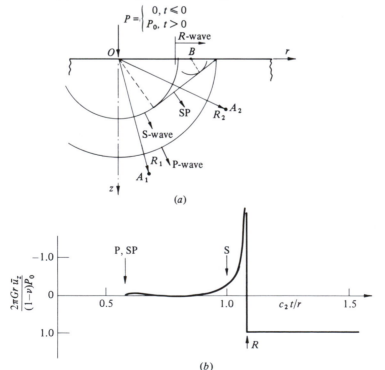

velocities c_1 and c_2. At point A_1 within the solid, distance $R_1 = (r_1^2 + z_1^2)^{1/2}$ from O, the material is unstressed until a time $t = R_1/c_1$, when the P-wave arrives, and it experiences a step radial displacement; at a later time $t = R_1/c_2$, the S-wave arrives imparting a step circumferential displacement. The magnitude of the displacement decreases as $1/R$ and of the stresses as $1/R^2$. The interaction of the P-wave with the free surface of the half-space initiates a weak disturbance – the 'head wave' or SP-wave – which propagates with velocity c_2 as shown in Fig. 11.2(a) and which influences slightly the displacements and stresses experienced by a subsurface point such as A_2. The interaction of the S-wave with the free surface gives rise to the Rayleigh wave which we have seen propagates with a velocity c_3 slightly less than the S-wave, and influences points which are on or close to the surface. The time-history of normal displacement $\bar{u}_z$ of points on the surface is shown in Fig. 11.2(b). It is evident that the predominant effect is that of the Rayleigh wave which decays as $1/R^{1/2}$, which is more slowly than either P or S waves. After the Rayleigh wave has passed, the surface is left with its static displacement under the action of a steady force P_0, given by equation (3.22b).

The effects of an impulsive line load and harmonically varying line and point loads have been analysed in a classic paper by Lamb (1904). We shall consider a point load

$$P(t) = P^* \cos \omega t \tag{11.8}$$

acting normally to the surface at the origin O. In the steady state, the wave system comprises P-waves moving with velocity c_1 and S-waves moving with velocity c_2 on spherical wavefronts centred at O. In addition Rayleigh waves radiate outwards on the surface of the half-space with velocity c_3.

Within the body, at a radial distance R from O which is large compared with the wavelength, the radial displacement u_R is entirely due to the pressure wave and is given by (Lamb, 1904; Miller & Pursey, 1954)

$$u_R = \frac{P^*}{2\pi GR} \frac{\cos \theta \, (\mu^2 - 2 \sin^2 \theta)}{F_0 \, (\sin \theta)} \cos (\omega t - k_1 R) \tag{11.9a}$$

The transverse displacement u_θ is due to the shear wave and is given by

$$u_\theta = \frac{i\mu^3 P^*}{2\pi GR} \frac{\sin 2\theta \, (\mu^2 \sin^2 \theta - 1)^{1/2}}{F_0 \, (\mu \sin \theta)} \sin (\omega t - k_2 R) \tag{11.9b}$$

In equations (11.9), $\theta = \cos^{-1}(z/R)$, $k_1 = \omega/c_1$, $k_2 = \omega/c_2$,

$$\mu = c_1/c_2 = \{2(1 - v)/(1 - 2v)\}^{1/2},$$
$$F_0(\zeta) = (2\zeta^2 - \mu^2)^2 - 4\zeta^2 (\zeta^2 - \mu^2)^{1/2}(\zeta^2 - 1)^{1/2}$$

On the surface at a distance r from O, which is again large compared with the wavelength, the displacements $\bar{u}_r$ and $\bar{u}_z$ are due to the Rayleigh wave and are given by

$$\bar{u}_r = \frac{P^*}{G} \left(\frac{k^3}{2\pi r} \right)^{1/2} F_r(v) \sin (\omega t - k_3 r - \pi/4) \qquad (11.10a)$$

$$\bar{u}_z = \frac{P^*}{G} \left(\frac{k^3}{2\pi r} \right)^{1/2} F_z(v) \cos (\omega t - k_3 r - \pi/4) \qquad (11.10b)$$

where $k_3 = \omega/c_3$; $F_r(v)$ and $F_z(v)$ are functions of Poisson's ratio (see Miller & Pursey, 1954). For $v = 0.25$, $F_r(v) = 0.125$, $F_z(v) = 0.183$.

Equations (11.9) and (11.10) are not accurate close to the origin but, in any case, the displacement and corresponding stresses become infinite at the point of application of a concentrated force (as R and r approach zero). The more realistic situation of a uniform pressure p acting on a circular area of radius a and oscillating with angular frequency ω has been analysed by Miller & Pursey (1954). This is the dynamic equivalent of the static problem discussed in §3.4(a). The wave motion at a large distance from the loaded circle (R, $r \gg a$) is the same as for a concentrated force $P = \pi a^2 p$ and the elastic displacements are given by equations (11.9) and (11.10). The mean normal displacement within the contact area $(\bar{u}_z)_m$ is of interest since it determines the 'receptance' of the half-space to an oscillating force. The receptance is defined as the ratio of the mean surface displacement $(\bar{u}_z)_m$ within the loaded area to the total load.† It is a complex quantity: the real part gives the displacement which is in-phase with the applied force; the imaginary part gives the displacement which is $\pi/2$ out-of-phase with the force.

If we write the inverse or reciprocal of the receptance in the form

$$\frac{P}{(\bar{u}_z)_m} = Ga \left\{ f_1 \cos \omega t - \left(\frac{\omega a}{c_2} \right) f_2 \sin \omega t \right\} \qquad (11.11)$$

it will be recognised as having the same form as the expression for the inverse receptance of a light spring in parallel with a viscous dashpot. The functions f_1 and f_2 depend upon Poisson's ratio and the frequency parameter $(\omega a/c_2)$. Values taken from Miller & Pursey (1954) are shown by the full lines in Fig. 11.3. In the range considered, f_1 and f_2 do not vary much with frequency so that, to a reasonable approximation, the elastic half-space can be modelled by a light spring in parallel with a dashpot. The energy 'dissipated' by the dashpot corresponds to the energy radiated through the half-space by wave motion. The stiffness of the spring may be taken to be independent of frequency and equal to the static stiffness of the half-space given by equation (3.29a); the

---

† An alternative quantity which is commonly used to give the same information is the 'impedance' which is the ratio of the force to the mean velocity of surface points in the loaded area.

spring and dashpot combination has a time constant $T = f_2 a/f_1 c_2 \approx 0.74 a/c_2$. In this way the power radiated through the half-space by wave motion can easily be calculated to be ($\nu = 0.25$)

$$\dot{W} = 0.074 P^2 \omega^2 / G c_2 \qquad (11.12)$$

Using equations (11.9) and (11.10) the partition of this energy between the different wave motions has been found by Miller & Pursey (1955). The pressure waves account for 7%, the shear waves for 26% and the surface waves for 67% of the radiated energy. If we note that the pressure and shear waves decay in amplitude (neglecting dissipation) with (distance)$^{-1}$ whilst the surface waves

Fig. 11.3. Receptance functions f_1 and f_2 for an elastic half-space: solid line – uniform pressure on circle radius a; large-dashed line – uniform pressure on strip width $2a$; chain line – uniform pressure on semi-infinite rod; small-dashed line – uniform displacement on circle radius a.

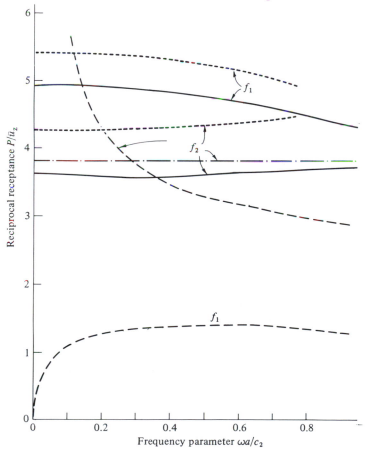

decay with (distance)$^{-1/2}$, it is clear that the predominant effect at some distance from the point of excitation is the surface wave. This explains why earthquakes can be damaging over such a large area.

The spring and dashpot model can be applied to other situations. A semi-infinite thin rod transmits one-dimensional longitudinal waves as described in §1. In view of its infinite length the rod has zero static stiffness in tension and compression. By equation (11.1) the force on the end of the rod is proportional to the *velocity* of the end. Thus under the action of an oscillating force the rod acts like a pure dashpot. The function $f_1 = 0$ and $f_2 = 3.84$. Miller & Pursey (1954) have also considered an elastic half-space loaded two-dimensionally by an oscillating pressure applied to a strip of width $2a$. In this case the functions f_1 and f_2 show larger variations with frequency (Fig. 11.3).

With an interest in the motion transmitted to the ground through the foundation of a vibrating machine, Arnold *et al.* (1955), Robertson (1966) and Gladwell (1968) studied the allied problem of a circular region on the surface of an elastic half-space which is oscillating with a uniform normal displacement. In this case the pressure distribution is not uniform. The receptance functions f_1 and f_2 computed by Gladwell are also plotted in Fig. 11.3. Not surprisingly they do not differ much from the case of a uniform pressure. When $\omega \to 0$, f_1 is given by the static displacement under a circular rigid punch (eq. (3.36)).

In this section we have considered the stresses and displacements in an elastic half-space in response to a sinusoidally oscillating pressure applied to a small circular region on the surface. In the language of the vibration engineer we have determined its linear dynamic response to harmonic excitation. In the next section, dealing with impact, we shall be concerned with the response of the half-space to a single pressure pulse. However, if the variation of the pulse strength with time $P(t)$ is known it can be represented by a continuous spectrum of harmonic excitation $F(\omega)$ by the transformation

$$F(\omega) = \left(\frac{2}{\pi}\right)^{1/2} \int_{-\infty}^{\infty} P(t)\, e^{i\omega t}\, dt \tag{11.13}$$

The response to harmonic excitation at a single frequency ω has been presented in this section. The response to a spectrum of harmonic excitation $F(\omega)$ can be found by superposition, i.e. by integration with respect to ω. In practice, the integration is seldom easy and requires numerical evaluation.

Finally we note that, although our discussion has been restricted to the dynamic response of a half-space to purely normal forces, behaviour which is qualitatively similar arises when tangential forces or couples are applied to the surface. For example, a light circular disc of radius a attached to the surface, in addition to a purely normal oscillation discussed above, can undergo three other modes of vibration: translation parallel to the surface, rocking about an axis lying in the surface, and twisting about the normal axis. Receptance functions for each of these modes are conveniently summarised by Gladwell (1968).

11.3   Contact resonance

In the previous section we saw that an elastic half-space responds to an oscillating force applied to the surface like a spring in parallel with a dashpot. If now a body of mass m is brought into contact with the half-space the resulting system comprises a mass, spring and dashpot, which might be expected to have a characteristic frequency of vibration and to exhibit resonance when subjected to an oscillating force.

We shall consider first the case of a rigid mass attached to the half-space over a fixed circular area of radius a. This is the problem investigated by Arnold *et al.* (1955). It has obvious application to ground vibrations excited by heavy machinery and also the vibration of buildings excited by earth tremors (Richart *et al.*, 1970). For motion normal to the surface, receptance of the half-space is given by (11.11), so that, denoting the displacement of the mass by $\bar{u}_z$, the equation of motion of the system when excited by an oscillating force $P \cos \omega t$ is:

$$m\ddot{\bar{u}}_z + (Ga^2 f_2/c_2)\dot{\bar{u}}_z + Gaf_1\bar{u}_z = P \cos \omega t \tag{11.14}$$

The frequency of free vibrations is $\omega_0(1 - \zeta^2)^{1/2}$, where the undamped natural frequency ω_0 is given by

$$\omega_0^2 = Gaf_1/m \tag{11.15a}$$

and the damping factor ζ by

$$\zeta = \tfrac{1}{2}(f_2/f_1)(\omega_0 a/c_2) = \tfrac{1}{2}(f_2/f_1^{1/2})(\rho a^3/m)^{1/2} \tag{11.15b}$$

A sharp resonance peak will be obtained if $\zeta \ll 1$. Now $\tfrac{1}{2}(f_2/f_1^{1/2}) \approx 1$, so that the damping factor due to wave propagation is small if the mass of the attached body is large compared with the mass of a cube of the half-space material of side a. In this case the resonant frequency is very nearly equal to ω_0, given by equation (11.15a). Resonance curves for different values of $(\rho a^3/m)$ are plotted for the different modes of vibration in Arnold *et al.* (1955).

We shall turn now to the situation where two non-conforming bodies are pressed into contact by a steady force P_0 and then subjected to an oscillating force $\Delta P \cos \omega t$. As in static contact stress theory we take the size of the contact area to be small compared with the dimensions of either body, in which case it follows that the parameter $(\rho a^3/m)$ must be small for both bodies. This means that the vibrational energy absorbed by wave motion is small. Hence, for either body, the damping term in equation (11.14) is negligible and the elastic stiffness term is given by the static stiffness $Gaf_1(\omega = 0)$. Since both bodies are deformable the effective 'contact spring' between them is the series combination of the stiffness of each body regarded as an elastic half-space. The mass of each body may be considered to be concentrated at its centroid. It is then a simple matter to calculate their frequency of *contact resonance.*

The frequency of contact resonance may be approached from another point of view. The relation between normal contact force and relative displacement of the two bodies is given by equation (4.23) for a circular contact area and by equation (4.26c) for an elliptical contact. Both may be written

$$P = K\delta^{3/2} \tag{11.16}$$

where the constant K depends upon the geometry and elastic constants of the two bodies. This relationship is nonlinear, but for small variations ΔP about a mean load P_0, the effective stiffness is given by

$$s = \frac{\mathrm{d}P}{\mathrm{d}\delta} = \tfrac{3}{2}(K^2 P_0)^{1/3} \tag{11.17}$$

If the bodies have masses m_1 and m_2 and are freely supported, the frequency of contact resonance is given by

$$\omega_0{}^2 = \frac{s(m_1 + m_2)}{m_1 m_2} \tag{11.18}$$

As we have seen, the effective damping arising from wave propagation is negligible but in practice there will be some damping due to elastic hysteresis as described in §6.4.

At resonance, when large amplitudes of vibration occur, the behaviour is influenced by the nonlinear form of the force–displacement relation (11.16). Under a constant mean load P_0 the effective stiffness decreases with amplitude, so that the resonance curve takes on the 'bent' form associated with a 'softening' spring (see Den Hartog, 1956). Thus the frequency at maximum amplitude is less than the natural frequency given by equation (11.18), which assumes small amplitudes. Under severe resonant conditions the two bodies may bounce out of contact for part of the cycle.

We have seen how contact resonance arises in response to an oscillating force. It also occurs in rolling contact in response to periodic irregularities in the profiles of the rolling surfaces (see Gray & Johnson, 1972). The vibration response of two discs rolling with velocity V to sinusoidal corrugations of wavelength λ on the surface of one of them is shown in Fig. 11.4. With the smaller corrugation the amplitude of vibration does not exceed the static compression, so that the surfaces are in continuous contact. A conventional resonance curve is obtained. With the larger corrugation the discs bounce out of contact at resonance and the resonance curve exhibits the 'jump' which is a feature of a highly nonlinear system.

11.4   Elastic impact

The classical theory of impact between frictionless elastic bodies is due to Hertz and follows directly from his statical theory of elastic contact (Chapter 4). The theory is quasi-static in the sense that the deformation is assumed to be restricted to the vicinity of the contact area and to be given by the statical theory: elastic wave motion in the bodies is ignored and the total mass of each body is assumed to be moving at any instant with the velocity of its centre of mass. The impact may be visualised, therefore, as the collision of two rigid railway trucks equipped with light spring buffers; the deformation is taken to be concentrated in the springs, whose inertia is neglected, and the trucks move as rigid bodies. The validity of these assumptions will be examined subsequently.

(a) Collinear impact of spheres

The two elastic spheres, of mass m_1 and m_2, shown in Fig. 11.5, are moving with velocities v_{z1} and v_{z2} along their line of centres when they collide at O. We shall begin by considering collinear impact in which $v_{x1} = v_{x2} = \omega_{y1} =$

Fig. 11.4. Contact resonance curves for rolling discs with one corrugated surface. Corrugation amplitude/static compression: circle – 0.30; cross – 0.55.

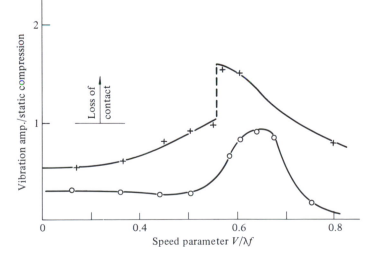

$\omega_{y2} = 0$. During impact, due to elastic deformation, their centres approach each other by a displacement δ_z. Their relative velocity is $v_{z2} - v_{z1} = d\delta_z/dt$ and the force between them at any instant is $P(t)$. Now

$$P = m_1 \frac{dv_{z1}}{dt} = -m_2 \frac{dv_{z2}}{dt}$$

hence

$$-\frac{m_1 + m_2}{m_1 m_2} P = \frac{d}{dt}(v_{z2} - v_{z1}) = \frac{d^2\delta_z}{dt^2} \tag{11.19}$$

The relationship between P and δ_z is now taken to be that for a static elastic contact given by equation (4.23), i.e.

$$P = (4/3)R^{1/2}E^*\delta_z^{3/2} = K\delta_z^{3/2} \tag{11.20}$$

where $1/R = 1/R_1 + 1/R_2$ and $1/E^* = (1 - v_1^2)/E_1 + (1 - v_2^2)/E$. Writing $1/m$ for $(1/m_1 + 1/m_2)$ we get

$$m\frac{d^2\delta_z}{dt^2} = -K\delta_z^{3/2} \tag{11.21}$$

Integrating with respect to δ_z gives

$$\tfrac{1}{2}\left\{V_z^2 - \left(\frac{d\delta_z}{dt}\right)^2\right\} = \tfrac{2}{5}\frac{K}{m}\delta_z^{5/2}$$

Fig. 11.5

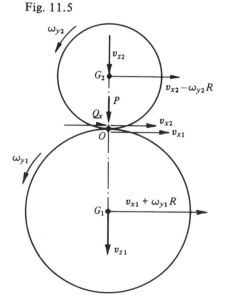

where $V_z = (v_{z2} - v_{z1})_{t=0}$ is the velocity of approach. At the maximum compression δ_z^*, $\mathrm{d}\delta_z/\mathrm{d}t = 0$, which gives

$$\delta_z^* = \left(\frac{5mV_z^2}{4K}\right)^{2/5} = \left(\frac{15mV_z^2}{16R^{1/2}E^*}\right)^{2/5} \tag{11.22}$$

The compression–time curve is found by a second integration, thus

$$t = \frac{\delta_z^*}{V_z} \int \frac{\mathrm{d}(\delta_z/\delta_z^*)}{\{1 - (\delta_z/\delta_z^*)^{5/2}\}^{1/2}} \tag{11.23}$$

This integral has been evaluated numerically by Deresiewicz (1968) and converted into a force–time curve in Fig. 11.6. After the instant of maximum compression t^*, the spheres expand again. Since they are perfectly elastic and frictionless, and the energy absorbed in wave motion is neglected, the deformation is perfectly reversible. The total time of impact T_c is, therefore, given by

$$T_c = 2t^* = \frac{2\delta_z^*}{V_z} \int_0^1 \frac{\mathrm{d}(\delta_z/\delta_z^*)}{\{1 - (\delta_z/\delta_z^*)^{5/2}\}^{1/2}} = 2.94\delta_z^*/V_z$$

$$= 2.87(m^2/RE^{*2}V_z)^{1/5} \tag{11.24}$$

The above analysis applies to the contact of spheres or to bodies which make elastic contact over a circular area. It can be adapted to bodies having general curved profiles by taking the parameter K in the static compression law from equation (4.26c) for the approach of two general bodies. The quasi-static impact of a rigid cone with an elastic half-space has been analysed by Graham (1973).

We can now examine the assumption on which the Hertz theory of impact is based: that the deformation is quasi-static. In §1, when discussing the impact of a thin rod, it was argued that the deformation in the rod would be quasi-

Fig. 11.6. Variation of compression δ_z and force P with time during a Hertz impact. Broken line – $\sin(\pi t/2t^*)$.

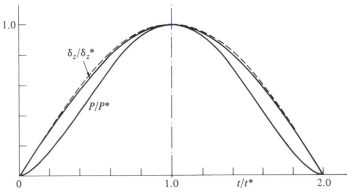

static if the duration of the impact was long enough to permit stress waves to traverse the length of the rod many times. Love (1952) suggested that the same criterion applies in this case. For like spheres, the time for a longitudinal wave to travel two ball diameters is $4R/c_0$. The time of impact, given by equation (11.24), can be expressed as $5.6 \, (R^5/c_0^4 V_z)^{1/5}$, so that the ratio of contact time to wave time $\approx (V_z/c_0)^{1/5}$. According to Love's frequently quoted criterion this quantity should be much less than unity for a quasi-static analysis of impact to be valid.

However, it now appears that Love's criterion, at least in the form stated, is not the appropriate one for three-dimensional colliding bodies. It clearly leads to logical difficulties when one of the bodies is large so that no reflected waves return to the point of impact! We shall now outline an alternative approach due to Hunter (1956), based on the work described in the last section. There it was shown that the dynamic response of an elastic half-space could be found with good approximation by regarding the half-space as an elastic spring in parallel with a dashpot; the energy 'absorbed' by the dashpot accounting for the energy radiated through the half-space by wave motion. Provided the time constant of the system is short compared with the period of the force pulse applied to the system, the force variation during the impact will be controlled largely by the spring, i.e. in a quasi-static manner, and the energy absorbed by the dashpot will be a small fraction of the total energy of impact. We will now find the condition for this to be so.

The force–time variation for a quasi-static elastic impact is given by equations (11.20) and (11.23) and is plotted in Fig. 11.6. It is not an explicit relationship but it is apparent from the figure that it can be approximated by

$$P(t) = P^* \sin \omega t = P^* \sin (\pi t/2t^*), \quad 0 \leqslant t \leqslant 2t^* \tag{11.25}$$

The spring–dashpot model of an elastic half-space has spring stiffness $s(\approx 5Ga)$ and time constant $T \approx 0.74a/c_2 \approx 1.2a/c_0$. When such a system is subjected to the force pulse expressed by equation (11.25), the energy absorbed by the dashpot is small and the response is dominated by the spring if the relaxation time T is short compared with the period of the pulse $2t^*$. If we now take a to be constant and equal to a^* the ratio of times may be written

$$\frac{T}{2t^*} \approx 0.4 \, \frac{a^* V_z}{\delta_z^* c_0} = 0.4 \, \frac{R V_z}{a^* c_0} \tag{11.26}$$

For quasi-static conditions to be approached this ratio must be much less than unity. For comparison with Love's criterion we consider two like spheres where $m_1 = m_2 = (4/3)\pi\rho R_1^3$ and $R = R_1/2$. Equation (11.26) then reduces to

$$T/2t^* \approx 0.3(V_z/c_0)^{3/5} \tag{11.27}$$

This is a much less restrictive condition than that put forward by Love: $T/2t^*$ is less than 1% provided $V_z < 0.002c_0$. As we shall see, however, a more severe restriction is placed on the velocity of impact for the above theory to be valid by the fact that most real materials cease to deform elastically at impact speeds very much less than those possible under the criterion of (11.27).

Although the impact of elastic spheres at practical speeds is virtually quasi-static, Thompson & Robinson (1977) have drawn attention to the behaviour immediately following first contact. Under quasi-static conditions the contact radius a is related to the indentation δ_z by $a^2 = \delta_z R$, so that a grows at a rate $\dot{a} = \dot{\delta}_z R/2a$, where $\dot{\delta}_z$ approximately equals the velocity of impact V_z. Thus, at first contact when a is vanishingly small, $\dot{a}$ can exceed the velocity with which elastic waves propagate on the surface. It turns out, however, that this so-called 'super-seismic phase' occupies a fraction of the total contact time of order V_z/c_1 which is insignificant.

(b) Oblique impact of spheres

If the spheres in Fig. 11.5 have a general coplanar motion then tangential velocities at the point of impact v_x and angular velocities ω_y are introduced. With frictionless surfaces the tangential and rotational motion is undisturbed by the impact. With friction, on the other hand, tangential tractions arise at the interface which influence the motion in an involved way. Denoting the resultant friction force by Q_x, the linear momentum in the tangential direction gives

$$Q_x = m_1 \frac{d}{dt}(v_{x1} + \omega_{y1}R_1) = -m_2 \frac{d}{dt}(v_{x2} - \omega_{y2}R_2) \qquad (11.28)$$

Now the moment of momentum of each sphere about the axis Oy is conserved, i.e.

$$\frac{d}{dt}\{m_1 v_{x1}R_1 + m_1(R_1^2 + k_1^2)\omega_{y1}\}$$

$$= \frac{d}{dt}\{-m_2 v_{x2}R_2 + m_2(R_2^2 + k_2^2)\omega_{y2}\} = 0 \qquad (11.29)$$

where k_1 and k_2 are radii of gyration of the spheres about their centres of mass. Eliminating ω_{y1} and ω_{y2} from (11.28) and (11.29) gives

$$Q_x = \frac{m_1}{1 + R_1^2/k_1^2}\frac{dv_{x1}}{dt} = -\frac{m_2}{1 + R_2^2/k_2^2}\frac{dv_{x2}}{dt} \qquad (11.30)$$

Writing $m_i/(1 + R_i^2/k_i^2) = m_i^*$ and $1/m^* = 1/m_1^* + 1/m_2^*$ we get

$$\frac{1}{m^*}Q_x = \frac{d}{dt}(v_{x1} - v_{x2}) = \frac{d^2\delta_x}{dt^2} \qquad (11.31)$$

where δ_x is the tangential elastic displacement between the two spheres at the point of contact. This equation governs tangential deformation at the contact in the way equation (11.19) governs normal compression. Deformation under the action of tangential forces, however, is complicated by micro-slip. If Q_x reaches its limiting value $\pm\mu P$ the surfaces will slip completely but if $Q_x < |\mu P|$ there may be no slip, but in general an annulus of micro-slip would be expected at the edge of the contact area where the pressure is low. The variations of tangential traction and micro-slip arising from simultaneous variations in tangential and normal forces have been studied by Mindlin & Deresiewicz and are discussed briefly in §7.3. The traction at any instant depends not only upon the values of P and Q, but upon the history of P and Q. This approach has been applied to oblique impact by Maw *et al.* (1976, 1981). The tangential tractions do not affect the normal motion if the materials of the two bodies are elastically similar (i.e. β as defined by eq. (5.3) is zero). However, we have seen in §5.4 that, even for dissimilar materials, the effect is small and may reasonably be neglected. The variation of contact size and contact pressure throughout the impact are thus given by the Hertz theory of impact, independently of friction forces.

The variations of tangential traction and micro-slip throughout the impact have been calculated step by step for different incident conditions. The elastic constants of the two bodies enter the calculation through the ratio of the tangential to normal compliance of a circular contact (see eqs. (7.43) and (7.44)). We define the stiffness ratio κ by

$$\frac{1}{\kappa} \equiv \frac{\dfrac{1-\nu_1/2}{G_1} + \dfrac{1-\nu_2/2}{G_2}}{\dfrac{1-\nu_1}{G_1} + \dfrac{1-\nu_2}{G_2}} \tag{11.32}$$

Thus κ is a material constant close to unity: for similar materials and $\nu = 0.3$, $\kappa = 0.824$. The incident conditions are specified by the non-dimensional parameter $\psi \equiv \kappa V_x/\mu V_z$ where $V_x = (v_{x1} - v_{x2})_{t=0}$ is the tangential velocity of approach before impact. Note that $\tan^{-1}(V_x/V_z)$ is the angle of incidence with which the surfaces approach each other at O. The behaviour during the impact depends upon a second parameter $\chi = \kappa m/2m^*$. For similar homogeneous spheres, with $\nu = 0.3$, $\chi = 1.44$. The variation in Q_x throughout the impact is shown in Fig. 11.7 for different incident conditions. For angles of incidence which are small compared with the angle of friction ($\psi \leqslant 1$) there is no slip at the start of the impact. With larger angles of incidence ($1 < \psi < 4\chi - 1$) the impact starts and finishes with complete slip; in between there is partial slip. At sufficiently high incidence ($\psi \geqslant 4\chi - 1$) sliding takes place throughout the complete time of impact.

Provided that the angle of incidence is not too large it is clear from Fig. 11.7 that the tangential force Q_x undergoes a reversal during the impact, whereas the normal force completes a half cycle only. In the last section we saw that two bodies in contact had a frequency of 'contact resonance' determined by the normal contact stiffness and their masses (eq. (11.18)). A similar behaviour would be expected in tangential motion. The ratio of the tangential to normal frequencies of contact resonance will be

$$\omega_t/\omega_n = (\kappa m/m^*)^{1/2} = (2\chi)^{1/2} \tag{11.33}$$

For solid spheres $\omega_t/\omega_n = 1.7$ which implies that the tangential force almost completes a full cycle during the time when the normal force goes through half a cycle.

The negative tangential force towards the end of the impact is responsible for a negative tangential rebound. Expressing the rebound velocity V_x' by the rebound parameter $\psi' = \kappa V_x'/\mu V_z$, the rebound conditions are plotted as a function of the incident parameter ψ in Fig. 11.8. The tangential rebound velocity V_x' is found to be mainly negative except when $\psi > 4\chi$. The classical rigid body theory of impact, which ignores contact deformation, predicts that the tangential rebound velocity V_x' is either positive, if slipping is continuous, or zero, if slipping ceases during the contact, as shown by the broken lines in Fig. 11.8. The negative tangential rebound velocities predicted in Fig. 11.8 have been substantiated by experiment (see Maw *et al.*, 1981). They are most

Fig. 11.7. Oblique impact of homogeneous solid spheres: the variation of tangential force throughout the impact. ($\nu = 0.3$, $\chi = 1.44$.)

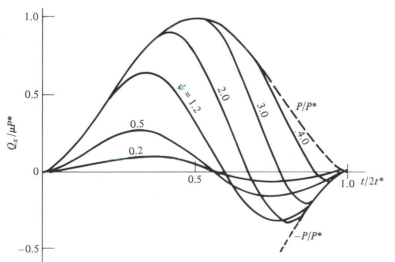

Fig. 11.8. Tangential velocities of incidence and rebound in the oblique impact of homogeneous solid spheres. Broken line – rigid body theory. A: $\nu = 0.3$, B: $\nu = 0.5$.

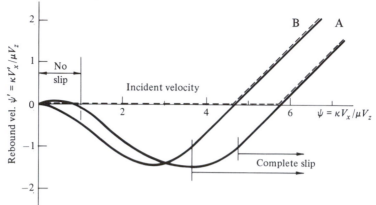

likely to arise when the coefficient of friction is large, for example, with dry rubber surfaces.

(c) Wave motion due to impact

Although the fraction of the impact energy which is radiated as elastic waves is generally very small, in some applications such as seismology it may be important. Tsai & Kolsky (1967) dropped steel balls onto a large block of glass under conditions of elastic impact and measured the radial strain in the surface waves. The variation of strain with time at a particular radial position is shown in Fig. 11.9. An analysis was made along the following lines. The force–time characteristic of Hertzian impact (Fig. 11.6) was assumed to apply and was

Fig. 11.9. Surface wave on a glass block produced by impact of a steel sphere (Tsai & Kolsky, 1967). A – circles joined by dashes – approximation calculation (1967); B – broken line – improved calculation (Tsai, 1968). Solid line – experimental.

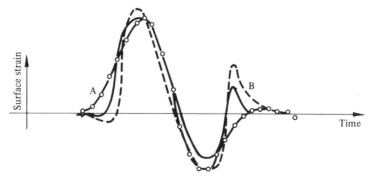

transformed into a continuous spectrum of harmonically varying forces $F(\omega)$, according to equation (11.13). The radial surface displacement $\bar{u}_r$ due to a single frequency of excitation is given by equation (11.10a). By substituting the force spectrum $F(\omega)$ for P, integrating with respect to ω and differentiating with respect to r, the time variation of radial strain $\partial\bar{u}_r/\partial r$ due to the impact pulse is obtained. The result of these calculations is shown by Curve A in Fig. 11.9. The agreement with the measured strains is fairly satisfactory. A refined analysis was then made by Tsai (1968) using the more exact expressions for radial displacement given by Miller & Pursey (1954) and taking into account the non-uniform pressure distribution and the variation in contact radius a during the impact. The result is shown by Curve B. The refinements in the analysis do not make a major difference to the result, but they do appear to account for the sharp secondary peak in the measured strain pulse.

More exact dynamic analyses of elastic impact have been made by Tsai (1971) for spherical bodies and by Bedding & Willis (1973) for the penetration of an elastic half-space by a rigid wedge and cone.

The discussion in this section so far has been concerned with collisions between compact bodies, in which stress wave effects account for only a small fraction of the energy of impact and do not influence the local deformation significantly. If one or both of the bodies is slender this conclusion no longer applies.

Let us return to the example of a thin rod, discussed in §1, this time struck on its end by a sphere moving with velocity V. The three-dimensional state of stress at the end of the rod demands some degree of approximation. A convenient approach is to choose a point H, just inside the end of the rod (Fig. 11.10(a)),

Fig. 11.10. (a) Impact of a sphere on the end of a slender rod; (b) Spring–dashpot model.

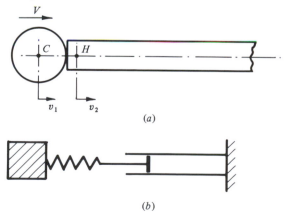

(a)

(b)

and assume that quasi-static deformation due to the impact takes place to the left of H and that one-dimensional elastic wave propagation takes place to the right of H. The choice of the location of H is somewhat arbitrary, but for most purposes it can be regarded as being adjacent to the end of the rod. If $P(t)$ is the contact force during impact, the momentum equation for the sphere gives the velocity of its centre C to be

$$v_1 = V - \frac{1}{m} \int_0^t P(t)\, dt \tag{11.34}$$

where m is the mass of the sphere. For the rod, from equation (11.1), the velocity of H is given by

$$v_2 = \frac{1}{A\rho c_0} P(t) \tag{11.35}$$

The approach δ_z of the centre of the sphere C and the point H is given by

$$\delta_z = \int_0^t (v_1 - v_2)\, dt = Vt - \frac{1}{m} \int_0^t dt \int_0^t P(t)\, dt$$

$$- \frac{1}{A\rho c_0} \int_0^t P(t)\, dt \tag{11.36}$$

The system can be modelled by a nonlinear spring representing the contact deformation in series with a dashpot which represents the wave motion in the rod (Fig. 11.10(*b*)). If the contact force–compression law is specified, for example by equation (11.20) for an elastic impact, equation (11.36) can be solved numerically to find the force–time history $P(t)$ and the dynamic stresses set up in the rod. Alternatively, if the dynamic strains in the rod are measured, equation (11.36) can be used to determine the force–deformation law at the point of impact (see Crook, 1952). For this technique to be satisfactory the impact must be complete before reflected waves in the rod return to the impact end, which requires the mass of the striker not to be too large compared with the total mass of the rod. At the other extreme, if the striker mass is too small v_2, given by (11.35), becomes negligible compared with v_1 and the rod behaves like a half-space. Davies (1948) has shown this state of affairs arises when the diameter of the sphere is less than half the diameter of the rod.

The approach outlined above can be applied to the longitudinal impact of two rods with rounded ends. A similar situation arises in the transverse impact of a beam by a striker; the local force–compression behaviour is determined by quasi-static considerations, but appreciable energy is transferred into bending vibrations of the beam (see Goldsmith, 1960; or Johnson, 1972).

11.5 Inelastic impact†

(a) Onset of yield

The Hertz theory of elastic impact has been presented in the foregoing section. An elastic-plastic material will reach the limit of elastic behaviour at a point beneath the surface when the maximum contact pressure p_0 at the instant of maximum compression reaches the value $1.60Y$, given by equation (6.9), where Y is the yield stress of the softer body. The maximum value p_0^* during elastic impact can be obtained by using equations (11.20) and (11.22) from which

$$p_0^* = \frac{3}{2\pi}\left(\frac{4E^*}{3R^{3/4}}\right)^{4/5}(\tfrac{5}{4}mV^2)^{1/5} \tag{11.37}$$

where $1/m = 1/m_1 + 1/m_2$, $1/R = 1/R_1 + 1/R_2$ and V is the relative velocity at impact.‡ Substituting the critical value of p_0 gives an expression for the velocity V_Y necessary to cause yielding:

$$\tfrac{1}{2}mV_Y^2 \approx 53R^3 Y^5/E^{*4} \tag{11.38}$$

In the case of a uniform sphere striking the plane surface of a large body, equation (11.38) reduces to

$$\frac{\rho V_Y^2}{Y} = 26(Y/E^*)^4 \tag{11.39}$$

where ρ is the density of the sphere. The impact velocity to cause yield in metal surfaces is very small; for a hard steel sphere striking a medium hard steel ($Y = 1000$ N/mm^2), $V_Y \approx 0.14$ m/s. It is clear that most impacts between metallic bodies involve some plastic deformation.

(b) Plastic impact at moderate speeds

In the last section we justified a quasi-static approach to finding the contact stresses during elastic impact, provided the impact velocity is small compared with the elastic wave speed. This condition remains valid when plastic deformation occurs, since the effect of plastic flow is to reduce the intensity of the contact pressure pulse and thereby to diminish the energy converted into elastic wave motion. At moderate impact velocities (up to 500 m/s, say) we can make use of our knowledge of inelastic contact stresses under static conditions (from Chapter 6) to investigate impact behaviour. We shall first consider normal impact.

---

† For general references see: Johnson (1972) or Zukas & Nicholas (1982).

‡ Since we are concerned here with normal impact only the suffix z will be omitted.

Up to the instant of maximum compression the kinetic energy is absorbed in local deformation, elastic and plastic, of the two colliding bodies, i.e.

$$\tfrac{1}{2}mV^2 = W = \int_0^{\delta^*} P \, d\delta \qquad (11.40)$$

where $1/m = 1/m_1 + 1/m_2$ and V is the relative velocity of impact. After the point of maximum compression the kinetic energy of rebound is equal to the work done during elastic recovery, thus

$$\tfrac{1}{2}mV'^2 = W' = \int_0^{\delta^*} P' \, d\delta' \qquad (11.41)$$

where primed quantities refer to the rebound. We wish to determine the maximum contact stress, the duration of the impact and the 'coefficient of restitution' (V'/V) in terms of the impact velocity V and the properties of the two bodies. We shall restrict the discussion to spherical profiles, but the analysis may be extended to more general profiles without difficulty.

It is clear from equations (11.40) and (11.41) that the impact behaviour is determined by the compliance relation $P(\delta)$ for the contact, in both loading and unloading. These relationships under static conditions have been discussed in §§6.3 & 6.4 (see Fig. 6.17). In the elastic range $(P \leqslant P_Y)$ loading and unloading are identical, expressed by equation (11.20). Yield initiates at a point beneath the surface and, as the plastic zone spreads, the mean contact pressure rises from $\sim 1.1Y$ to $\sim 3Y$ when the fully plastic condition is reached. Thereupon, in the absence of strain hardening, the contact pressure remains approximately constant, referred to as the flow pressure or yield pressure.

Unfortunately the compliance relationship for an elastic-plastic contact is not precisely defined, so that a theory of elastic-plastic impact is necessarily approximate. Since most impacts between metal bodies result in a fully plastic indentation we can concentrate on this regime. In our static analysis we assumed (*a*) that the total (elastic and plastic) compression δ was related to the contact size by: $\delta = a^2/2R$, i.e. neither 'pile-up' nor 'sinking in' occurs at the edge of the indentation, and (*b*) that the mean contact pressure p_m is constant and equal to 3.0Y. These assumptions led to the compliance relation (6.41), which gives a fair prediction of the experimental results, as shown in Fig. 6.17. Making the same assumptions here and using equation (11.40) gives

$$\tfrac{1}{2}mV^2 = \int_0^{a^*} \pi a^2 p_d(a/R) \, da = \pi a^{*4} p_d/4R \qquad (11.42)$$

where p_d is used to denote the mean contact pressure during *dynamic* loading. We note that the quantity $\pi a^{*4}/4R$ is the apparent volume of material v_a displaced by an indenter of radius R. With a material which strain-hardens according to a power law with index n (eq. (6.73)) it has been shown by Mok & Duffy (1965) that the right-hand side of equation (11.42) is multiplied by the factor $4n/(4n + 1)$ and p_d is the dynamic pressure at the instant of maximum compression.

Taking the rebound to be elastic, the energy of rebound W' is given by substituting equation (6.45) for the compliance relation into equation (11.41), where $P^*(= \pi a^{*2} p_d)$ is the compressive force between the bodies at the start of the rebound. Using equation (4.22) to eliminate the radii, the kinetic energy of rebound can be expressed in terms of the size of the indentation by

$$\tfrac{1}{2}mV'^2 = W' = \frac{3P^{*2}}{10a^*E^*} = \tfrac{3}{10}\pi^2 a^{*3} p_d^2 / E^* \tag{11.43}$$

Eliminating a^* from equations (11.42 and 43) gives an expression for the coefficient of restitution:

$$e^2 \equiv \frac{V'^2}{V^2} = \frac{3\pi^{5/4} 4^{3/4}}{10} \left(\frac{p_d}{E^*}\right)\left(\frac{\tfrac{1}{2}mV^2}{p_d R^3}\right)^{-1/4} \tag{11.44}$$

or, by writing $p_d \approx 3.0 Y_d$ where Y_d is the dynamic yield strength,

$$e \approx 3.8 (Y_d/E^*)^{1/2} (\tfrac{1}{2}mV^2/Y_d R^3)^{-1/8} \tag{11.45}$$

It is clear from this analysis that the coefficient of restitution is not a material property, but depends upon the severity of the impact. At sufficiently low velocities ($V < V_Y$ given by eq. (11.38)) the deformation is elastic and e is very nearly equal to unity. The coefficient of restitution gradually falls with increasing velocity. When a fully plastic indentation is obtained our theory suggests that e is proportional to $V^{-1/4}$. Some experimental results taken from Goldsmith (1960), shown in Fig. 11.11, illustrate this behaviour.

Fig. 11.11. Measurements of the coefficient of restitution of a steel ball on blocks of various materials (from Goldsmith, 1960). Cross – hard bronze; circle – brass; triangle – lead. Lines of slope $-\tfrac{1}{4}$.

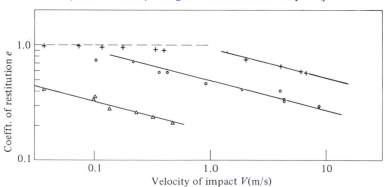

The coefficient of restitution is also very dependent upon the hardness of the material: according to equation (11.45) it is proportional to $Y_d^{5/8}$. Some experiments by Tabor (1948) with different materials are compared with equation (11.45) in Fig. 11.12. The general trend of the experiments follows the theory†, but the measured values of e are somewhat low.

Tabor (1948) and Crook (1952) used impact experiments to deduce the dynamic yield pressure p_d. As expected they found that the dynamic pressure was greater than the static yield pressure p_m by a factor which was somewhat larger for soft metals whose yield stress is sensitive to rate of strain (see Table 11.2). Furthermore it was found that the contact pressure does not remain constant throughout the period of plastic deformation, but falls, as the striker decelerates, to a value which is closer to the static pressure at the start of the rebound (denoted by p_r in Table 11.2). This fact accounts for the observed coefficients of restitution being lower than those predicted by the simple theory, which is based on the yield pressure being maintained constant up to the instant of closest approach.

Fig. 11.12. Variation of coefficient of restitution with dynamic hardness p_d. Solid circle – steel; triangle – Al alloy; square – brass; open circle – lead.

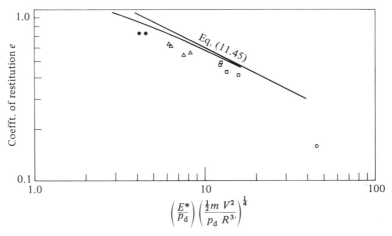

$$\left(\frac{E^*}{p_d}\right)\left(\frac{\tfrac{1}{2}m\,V^2}{p_d\,R^{3'}}\right)^{\tfrac{1}{4}}$$

† Tabor (1948) develops a theory from a slightly different premise. Instead of taking $\delta = a^2/2R$, he assumes that the energy dissipated in plastic deformation $(W - W') = p_d v_r$, when v_r is the *residual* volume of the indentation after rebound. This assumption modifies equation (11.42) to read

$$\tfrac{1}{2}m(V^2 - \tfrac{3}{8}V'^2) = p_d v_a = \pi a^{*4} p_d/4R \qquad (11.42a)$$

where V' is given by equation (11.43). The influence upon the coefficient of restitution is shown in Fig. 11.12.

The total time of impact is made up of two parts: the time of plastic indentation t_p and the time of elastic rebound t'. Making the same assumptions as before, that the flow pressure p_d is constant and that the compression $\delta = a^2/2R$, the indentation time may be easily calculated. The equation of relative motion of the two bodies is

$$m \frac{d^2\delta}{dt^2} = -\pi a^2 p_d = -2\pi R p_d \delta$$

where $1/m = 1/m_1 + 1/m_2$ and $1/R = 1/R_1 + 1/R_2$. The solution to this equation gives

$$t_p = \left(\frac{\pi m}{8Rp_d}\right)^{1/2} \tag{11.46}$$

which is independent of the velocity of impact. For a 10 mm diameter steel ball impinging on softer metals t_p has values in the range 10^{-4}–10^{-5} s. Assuming the rebound to be elastic and governed by the Hertz theory, the rebound time t' can be found from equations (11.45) and (11.24) with the result:

$$t' = 1.2 e t_p \tag{11.47}$$

where the coefficient of restitution e is given by (11.45) and t_p by (11.46). As the impact becomes more plastic, e falls and the rebound time t' becomes a smaller proportion of the total time of impact $(t_p + t')$.

During an oblique plastic impact friction forces between the projectile and the target are called into play and an elongated crater is produced. The oblique impact of a rigid sphere with a plastic solid has been analysed by Rickerby & Macmillan (1980) on very much the same basis as the theory of normal impact described above. The surface of the plastically deformed target is assumed to remain flat outside the crater and the penetration of the sphere is assumed to be resisted by a constant dynamic flow pressure p_d together with a frictional traction μp_d. Step-by-step calculations of the motion of the sphere, leading to the volume of the crater and the loss of kinetic energy of the projectile are well supported by experiment (Hutchings *et al.*, 1981)

Table 11.2

Metal	p_d/p_m	p_r/p_m
Steel	1.28	1.09
Brass	1.32	1.10
Al alloy	1.36	1.10
Lead	1.58	1.11

The simple theories of plastic impact outlined above are based on the assumption that the maximum penetration δ^* is given approximately by $a^{*2}/2R$. For this to be the case a^*/R must be less than about 0.5 so that, by equation (11.42), $(\frac{1}{2}mV^2/p_d R^3)$ must be less than about 0.05. For a steel sphere striking a steel surface this requires V to be less than 100 m/s.

(c) High speed impact

At higher speeds, associated with bullets and, in the extreme, with meteorites, the permanent deformation is much greater and the nature of the impact phenomenon changes in ways which depend upon the mechanical properties of both projectile and target. In addition, the energy dissipated during the impact produces a local temperature rise which can appreciably influence the material properties.

Johnson (1972) has suggested that the non-dimensional parameter $(\rho V^2/Y_d)$ provides a useful guide for measuring the regime of behaviour for the impact of metals. Table 11.3 is adapted from Johnson.

For illustrative purposes we shall consider the impact of a hard sphere of density ρ_1, impacting a massive block of density ρ and dynamic yield strength Y_d. Taking $p_d = 3Y_d$ the parameter $(\frac{1}{2}mV^2/p_d R^3)$ may then be written

$$\left(\frac{\frac{1}{2}mV^2}{p_d R^3}\right) = 0.72 \left(\frac{\rho_1}{\rho}\right) \left(\frac{\rho V^2}{Y_d}\right) \approx \frac{\rho V^2}{Y_d}$$

since ρ_1/ρ is not going to differ much from unity compared with the variations in $(\rho V^2/Y_d)$ which appear in Table 11.3. For most metals the ratio of elastic modulus to yield stress E^*/Y_d is 100 or more, so that for purely elastic deformation, by equation (11.39), $(\rho V^2/Y_d)$ is generally less than 10^{-6}. When the elastic limit is first exceeded the plastic zone is contained beneath the surface, but fully plastic indentations are produced when the parameter $(aE^*/Y_d R)$

Table 11.3

Regime	$\rho V^2/Y_d$	Approx. velocity V (m/s)
Elastic	$<10^{-6}$	<0.1
Fully plastic indentation	$\sim 10^{-3}$	~ 5
Limits of shallow indentation theory	$\sim 10^{-1}$	~ 100
Extensive plastic flow, beginning of hydrodynamic behaviour	~ 10	$\sim 1\,000$
Hypervelocity impact	$\sim 10^3$	$\sim 10\,000$

exceeds about 30 (from Fig. 6.15), which corresponds to $(\rho V^2/Y_d) \approx 10^{-3}$. Between $(\rho V^2/Y_d) = 10^{-3}$ and 10^{-1} the impact is reasonably described by the quasi-static shallow indentation theory presented above. In the velocity range discussed so far heating effects are negligible.

A further increase in impact speed leads to more extensive plastic deformation. The plastic strains are large and shear heating reduces the dynamic yield strength of the material. If the projectile is hard compared with the target the crater diameter increases to more than that of the projectile and penetrations greater than one diameter are obtained. This is the typical range of bullet speeds. When $(\rho V^2/Y_d)$ approaches unity the nature of the deformation changes and can no longer be regarded as quasi-static. Under these circumstances the inertia stresses associated with the local plastic deformation are comparable in magnitude with the yield stress of the material which resists deformation. Inertia stresses become important in the plastically deforming zone because of the very high rates of strain which are occurring there. In the surrounding elastically deforming material inertia effects remain small. The parameter $(\rho V^2/Y_d)$ can be interpreted as the ratio of the 'stagnation pressure' of the moving projectile, conceived as a fluid jet, to the strength of the target in shear. When this ratio appreciably exceeds unity, the inertia of the deforming material becomes more important than its yield strength, so that it behaves more like an ideal fluid than a plastic solid. Theoretical analyses of high speed impact have been made on this basis by Bjork and others (see Kornhauser, 1964) with moderate success.

We now write

$$\frac{\rho V^2}{Y_d} = \left(\frac{E}{Y_d}\right)\left(\frac{\rho V^2}{E}\right) = \frac{E}{Y_d}\left(\frac{V}{c_0}\right)^2$$

Since E/Y_d is greater than 100, it is clear that $(\rho V^2/Y_d)$ will exceed unity and fluid-like behaviour will develop *before* (V/c_0) approaches unity and dynamic effects occur in the bulk of the solid. However when $(\rho V^2/Y_d)$ reaches values around 10^3, (V/c_0) approaches or exceeds unity, and the impact sets up intense shock waves in the material. This is the region of hypervelocity impact normally associated with meteorites and laser beams. The heat liberated may be sufficient to melt or vaporise some of the projectile and target. A fine curtain of spray is ejected from the crater at a speed in excess of the impact velocity. A larger shallow crater with a pronounced lip is produced; if the projectile is ductile, it mushrooms on impact and turns itself inside out. This behaviour has been reproduced by Johnson *et al.* (1968) using a plasticine projectile and target, whose low yield strength enables hypervelocity impact conditions to be obtained at a velocity less than 1000 m/s. For further information on hypervelocity impact the reader is referred to Kornhauser (1964).

(d) Impact of viscoelastic solids

So far in this section we have described the inelastic behaviour of materials in terms of plastic flow, characterised by a dynamic yield stress Y_d. This is appropriate for metals but it is not a good model for polymeric materials, including rubber, which are better described in terms of viscoelasticity. The quasi-static impact of a projectile with a linear viscoelastic solid can be analysed by the methods described briefly in §6.5.

Provided the energy dissipated at impact is a fairly small fraction α of the kinetic energy of impact, a rough and ready estimate of the coefficient of restitution may be made from measurements of the energy dissipated in a cyclic strain experiment whose period is comparable with the time of impact. It is usual to express the energy dissipation in cyclic strain by the loss tangent, $\tan \phi$, where ϕ is the phase angle between the cyclic stress and strain. The loading and unloading during impact correspond roughly to a half cycle, whereupon the coefficient of restitution is given by

$$e = (1 - \alpha)^{1/2} = (1 - \pi \tan \phi)^{1/2} \tag{11.48}$$

To take a specific example, a Maxwell material (defined in §6.5) strained at a frequency ω has a loss tangent $= 1/\omega T$, where T is the time constant of the material. If T_c is the time of impact, we can take ω as π/T_c, whereupon

$$e = (1 - \alpha)^{1/2} \approx 1 - \tfrac{1}{2}\alpha = 1 - \tfrac{1}{2}(T_c/T) \tag{11.49}$$

provided that α is small, i.e. $T_c/T \ll 1$. Since the impact is predominantly elastic, T_c can be taken to be the elastic impact time given by equation (11.24).

In order to carry out a more exact analysis of a rigid sphere of mass m striking a viscoelastic half-space we have to make use of the results due to Ting (1966) outlined in §6.5. A general incompressible linear viscoelastic material has a creep compliance $\Phi(t)$ and a relaxation function $\Psi(t)$. Different equations govern the loading and unloading parts of the impact process. During loading ($0 < t < t^*$) the penetration $\delta(t)$ is related to the contact size $a(t)$ by the elastic equation:

$$\delta(t) = a^2(t)/R \tag{11.50}$$

The sphere retards under the action of the contact force $P(t)$, which from equation (6.60) is given by

$$-m\ddot{\delta}(t) = P(t) = \frac{8}{3R} \int_0^t \Psi(t - t') \frac{d}{dt'} a^3(t') \, dt' \tag{11.51}$$

The variation of force and penetration with time during loading are obtained by the simultaneous solution of equations (11.50) and (11.51). The maximum contact size coincides with the maximum penetration when $t = t^*$.

During the rebound ($t > t^*$), $a(t)$ is decreasing, whereupon $P(t)$ and $\delta(t)$ depend upon the time t_1 during loading, at which the contact size $a(t_1)$ was equal to the current contact size $a(t)$. The penetration is given by

$$\delta(t) = a^2(t)/R$$
$$- \int_{t^*}^{t} \Phi(t-t') \frac{d}{dt'} \left[\int_{t_1}^{t'} \Psi(t'-t'') \frac{d}{dt''} \{a^2(t'')\} dt'' \right] dt' \quad (11.52)$$

and the contact force by

$$-m\ddot{\delta}(t) = P(t) = \frac{8}{3R} \int_{0}^{t_1} \Psi(t-t') \frac{d}{dt'} a^3(t') dt' \quad (11.53)$$

The simultaneous solution to equations (11.52) and (11.53) gives the variation of force and displacement during the rebound.

A solution in closed form to these equations has been obtained by Hunter (1960) for the simple case of a Maxwell material in which the dissipation is small (i.e. in which the time of impact T_c is short compared with the relaxation time T of the material). The coefficient of restitution was found to be given by

$$e \approx 1 - (4/9)(T_c/T) \quad (11.54)$$

This result is very close to the approximate value obtained in equation (11.49) from the energy loss in cyclic strain.

For more complex materials, or when T_c/T is no longer small, equations (11.50)–(11.53) must be solved numerically step by step. This has been done by Calvit (1967) using creep and relaxation functions estimated from cyclic strain tests on perspex. Values of the coefficient of restitution and the time of impact obtained by experiment were both found to be somewhat lower than calculated.

11.6 Travelling loads – high speed sliding and rolling

In the discussion of sliding and rolling contact in Chapters 7 and 8 it was assumed that the velocity of the point of contact over the surface was sufficiently slow for the deformation to be quasi-static. This is true for most engineering purposes but, if the velocity approaches the speeds of elastic wave propagation, the inertia of the material plays a part and modifies the contact stresses. By analogy with a body moving through a fluid, we can identify three regimes: 'subsonic', 'transonic' and 'supersonic'† depending upon the ratio of the velocity to the elastic wave speed. On the surface of an elastic solid the behaviour is complicated by the fact that there are three wave speeds involved: dilatational c_1, shear c_2 and surface c_3. In this section we shall merely outline

† Referred to by some authors as 'subseismic', etc.

the main features of the dynamic effects for two-dimensional deformation (plane strain) only.†

(a) Moving line load on an elastic half-space

The fundamental problem concerns the stresses and deformation set up in an elastic half-space by a concentrated line force P per unit length moving with velocity V over the surface. It is the dynamic equivalent of the static problem considered in §2.2. An analysis using the complex variable method, which covers the subsonic, transonic and supersonic regimes, has been made by Cole & Huth (1958). They only investigate steady-state solutions, for which an Eulerian coordinate system which moves with the load can be used. Mach numbers M_1 and M_2 are defined as

$$M_1 = V/c_1 \quad \text{and} \quad M_2 = V/c_2$$

Under 'subsonic' conditions, when M_1 and $M_2 < 1$, we define:

$$\beta_1 = (1 - M_1^2)^{1/2}, \quad \beta_2 = (1 - M_2^2)^{1/2}$$

and

$$N = (2 - M_2^2)^2 - 4\beta_1\beta_2$$

Cole & Huth show that the surface displacement $\bar{u}_z$ at a point distance x from the load is given by

$$\bar{u}_z = \frac{2(1 + \nu)P}{\pi E} \frac{\beta_1 M_2^2}{N} \ln |x| + C \tag{11.55}$$

where C is a constant determined by the choice of the datum for displacements.

At low velocity, $V \to 0$, $\beta_1 \to 1$ and $N \to 2(M_1^2 - M_2^2) = -M_2^2/(1 - \nu)$. Thus the expression for the surface displacement reduces to

$$\bar{u}_z = -\frac{2(1 - \nu^2)P}{\pi E} \ln |x| + C \tag{11.56}$$

which is the static result obtained in equation (2.19). With an increase in velocity the logarithmic shape of the surface remains unchanged, but the magnitude of the displacement increases by the factor $\{\beta_1 M_2^2/(1 - \nu)N\}$ approaching an infinite value as N approaches zero. Reference to equation (11.7) will confirm that this situation arises when the velocity V coincides with the speed c_3 of Rayleigh surface waves. Then $M_2 = \alpha$ and $M_1 = \alpha c_2/c_1$. It is not surprising that an undulation which travels freely along the surface at speed c_3 increases without limit if forced by a load moving at the same speed. Cole & Huth have obtained expressions for the stress components beneath

---

† The three-dimensional problem of a moving point force on an elastic half-space has been considered by Eason (1965).

the surface as a function of the speed and elastic properties of the surface. The distributions of normal stress $\sigma_z(x)$ along a line at depth z for $M_2 = 0$, 0.5 and 0.9 are shown in Fig. 11.13. The amplification of the stress with speed as V approaches $c_3 (M_2 \to 0.93)$ is clearly shown, but for $M_2 < 0.5$ the difference from the static stress distribution is small. At high subsonic speeds the normal acceleration $\ddot{u}_z$ leads to the stress σ_z becoming tensile at some distance on either side of the load. In the subsonic regime the normal surface displacements are symmetrical ahead of and behind the load so that no net work is done by the moving load.

If we turn now to the completely 'supersonic' case where V is greater than the largest of the wave speeds, then both M_1 and $M_2 > 1$. We now write

$$\beta_1' = (M_1^2 - 1)^{1/2}, \quad \beta_2' = (M_2^2 - 1)^{1/2}$$

and

$$N' = (2 - M_2^2) + 4\beta_1'\beta_2'$$

The solution for the displacements and stresses is completely different from the 'subsonic' case. The surface displacement may be written

$$\bar{u}_z = \begin{cases} 0, & x < 0 \\ \dfrac{2(1+\nu)P}{E} \dfrac{\beta_1' M_2^2}{N'}, & x > 0 \end{cases} \qquad (11.57)$$

Ahead of the load the surface is undisturbed; behind the load, it is uniformly depressed by an amount which depends upon the speed. The stresses in the half-space are zero everywhere except along the lines of two 'shock waves' propagating from the point of application of the load. At the wave front the stress is theoretically infinite. The shock waves travel at velocities c_1 and c_2

Fig. 11.13. Normal stress σ_z due to a line load P moving subsonically ($V < c_2$) over the surface of an elastic half-space, for $M_2 = 0$, 0.5 and 0.9. ($\nu = \frac{1}{3}$, $M_2/M_1 = 2$.)

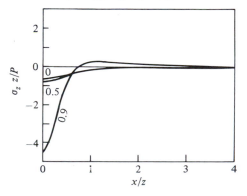

and hence make angles of $\cot^{-1}(\beta_1')$ and $\cot^{-1}(\beta_2')$ with the surface as shown in Fig. 11.14. This process is no longer conservative; the moving load does work as it permanently depresses the surface and strain energy is steadily radiated away by the shock waves.

The 'transonic' regime is complicated. In the small range of speed $c_3 < V < c_1$, N changes sign so that a downward force on the surface gives rise to an upward displacement in the vicinity of the force. For $c_2 < V < c_1$, N is complex and the stresses and deformation are a combination of a shock wave travelling at speed c_2 and a 'subsonic' pattern associated with c_1. For details consult Cole & Huth (1958).

(b) High speed rolling or sliding of a cylinder

We shall now consider the stresses and deformation set up by a long rigid, frictionless cylinder which slides or rolls over the surface of an elastic half-space with velocity V perpendicular to its axis (Craggs & Roberts, 1967). We have seen that in the 'subsonic' regime the deformation of the surface by a moving line load is similar in shape to that produced by a stationary load; the dynamic effect is to amplify the displacements by the speed factor $\{\beta_1 M_2^2/(1-\nu)N\}$ where β_1, M_2 and N are defined above. Thus the half-space deforms as though its rigidity were reduced by the same factor. Now the contact pressure distribution with the moving cylinder can be built up by the super-position of concentrated loads, such that the resultant deformation in the contact zone matches the profile of the cylinder. It follows that the pressure distribution will be similar to the static (Hertz) case, and the contact width will be increased by:

$$a(V) = a(0)\{\beta_1 M_2^2/(1-\nu)N\}^{1/2} = \left\{ \frac{4(1+\nu)PR}{\pi E} \frac{\beta_1 M_2^2}{N} \right\}^{1/2} \qquad (11.58)$$

Fig. 11.14. Line load moving supersonically ($V > c_1$) over the surface of an elastic half-space showing shock waves. ($\nu = 0.25$.)

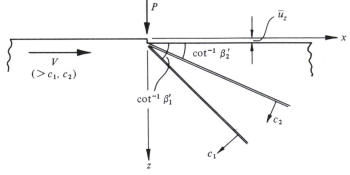

For a given load P, therefore, the maximum contact pressure $p_0(V)$ is reduced by the same factor. It is clear from Fig. 11.13 that the subsurface stresses do not follow a distribution which is similar to the static case. They may be deduced from the results given by Craggs & Roberts (1967).

It is immediately apparent from equation (11.58) that the indentation of the half-space by the cylinder becomes excessively large when V approaches the Rayleigh wave speed ($N \to 0$). Above the Rayleigh wave speed ($c_3 < V < c_2$) the change in sign of the surface displacements makes it impossible for the cylinder to make contact with the surface of the half-space along a continuous arc. In fact no physically acceptable steady-state solution appears to exist in the range $c_3 < V < c_1$ (Craggs & Roberts, 1967).

In the 'supersonic' regime ($V > c_1$), however, a simple solution can be found from the results for a line load. The surface is undisturbed until it meets the cylinder at $x = -a$. By equation (11.57) each increment of pressure $p \, dx$ depresses the surface by an amount

$$d\bar{u}_z = \frac{2(1 + \nu)}{E} \frac{\beta_1' M_2^2}{N'} p \, dx$$

Thus the pressure is proportional to the slope of the profile $d\bar{u}_z/dx$, which is triangular. The surface leaves the cylinder at its lowest point at a depth d given by

$$d = \frac{a^2}{2R} = \frac{2(1 + \nu)P}{E} \frac{\beta_1' M_2^2}{N'} \tag{11.59}$$

as shown in Fig. 11.15. Trains of stress waves are propagated from the arc of contact at velocities c_1 and c_2.

Fig. 11.15. Frictionless cylinder rolling or sliding supersonically ($V > c_1$) over the surface of an elastic half-space.

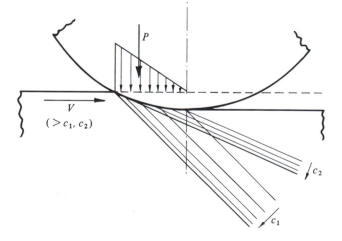

12

Thermoelastic contact

12.1   Introduction

Classical elastic contact stress theory concerns bodies whose temperature is uniform. Variation in temperature within the bodies may, of itself, give
rise to thermal stresses but may also change the contact conditions through
thermal distortion of their surface profiles. For example if two non-conforming
bodies, in contact over a small area, are maintained at different temperatures,
heat will flow from the hot body to the cold one through the 'constriction'
presented by their contact area. The gap between their surfaces where they do
not touch will act more or less as an insulator. The interface will develop an
intermediate temperature which will lie above that of the cold body, so that
thermal expansion will cause its profile to become more convex in the contact
region. Conversely the interface temperature will lie below that of the hot body,
so that thermal contraction will lead to a less convex or a concave profile. Only
if the material of the two bodies is similar, both elastically and thermally, will
the expansion of the one exactly match the contraction of the other; otherwise
the thermal distortion will lead to a change in the contact area and contact
pressure distribution. This problem will be examined in §4 below.

A somewhat different situation arises when heat is generated at or near to
the interface of bodies in contact. An obvious example of practical importance
is provided by frictional heating at sliding contacts. Also inelastic deformation
in rolling contact liberates heat beneath the surface which, with poorly conducting materials, can lead to severe thermal stresses. The passage of a heavy electric
current between non-conforming surfaces in contact leads to a high current
density and local heating at the contact constriction.

The analysis of a thermoelastic contact problem consists of three parts:
(i) the analysis of heat conduction to determine the temperature distribution
in the two contacting bodies; (ii) the analysis of the thermal expansion of the

bodies, to determine the thermal distortion of their surface profiles; (iii) the isothermal contact problem to find the contact stresses resulting from the deformed profiles. In the simplest cases these three aspects are uncoupled and the analysis can proceed in the above sequence. In many cases these aspects are not independent. Where heat is generated by sliding friction, for example, the distribution of heat liberated at the interface, which governs the temperature distribution, is proportional to the distribution of contact pressure, which itself depends on the thermoelastic distortion of the solids. Nevertheless our discussion of thermoelastic contact will follow that sequence.

Readers of this book will have already appreciated the advantages, when calculating elastic deformations, of representing bodies in contact, whatever their actual profile, by an elastic half-space bounded by a plane surface. The same idealisation is also helpful in calculating the temperature distributions in thermal problems. It may be justified in the same way; the temperature *gradients* which give rise to thermal stress and distortion are large only in the vicinity of the contact region where the actual surfaces of the bodies are approximately plane. Widespread changes in the temperature of the bulk of the bodies lead only to overall and approximately uniform expansion or contraction which neither introduces thermal stresses nor significantly changes the profile in the contact zone.

12.2 Temperature distributions in a conducting half-space

The theory of heat conduction in solids is not the concern of this book. A full account of the theory and the analysis of most of the problems we require are contained in the book by Carslaw & Jaeger (1959). Only the results will be summarised here. We are interested in the flow of heat into a half-space through a restricted area of the surface. We shall start with the temperature distribution in the half-space due to a 'point source' of heat located at the surface. Since the conduction equations are linear the temperature distribution due to any distribution of heat supplied to the surface can be found by the superposition of the solution for 'point sources' in the same way as elastic stress distributions due to surface tractions were found from 'point force' solutions in Chapters 2 and 3.

The half-space is taken to be uniform with conductivity k, density ρ, specific heat capacity c and thermal diffusivity $\kappa \, (= k/\rho c)$.

(a) Instantaneous point source

A quantity of heat H is liberated instantaneously at time $t = 0$ at the origin O on the surface of a half-space, whose temperature is initially uniform and equal to θ_0. The temperature at subsequent times at a point situated a radial distance R from O is given by (C & J §10.2)†

$$\theta - \theta_0 = \frac{H}{4\rho c(\pi \kappa t)^{3/2}} \exp(-R^2/4\kappa t) \tag{12.1}$$

At any point in the solid the temperature rises rapidly from θ_0 to a maximum value when $t = R^2/6\kappa$ and slowly decays to θ_0 as the heat diffuses through the solid.

(b) Instantaneous line source

The treatment of two-dimensional problems is facilitated by the use of a line source in which H units of heat per unit length are instantaneously liberated on the surface of a half-space along the y-axis. The temperature distribution is cylindrical about the y-axis and at a distance R is given by (C & J §10.3)

$$\theta - \theta_0 = \left(\frac{H}{2\pi k t}\right) \exp(-R^2/4\kappa t) \tag{12.2}$$

(c) Continuous point source

If heat is supplied to the half-space at O at a steady rate $\dot{H}$, the temperature at a distance R from O may be found by integrating (12.1) with respect to time. It varies according to

$$\theta - \theta_0 = (\dot{H}/2\pi kR) \operatorname{erfc}(R/4\kappa t) \tag{12.3}$$

where $\operatorname{erfc}(\chi) = 1 - \operatorname{erf}(\chi) = 1 - (2/\pi^{1/2}) \int_0^\chi \exp(-\xi^2)\, d\xi$.

After a sufficient time has elapsed, a steady state is reached in the neighbourhood of the source (i.e. where $R \ll 4\kappa t$) in which the temperature is given by

$$\theta - \theta_0 = \dot{H}/2\pi kR \tag{12.4}$$

The infinite temperature at $R = 0$ is a consequence of assuming that the heat is introduced at a point.

(d) Distributed heat sources (C & J §10.5)

In reality heat is introduced into the surface of a solid over a finite area. Assuming the remainder of the surface to be perfectly insulated, the temperature within the solid may be found by the superposition of point or line sources. If heat is supplied at a steady rate $\dot{h}$ per unit area, then equations (12.3) and (12.4) can be used.

Suppose we wish to find the steady-state distribution of temperature throughout the surface of a half-space when heat is supplied steadily to a small area A of the surface. For a single source equation (12.4) applies, in which we

† Carslaw & Jaeger (1959).

shall denote by $\bar{\theta}$ temperature at a point on the surface distance r from the source. The reader of this book will recognise that equation (12.4) is analogous to equation (3.22b) which expresses the normal displacement $\bar{u}_z$ of a point on the surface of an elastic half-space due to a point force P acting at a distance r, viz.

$$\bar{u}_z = \frac{1-\nu^2}{\pi E}\frac{P}{r}$$

This analogy may be used to determine the steady-state distribution of surface temperature due to a distributed heat supply. For example, the temperature due to uniform supply of heat to a circular area of radius a is analogous to the displacement produced by a uniform pressure. If $(1-\nu^2)/E$ is replaced by $\frac{1}{2}k$ and p by $\dot{h}$, the surface temperature distribution is given by equations (3.29). The maximum temperature at the centre of the circle is

$$\bar{\theta}_{max} - \theta_0 = \dot{h}a/k \tag{12.5a}$$

and the average temperature over the heated circle is

$$\bar{\theta}_{max} - \theta_0 = 8\dot{h}a/3\pi k \tag{12.5b}$$

Similarly the temperature at the surface of a uniformly heated polygonal region may be found from the results of §3.3.

The same analogy is useful when it is required to find the distribution of heat supplied to a small area of the surface which would maintain a steady prescribed temperature distribution in that area. For example, consider a half-space in which a circular area of the surface, of radius a, is maintained at a steady uniform temperature $\bar{\theta}_c$. The temperature far away is θ_0 and the surface of the half-space outside the circle is insulated. The analogous elastic problem concerns the indentation of an elastic half-space by a rigid circular punch which imposes a uniform displacement of the surface $\bar{u}_z$. The pressure under the punch is given by equations (3.34) and (3.36), from which the required distribution of heat supply may be deduced to be

$$\dot{h} = \frac{2k(\bar{\theta}_c - \theta_0)}{\pi(a^2 - r^2)^{1/2}} \tag{12.6}$$

(e) Moving heat sources (C & J §10.7)

In order to investigate the temperature produced by frictional heating in sliding contact we need to examine the temperature produced in a half-space by a heat source which moves on the surface. If we are dealing with the steady state it is convenient to fix the heat source and imagine the half-space moving beneath it with a steady velocity V parallel to the x-axis. The temperature field is then a function of position but not of time.

We shall examine the two-dimensional case of an infinitely long source of heat parallel to the y-axis, uniformly distributed over the strip $-a \leqslant x \leqslant a$. Referring to Fig. 12.1, the distributed source is regarded as an array of line sources of strength $\dot{h}$ ds. The element of material at (x, z) at time t was located at $(x - Vt', z)$ at an earlier instant $(t - t')$. The heat liberated by the line source at s in the time interval dt' is $\dot{h}$ ds dt', whereupon the steady temperature of an element located instantaneously at x is found by the integration of equation (12.2) from $t = -\infty$ to the current instant $t = 0$.

$$\theta(x, z) - \theta_0$$

$$= \frac{\dot{h}}{2\pi k} \int_{-a}^{a} ds \int_{-\infty}^{0} \left[\exp \left\{ -\frac{(x - s - Vt')^2 + z^2}{4\kappa t'} \right\} \frac{dt'}{t'} \right]$$

(12.7)

The maximum temperature occurs on the surface ($z = 0$) for which equation (12.7) can be written in the form

$$\bar{\theta} - \theta_0 = \frac{\dot{h}a}{kL^{1/2}} F(L, X)$$
(12.8)

where $L = (Va/2\kappa)$ and $X = (Vx/2\kappa)$. The integrals have been evaluated by Jaeger (1942). Surface temperature distributions are shown in Fig. 12.2(a). The maximum temperature occurs towards the rear of the heated zone which has had the longest exposure to heat. Maximum and average temperatures for the heated zone are plotted against the speed parameter L in Fig. 12.2(b). The parameter L, known as the Peclet number, may be interpreted as the ratio of the speed of the surface to the rate of diffusion of heat into the solid. At large values of L (>5), the heat will diffuse only a short distance into the solid in the time taken for the surface to move through the heated zone. The heat flow will then be approximately perpendicular to the surface at all points. The

Fig. 12.1

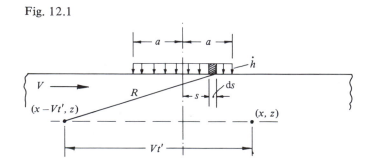

Fig. 12.2. Surface temperature rise due to a uniform moving line heat
source. (*a*) Temperature distribution; (*b*) Maximum and average
temperatures as a function of speed. A – Band source (max); B – square
source (max); C – square source (mean).

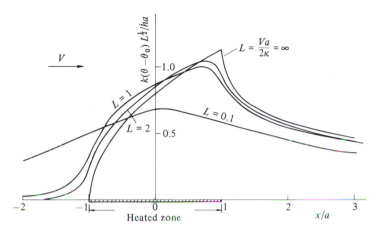

(*a*)

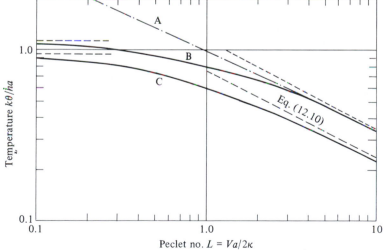

(*b*)

temperature of a surface point is then given by (C & J §2.9)

$$\bar{\theta} - \theta_0 = \frac{2\dot{h}(\kappa t)^{1/2}}{\pi^{1/2}k} = \frac{\dot{h}a}{k}\left\{\frac{2}{\pi}\left(\frac{2\kappa}{Va}\right)(1+x)\right\}^{1/2}, \quad -a \leqslant x \leqslant a \qquad (12.9)$$

The mean temperature for a band source is then

$$\bar{\theta}_{\text{mean}} - \theta_0 \approx \frac{4}{3\pi^{1/2}}\frac{\dot{h}a}{k}L^{-1/2} \qquad (12.10)$$

Since equation (12.9) is based on one-dimensional heat flow into the solid, it applies to a uniform source of any planform. Thus the mean temperature for a square source of side $2a$ is also given by expression (12.10).

At very low speeds ($L < 0.1$) the temperature distribution becomes symmetrical and similar to that for a stationary source. In the case of an infinitely long band source no steady-state temperature is reached, but the square source reaches a maximum temperature of $\theta_0 + 1.12\dot{h}a/k$ at the centre and a mean temperature in the heated zone of $\theta_0 + 0.946\dot{h}a/k$.

12.3 Steady thermoelastic distortion of a half-space

The equations which express the stresses and displacements in an elastic half-space due to an arbitrary steady distribution of temperature have been derived by various authors (e.g. Boley & Weiner, 1960). If the surface $z = 0$ is stress-free, then it can be shown that all parallel planes are stress-free, i.e. $\sigma_z = \tau_{yz} = \tau_{xz} = 0$ throughout. The normal displacement at depth z is given by (see Williams, 1961)

$$u_z = -(1+\nu)\alpha \int \theta \, dz \qquad (12.11)$$

Williams has expressed the thermoelastic stresses and displacements in an elastic half-space in terms of two harmonic functions and Barber (1975) has used this formulation to derive some general results which are useful in thermoelastic contact problems:

(i) If heat is supplied at a rate $\dot{h}$ per unit area to the free surface of a half-space, the surface distorts according to:

$$\frac{\partial^2 \bar{u}_z}{\partial x^2} + \frac{\partial^2 \bar{u}_z}{\partial y^2} = c\dot{h}(x,y) \qquad (12.12a)$$

With circular symmetry this equation becomes

$$\frac{1}{r}\frac{d}{dr}\left(r\frac{d\bar{u}_z}{dr}\right) = c\dot{h}(r) \qquad (12.12b)$$

where $c = (1 + \nu)\alpha/k$ is referred to as the 'distortivity' of the material.

In a two-dimensional situation $\partial^2 \bar{u}_z/\partial y^2 = 0$, so that equation (12.12a) implies that the curvature of the surface at any point is directly proportional to the rate of heat flow at that point; convex when heat is flowing into the surface and concave when it is flowing out. An insulated surface which is initially flat will remain flat. Thus uniform heating of a half-space over a long narrow strip will distort the surface such that it has a constant convex curvature within the strip and has plane inclined surfaces outside the strip. This general theorem is due to Dundurs (1974) (see also Barber, 1980a).

(ii) If the surface of a half-space, $z = 0$, is heated so that it has a distribution of surface temperature $\bar{\theta}(x, y)$, then the surface stress required to maintain the surface flat, such that $\bar{u}_z = 0$, is given by

$$\bar{\sigma}_z = -\tfrac{1}{2}\{cE/(1 - v^2)\}\bar{\theta}(x, y) \tag{12.13}$$

The application of an equal and opposite traction will free the surface from traction and allow it to distort. It follows from equation (12.13) that the surface displacements are those which would be caused by a surface pressure $p(x, y)$ proportional to the distribution of surface temperature $\bar{\theta}(x, y)$. In this way the steady thermal distortion of a half-space can be found by the methods discussed earlier in this book if the temperature distribution over the whole surface is known. In most contact situations, however, it is usual to assume that no heat is transferred across a non-contacting surface; the boundary conditions, therefore, are more appropriately expressed in terms of heat flux rather than temperature.

We will now look at a number of particular cases.

(a) Point source of heat

The temperature distribution due to a continuous point source of heat at point O on the surface is given by equation (12.4). The normal displacement of a point on the surface distance r from O then follows from equation (12.11)

$$u_z = -(1 + v)\alpha \int \dot{H}/\{2\pi k(r + z^2)^{1/2}\}\,dz$$

$$\bar{u}_z = -(c\dot{H}/2\pi) \ln (r_0/r) \tag{12.14}$$

where r_0 is the position on the surface where $\bar{u}_z = 0$. Since heat is being injected continuously into the solid and the surface is assumed to be insulated except at O, the expansion of the surface given by equation (12.14) increases without limit as the datum for displacements is taken at an increasing distance from the source.

(b) Uniform heating of a circular region

When heat is supplied steadily over a small area of the surface, the surface distortion can be found by superposition of the point source solutions (Barber, 1971a). For example, heating a thin annulus of radius a gives rise to surface displacements:

$$\bar{u}_z = -(c\dot{H}/2\pi) \ln (r_0/a), \quad r \leq a \tag{12.15a}$$

which is constant and

$$\bar{u}_z = -(c\dot{H}/2\pi) \ln (r_0/r), \quad r > a \tag{12.15b}$$

which is the same as for a point source at the centre. From these results we may easily proceed to uniform heating of a circular area, radius a, whence

$$\bar{u}_z = \begin{cases} -(c\dot{H}/4\pi)\{2 \ln (r_0/a) + (1 - r^2/a^2)\}, & r \leq a \tag{12.16a} \\ -(c\dot{H}/2\pi) \ln (r_0/r), & r > a \tag{12.16b} \end{cases}$$

This distorted surface is shown in Fig. 12.3. Alternatively the results expressed by equations (12.16) could have been obtained directly by integrating equation (12.12b) with $\dot{h} = \dot{H}/\pi a^2$ for $r \leq a$ and $\dot{h} = 0$ for $r > a$.

(c) Circular region at uniform temperature

To maintain a circular region of the surface at a uniform temperature θ_c which is different from the temperature at a distance θ_0 requires a supply of heat per unit area $\dot{h}(r)$ distributed according to equation (12.6). Hence substituting $2\pi r \dot{h}(r) \, dr$ for $\dot{H}$ in equation (12.13) and integrating from $r = 0$ to $r = a$ gives (Barber, 1971b)

$$\bar{u}_z = \begin{cases} -\dfrac{2}{\pi} cka(\theta_c - \theta_0) \left[\ln \left(\dfrac{r_0}{a} \right) - \ln \{1 + (1 - r^2/a^2)^{1/2}\} \right. \\ \qquad \left. + \left(1 - \dfrac{r^2}{a^2} \right)^{1/2} \right], r \leq a \tag{12.17a} \\ \\ -\dfrac{2}{\pi} cka(\theta_c - \theta_0) \ln (r_0/r), \quad r > a \tag{12.17b} \end{cases}$$

This distortion is also illustrated in Fig. 12.3.

(d) Moving heat source

We wish to find the thermal distortion of the surface due to the moving heat source shown in Fig. 12.1. Barber has shown that a concentrated line source $\dot{H}$, moving with velocity V causes a displacement of the surface at a point distance ξ ahead of the source given by

$$\bar{u}_z = -(2c\kappa \dot{H}/V) \exp (-X^2) I_0(X^2) \tag{12.18a}$$

where $X = (V\xi/2\kappa)^{1/2}$ and I_0 is a modified Bessel function.† At all points

---

† See Abramowitz & Stegun, *Handbook of Mathematical Functions*, Dover (1965) for definitions and tabulation of modified Bessel functions I_0 and I_1, and also for integration of equation (12.18a).

Fig. 12.3. Thermoelastic distortion of a half-space heated over a circular area radius a. Curve A – uniform-heat input, eqs. (12.16); curve B – uniform temperature, eqs. (12.17).

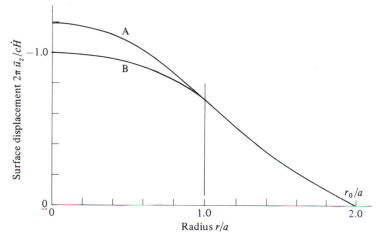

behind the source the displacement is constant and given by

$$\bar{u}_z = -2c\kappa\dot{H}/V \qquad\qquad (12.18b)$$

The thermal distortion due to a uniform rate of heating $\dot{h}$ over the strip $-a \leqslant x \leqslant a$ can be found by superposition. The displacement $\bar{u}_z(x)$ of a surface point is found by putting $\xi = s - x$ in equation (12.18a) when ξ is positive,

Fig. 12.4. Thermoelastic distortion of a half-space caused by a moving heat source.

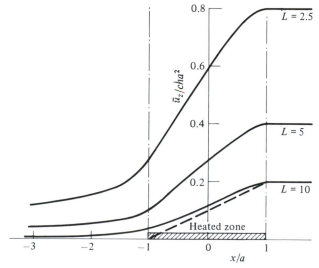

i.e. when the point in question is ahead of the element $\dot{h}\,ds$ of the heat source, and by using equation (12.18b) when ξ is negative. In this way the distorted surface shapes shown in Fig. 12.4 have been found as a function of the Peclet number L defined above. At the leading edge of the heat source ($x = -a$) the thermal displacement varies with speed according to

$$\bar{u}_z(-a) = -2c\dot{h}a^2\,e^{-2L}\{I_0(2L) + I_1(2L)\}/L \tag{12.19a}$$

$$\approx -2c\dot{h}a^2(\pi L^3)^{-1/2} \tag{12.19b}$$

for large values of L. At the trailing edge the displacement is

$$\bar{u}_z(a) = -2c\dot{h}a^2/L \tag{12.19c}$$

With increasing speed conduction ahead of the source becomes less effective so that the displacement at the leading edge falls. At high speed the displacement increases almost in proportion to the distance from the leading edge.

12.4 Contact between bodies at different temperatures

We shall consider first the situation in which a hot body, at temperature θ_1, is pressed into contact with a cooler one at temperature θ_2, and will restrict the discussion to frictionless surfaces and to a circular contact area of radius a.

If we assume, in the first instance, that the bodies make perfect thermal contact at their interface and that each body is effectively a half-space, then the heat conduction problem is straightforward. The temperature of the interface θ_c is uniform and the distribution of heat flux across the interface is given by equation (12.6), i.e.

$$\dot{h}(r) = \frac{2k_1(\theta_1 - \theta_c)}{\pi(a^2 - r^2)^{1/2}} = \frac{2k_2(\theta_c - \theta_2)}{\pi(a^2 - r^2)^{1/2}}$$

from which it may be seen that θ_c divides the difference in temperature between the two bodies in the inverse proportion to their thermal conductivities k_1 and k_2. The total heat flux is thus

$$\dot{H} = 4k_1a(\theta_1 - \theta_c) = 4k_2a(\theta_c - \theta_2) = 4ka(\theta_1 - \theta_2) \tag{12.20}$$

where $k = k_1k_2/(k_1 + k_2)$.

The distortion of the surface of an elastic half-space by this distribution of heat flux is given by equations (12.17). Since heat is flowing into the cooler body it develops a bulge in which the displacements are proportional to the value of its distortivity $c_2\,(=\alpha_2(1 + \nu_2)/k_2)$, while the warmer body develops a hollow of similar shape whose depth is proportional to c_1. If $c_2 = c_1$ the bulge in the cooler body will just fit the hollow in the warmer body and the contact stresses due to the external load will not be influenced by the existence of the heat flux.

This is clearly the case with identical materials. When the two materials are different the heat flux will give rise to an additional ('thermal') contact pressure such as to suppress the mismatch in the distorted profiles of the two surfaces. The required pressure, acting over the circle $r \leqslant a$, is that which (by equation (12.17)) would produce combined displacements of both surfaces given by

$$(\bar{u}_z)_1 + (\bar{u}_z)_2 = \frac{1}{2\pi}(c_2 - c_1)\dot{H}$$

$$\times \left[\ln(r_0/a) - \ln\{1 + (1 - r^2/a^2)^{1/2}\} + (1 - r^2/a^2)^{1/2}\right] \tag{12.21}$$

It can be found by the methods of Chapter 3, with the result (Barber, 1973)

$$p'(r) = (c_2 - c_1)\frac{\dot{H}E^*}{2\pi^2 a}\left[\frac{\pi^2}{8} - \chi_2\left\{\frac{a - (a^2 - r^2)^{1/2}}{a + (a^2 - r^2)^{1/2}}\right\}\right] \tag{12.22}$$

where $\chi_2(x)$ is Legendre's chi function†, which is defined by

$$\chi_2(x) = \frac{1}{2}\int_0^x \ln\left(\frac{1+s}{1-s}\right)\frac{ds}{s} = \sum_{m=1}^{m=\infty} \frac{x^{2m-1}}{(2m-1)^2}$$

This pressure distribution is plotted in Fig. 12.5. It corresponds to a total load

$$P' = (1/2\pi)(c_2 - c_1)\dot{H}E^*a = (2/\pi)k(c_2 - c_1)(\theta_1 - \theta_2)E^*a^2 \tag{12.23}$$

The differential thermal expansion of the two bodies (12.21) has been annulled by the application of the 'thermal' pressure distribution given by equation (12.22). To find the net pressure we must add a Hertz contact pressure to account for the isothermal elastic deformation, viz.:

$$p''(r) = (2E^*a/\pi R)\{1 - (r/a)^2\}^{1/2} \tag{12.24}$$

and the corresponding load

$$P'' = 4E^*a^3/3R \tag{12.25}$$

The total load P, thermal plus isothermal, is thus $P' + P''$, which leads to the relationship:

$$\beta(a/a_0)^2 + (a/a_0)^3 = 1 \tag{12.26}$$

where $a_0 = (3RP/4E^*)^{1/3}$ is the contact radius under isothermal (Hertz) conditions and $\beta = (3kR/2\pi a_0)(c_2 - c_1)(\theta_1 - \theta_2)$. This relationship for $\beta > 0$ is plotted on the right-hand side of Fig. 12.6. As expected an increase in temperature difference or in differential distortivity causes the relative curvature of the two surfaces to increase and the contact area to decrease. In the case of two

† L. Lewin, *Dilogarithms and Associated Functions*, MacDonald, London, 1958.

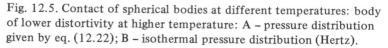

Fig. 12.5. Contact of spherical bodies at different temperatures: body of lower distortivity at higher temperature: A – pressure distribution given by eq. (12.22); B – isothermal pressure distribution (Hertz).

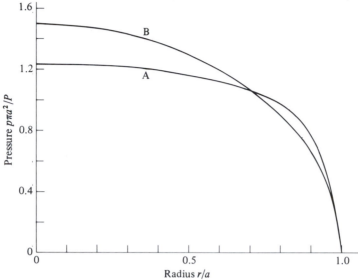

Fig. 12.6. Contact of spherical bodies at different temperatures. Exact solutions with annulus of imperfect contact $b < r < a$. Broken line – approximate solution (eq. (12.26)) assuming perfect contact throughout.

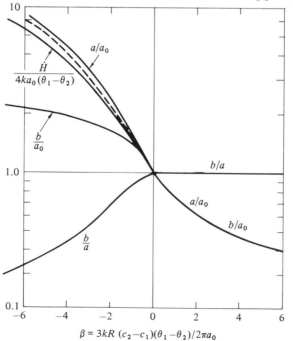

nominally flat surfaces ($R \rightarrow \infty$) the isothermal component of pressure is zero. Any slight departure from flatness will cause the heat transfer to be concentrated at the point of closest approach of the two surfaces. Consequent thermal expansion will cause a circular area of intimate contact to develop and the surrounding surfaces to separate. The size of the contact zone is then given by equation (12.23).

The above analysis of the contact of two dissimilar bodies whose temperatures are different is perfectly satisfactory provided that the product $(c_2 - c_1) \times (\theta_1 - \theta_2)$ is positive. If the situation is changed, such that the body of higher distortivity has the higher temperature, a new feature is introduced through the change in sign of the thermal pressure $p'(r)$. It may be seen from Fig. 12.5 that this pressure distribution falls to zero very steeply as $r \rightarrow a$ compared with the Hertz pressure. It can be shown that $p'(r)$ always exceeds $p''(r)$ in this region so that, when a negative (tensile) thermal pressure is added to the isothermal pressure, an annulus of *tensile* traction is found, however small the temperature difference between the surfaces. This suggests that the surfaces would peel apart at the edge, but it is not possible for them to do so and to still maintain equilibrium with the applied load. It must be concluded that there is no solution to the problem in the form posed above. Barber (1978) has shown that the paradox arises from the thermal boundary conditions which assume 'perfect contact' within the contact area, i.e. no discontinuity in temperature across the interface, and perfect insulation outside the contact. The difficulty can be removed by the introduction of an additional state of 'imperfect contact' in which the displacements (elastic and thermal) are such that the surfaces just touch and conduct some heat, but the contact pressure is zero. This state is achieved by a jump in temperature across the interface. These boundary conditions follow from the fact that, in reality, the change from perfect insulation to perfect conduction will not be discontinuous when surfaces come into contact. When the separation is sufficiently small heat can be transferred by radiation or conduction through the intervening gas; further, contaminant films and the inevitable roughness of real surfaces give rise to a thermal resistance at the interface which might be expected to depend inversely upon the contact pressure. The argument may be appreciated in its simplest form by reference to the one-dimensional model shown in Fig. 12.7(a). A rod of length l, Young's modulus E and coefficient of thermal expansion α is placed between two rigid conducting walls A and B, at temperatures θ_A and θ_B. The rod is attached to A and initially, when $\theta_A = \theta_B$, there is a small gap $g = g_0$ between the end of the rod and wall B. If θ_A is raised above θ_B, in the steady state the rod will acquire the temperature θ_A and the gap will close such that

$$g = g_0 - \alpha l(\theta_A - \theta_B) \tag{12.27}$$

This expression is only valid if $g > 0$, i.e. if

$$\theta_A - \theta_B \leqslant g_0/\alpha l \qquad (12.28)$$

If the rod expands to make 'perfect contact' with the wall B, so that its temperature varies linearly from θ_A to θ_B, its unrestrained expansion would be $\frac{1}{2}\alpha l(\theta_A - \theta_B)$, but the actual expansion cannot exceed g_0, so that a pressure will develop on the end of the rod given by

$$p = \frac{1}{2}E\alpha(\theta_A - \theta_B) - Eg_0/l \qquad (12.29)$$

The contact pressure must be positive so that

$$\theta_A - \theta_B \geqslant 2g_0/\alpha l \qquad (12.30)$$

There is thus a range of temperature difference

$$g_0/\alpha l < (\theta_A - \theta_B) < 2g_0/\alpha l$$

in which no steady-state solution is possible. This is a similar state of affairs to

Fig. 12.7. Contact of an elastic rod between two rigid walls at different temperatures ($\theta_A > \theta_B$). (*a*) The system. (*b*) The thermal resistance $f(R)$ as a function of the gap g or the contact pressure p.

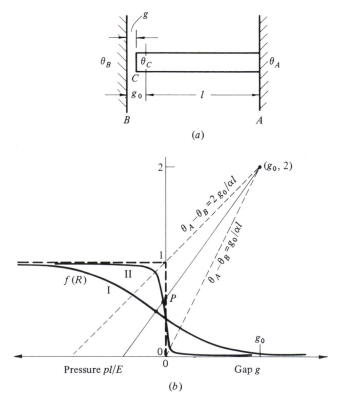

(*a*)

(*b*)

that encountered in the contact of spheres discussed above. To resolve the paradox we introduce a thermal resistance $R(g)$ which varies continuously with the gap, becoming very large as g becomes large. As we have seen negative gaps cannot exist and we should replace the unrestrained 'gap' $-g$ by a contact pressure $p = -Eg/l$. The resistance $R(p)$ will decrease as p increases. The temperature of the free end of the rod is denoted by θ_C so that, by equating the heat flux along the rod to that across the gap, we get

$$Sk(\theta_A - \theta_C)/l = (\theta_C - \theta_B)/R$$

where S is the cross-sectional area of the rod. Thus

$$\theta_A - \theta_C = (\theta_A - \theta_B)f(R) \tag{12.31}$$

where $f(R) = (1 + SkR/l)^{-1}$. This function is represented by curve I in Fig. 12.7(b). For large positive gaps R is large, hence $f(R)$ approaches zero; at high contact pressure (negative unrestrained gap), R is small and $f(R)$ approaches unity, but its precise form is unimportant.

The expression for the gap now becomes

$$g = g_0 - \tfrac{1}{2}\alpha l(\theta_A + \theta_C - 2\theta_B)$$
$$= g_0 - \alpha l(\theta_A - \theta_B) + \tfrac{1}{2}\alpha l(\theta_A - \theta_C) \tag{12.32}$$

Eliminating $(\theta_A - \theta_C)$ from equations (12.31) and (12.32) we find

$$f(R) = 2 + 2(g - g_0)/\alpha l(\theta_A - \theta_B) \tag{12.33}$$

This equation plots in Fig. 12.7(b) as a straight line which passes through the point $(g_0, 2)$, and whose gradient is inversely proportional to the temperature difference $(\theta_A - \theta_B)$. Where the line intersects the curve of $f(R)$ gives the steady solution to the problem: it determines the gap g if the point of intersection is to the right of 0 and the pressure p if the point of intersection is to the left of 0. Note that a point of intersection exists for a line of any gradient, hence a solution can be found for all values of $(\theta_A - \theta_B)$.

If we now make the resistance curve $f(R)$ increasingly sensitive to the gap and the contact pressure, as shown by curve II in Fig. 12.7, in the limit it takes the form of a 'step', zero to the right of 0 and unity to the left. More significantly it has a vertical segment between 0 and 1 when $g = 0$. An intersection with the straight line given by equation (12.33) is still possible in the range $g_0/\alpha l < (\theta_A - \theta_B) < 2g_0/\alpha l$, as indicated by the point P in Fig. 12.7(b). Both the gap and the contact pressure are zero at this point; the temperature of the end of the rod θ_C is intermediate between θ_A and θ_B, given by putting $g = 0$ in equation (12.32), and some heat flows across the interface. These are the boundary conditions referred to by Barber (1978) as 'imperfect contact' and investigated further by Comninou & Dundurs (1979).

Returning to the contact of spheres when the heat flow is such that β is negative, the existence of tensile stresses as $r \to a$ when perfect contact is assumed suggests that the contact area will be divided into a central region ($r \leqslant b$) of perfect contact surrounded by an annulus ($b < r \leqslant a$) of imperfect contact. Barber (1978) has analysed this situation with results which are shown in Fig. 12.6 for negative values of β. The variation in contact radius (a/a_0) given by equation (12.26), which assumes perfect contact throughout, is also shown for comparison. With increasing (negative) temperature difference the contact size grows as the thermal distortion makes the surfaces more conforming. The exact variation is not very different from that predicted by equation (12.26). The radius b of the circle of perfect contact, within which the contact pressure is confined, also grows but more slowly. It is shown in Fig. 12.6. as a ratio of the isothermal contact radius a_0 and also as a ratio of the actual radius a. The mean contact pressure falls, therefore, but not to the extent which would be expected if perfect contact prevailed throughout. With perfect contact the heat flux through the contact $\dot{H}$ is proportional to the contact radius a, so that the influence of thermal distortion on $\dot{H}$ is expressed by the approximate curve of a/a_0 against β given by equation (12.26) and shown dotted in Fig. 12.6. The exact variation of heat flux is also shown. The effect of an annulus of imperfect contact upon the heat flux is not large; the reduction in conductivity of the interface is offset to some extent by the increase in the size of the contact. The analogous problems of two-dimensional contact of cylindrical bodies and of nominally flat wavy surfaces have been solved by Comninou *et al.* (1981) and Panek & Dundurs (1979).

When contact is made between a flat rigid punch and an elastic half-space which is hotter than the punch, at first sight a hollow would be expected to form in the half-space so that contact would be lost from the centre of the punch. This cannot happen, however, since by equation (12.12b) the surface can only become concave if heat is flowing from it, whereas no heat flows if there is no contact. This is another situation, investigated by Barber (1982), in which a state of imperfect contact exists, this time in a central region of the punch.

A basic feature of Fig. 12.6 calls for comment: for a given temperature difference between the bodies, the heat transfer from the body of lower distortivity into that of higher distortivity ($\beta < 0$) is greater than the heat transfer in the opposite direction ($\beta > 0$). This phenomenon has been called 'thermal rectification' and is frequently observed when heat is transferred between dissimilar solids in contact. The above theory, with modifications to allow for the geometry of the experimental arrangement, has shown reasonable agreement with measurements of heat transfer between rods having rounded ends in contact (see Barber, 1971b).

12.5   Frictional heating and thermoelastic instability

In the sliding contact of nominally flat surfaces heat is liberated by friction at the interface at a rate

$$\dot{h} = \mu V p \tag{12.34}$$

where V is the sliding velocity and μ the coefficient of friction. If the pressure p is uniform then the heat conducted to the surfaces will be uniform and so will the surface temperature. It has been frequently observed with brake blocks, for example, that the stationary surface develops 'hot spots' where the temperature is much in excess of its expected mean value. This phenomenon was investigated by Barber (1969). He showed that initial small departures from perfect conformity concentrated the pressure and hence the frictional heating into particular regions of the interface. These regions expanded above the level of the surrounding surface and reduced the area of real contact, as described in the previous section, thereby concentrating the contact and elevating the local temperature still further. This process has come to be called 'thermoelastic instability' and has been studied in detail by Burton (1980). If sliding continues the expanded spots, where the pressure is concentrated, wear down until contact occurs elsewhere. The new contact spots proceed to heat, expand and carry the load; the old ones, relieved of load, cool, contract and separate. This cyclic process has been frequently observed in the sliding contact of conforming surfaces. The scale of the hot spots is large compared with the scale of surface roughness and the time of the cycle is long compared with the time of asperity interactions. The essential mechanism of thermoelastic instability may be appreciated by the simple example considered below.

Two semi-infinite sliding solids having nominally flat surfaces, which are pressed into contact with a mean pressure $\bar{p}$, are shown in Fig. 12.8. To avoid the transient nature of heat flow into a moving surface, the moving surface will be taken to be perfectly flat, and non-conducting. The stationary solid has a distortivity c and its surface has a *small* initial undulation of amplitude Δ and wavelength λ. In the present example, where the mating surface is non-conducting, it is immaterial whether the undulations are parallel or perpendicular to the direction of sliding. The isothermal pressure required to flatten this waviness is found in Chapter 13 (eq. (13.7)), to be

$$p'' = (\pi E^* \Delta / \lambda) \cos (2\pi x / \lambda) \tag{12.35}$$

The steady thermal distortion of the surface is given by

$$\frac{d^2 \bar{u}_z}{dx^2} = c\dot{h} = c\mu V p(x) \tag{12.36}$$

It is clear that the initial sinusoidal undulation of wavelength λ is going to result in a fluctuation of pressure at the same wavelength, which may be expressed by

$$p(x) = \bar{p} + p^* \cos(2\pi x/\lambda) \qquad (12.37)$$

We are concerned here only with the fluctuating components of pressure and heat flux which, when substituted in equation (12.36) and integrated, give the thermal distortion of the surface to be

$$\bar{u}_z = -(c\mu V p^* \lambda^2/4\pi^2) \cos(2\pi x/\lambda) \qquad (12.38)$$

The thermal pressure $p'(x)$ required to press this wave flat can now be added to the isothermal pressure given by (12.35) to obtain the relationship

$$p^* = \frac{\pi E^*}{\lambda} (\Delta + c\mu V p^* \lambda^2/4\pi^2)$$

whereupon

$$\frac{p^*}{\bar{p}} = \frac{\pi E^* \Delta/\lambda \bar{p}}{1 - c\mu V E^* \lambda/4\pi} \qquad (12.39)$$

As the sliding velocity approaches a critical value V_c given by

$$V_c = 4\pi/c\mu E^* \lambda \qquad (12.40)$$

the fluctuations in pressure given by equation (12.39) increase rapidly in magnitude (Fig. 12.8(c)).

When the fluctuation in pressure p^* reaches the mean pressure $\bar{p}$ the surfaces will separate in the hollows of the original undulations and the contact will concentrate at the crests (Fig. 12.8(d)). A simple treatment of this situation

Fig. 12.8. Mechanism of thermoelastic instability. Thermal expansion causes small initial pressure fluctuations to grow when the sliding speed approaches a critical value V_c. At high speed contact becomes discontinuous which further increases the non-uniformity of pressure.

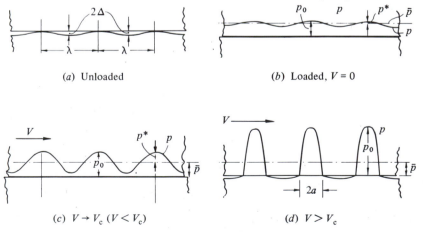

(a) Unloaded

(b) Loaded, $V = 0$

(c) $V \rightarrow V_c$ ($V < V_c$)

(d) $V > V_c$

may be carried out by assuming that the pressure in the contact patch, $-a \leqslant x \leqslant +a$, is Hertzian, i.e. $p(x) = p_0 \{1 - (x/a)^2\}^{1/2}$, where

$$p_0 = aE^*/2R \tag{12.41}$$

The curvature of $1/R$ of the distorted surface at $x = 0$ is given by

$$1/R = c\mu V p_0 + 4\pi^2 \Delta/\lambda^2 \tag{12.42a}$$

$$\approx c\mu V p_0 \tag{12.42b}$$

if the initial undulation is small compared with the subsequent thermal distortion. Thus equation (12.42) gives†

$$a \approx 2/c\mu VE^* \tag{12.43}$$

The transition from continuous to discontinuous contact at the interface takes place when V approaches V_c given by (12.40). Putting $V = V_c$ in (12.43) then gives an approximate expression for the critical contact size:

$$a_c \approx \lambda/2\pi \tag{12.44}$$

The non-uniform pressure distribution leads directly to non-uniform heat input and to a non-uniform distribution of surface temperature. The temperature distribution can be found using the analogy with the surface displacements produced by a pressure which is proportional to the heat flux at the surface. Below the critical speed, while the surfaces are in continuous contact, the pressure fluctuations are sinusoidal with an amplitude p^* given by equation (12.39). It follows that the fluctuations in heat flux and temperature will also be sinusoidal with amplitudes h^* and θ^*. From the analogy mentioned above we find

$$\theta^* = \lambda h^*/2\pi k = \mu V\lambda p^*/2\pi k \tag{12.45}$$

Above the critical speed, the surfaces are in discontinuous contact. The surface displacements and contact pressures where a wavy surface is in discontinuous contact with a plane are given in §13.2. From those results it may be deduced that the temperature difference between the centre of a contact patch and the centre of the trough is given by

$$\bar{\theta}(0) - \bar{\theta}(\lambda/2) \approx \frac{\mu V P}{\pi k} \frac{\sin \psi}{\psi} \left\{ \frac{1 - \cos \psi}{\sin^2 \psi} + \ln \left(\frac{1 + \cos \psi}{\sin \psi} \right) \right\} \tag{12.46}$$

where $\psi = \pi a/\lambda$.

The mechanism of thermoelastic instability may now be described with reference to the above example (Figs. 12.8 and 12.9). In static contact any waviness of the surfaces in contact will give rise to a non-uniform distribution

---

† A more exact treatment which matches the pressure and distortion throughout the contact patch has been carried out by Burton & Nerlikar (1975) for multiple contacts; for a single contact patch Barber (1976) finds $a = 2.32/c\mu VE^*$.

of contact pressure. At low sliding speeds the variations in pressure from the steady mean value are augmented by thermoelastic distortion according to equation (12.38). When the velocity reaches a critical value V_c given by equation (12.40) the amplitude of the fluctuation increases very rapidly and, if they have not already done so, the surfaces separate at the positions of the initial hollows. Contact is discontinuous and the size of the contact patches shrinks to a width about 1/3 of the original wavelength (eq. (12.44)). Further increase in speed results in a stable decrease in the contact patch size according to equation (12.43). The sudden rise in pressure and drop in contact area at the critical speed are accompanied by a sharp rise in temperature fluctuation, as shown in Fig. 12.9.

Real surfaces, of course, will have a spectrum of initial undulations. Equation (12.39) suggests that the pressure variation grows in proportion to the ratio Δ/λ, i.e. to the slope of the undulations. The critical velocity V_c, however, is independent of the amplitude and inversely proportional to the wavelength. This suggests that long wavelength undulations will become unstable before the short ones and thereby dominate the process. The size of the body imposes an upper limit to the wavelength and hence a lower limit to the critical speed. The undulations in real surfaces are two-dimensional having curvature in both directions. Following the onset of instability, the same reasoning that led to

Fig. 12.9. Variation of the contact width a and the amplitude of pressure and temperature fluctuations with sliding speed.

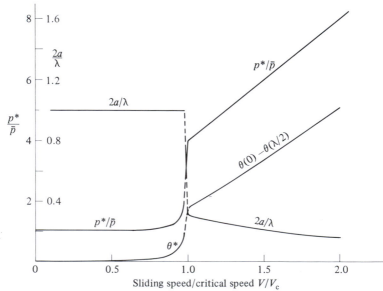

equation (12.43) gives the radius of a discrete circular contact area to be†

$$a \approx \pi/c\mu VE^*$$ (12.47)

Another way of expressing the influence of the size of the body is to say that, if the nominal contact area has a diameter less than $2a$ given by (12.47), the situation will be stable.

The above analysis simplifies the real situation in two important ways: (i) the thermoelastic solutions employed refer to the steady state, whereas the unstable variation in contact pressure and area is essentially a transient process, and (ii) both surfaces will be conducting and deformable to a greater or lesser extent. To investigate these effects Dow & Burton (1972) and Burton *et al.* (1973) have studied the stability of small sinusoidal perturbations in pressure between two extended sliding surfaces in continuous contact. The equation of unsteady heat flow was used. They show first that a pair of identical materials is very stable; however high the sliding speed an impractical value of the coefficient of friction (>2) would be required to cause instability. When the two materials are different a thermal disturbance, comprising a fluctuation in pressure and temperature, moves along the interface at a velocity which is different from that of either surface. An appreciable difference in the thermal conductivities of the two materials, however, leads to the disturbance being effectively locked to the body of higher conductivity; most of the heat then passes into that surface. In the limit we have the situation analysed above where one surface is non-conducting. The critical velocity then approaches that given by equation (12.40). Some heat is, in fact, conducted to the mating surface, at a rate given by equation (12.10) which reduces the heat causing thermoelastic deformation of the more conducting surface and thereby increases the critical velocity above that given by (12.40).

When the contact is discontinuous the analysis of transient thermoelasticity becomes more difficult. Some basic cases of the distortion of a half-space due to transient heating of a small area of the surface have been investigated by Barber (1972) and Barber & Martin-Moran (1982). These results have been used to investigate the transient shrinking of a circular contact area due to frictional heating when the moving surface is an insulator. The stationary conducting surface is assumed to have a slight crown so that before sliding begins there is an initial contact area of radius a_0. During sliding, in the steady state, the contact area shrinks to a radius a_∞. In this analysis the simplifying assumptions which we have used previously are applied: the pressure distribution is Hertzian and the curvature due to thermoelastic distortion is matched at the

---

† More exactly, for a single contact patch, Barber (1976) obtains $a = 1.28\pi/c\mu VE^*$

origin only. With these assumptions a_∞ is given by equation (12.47). Barber (1980*b*) shows that the contact radius shrinks initially at a uniform rate $1.34\kappa/a_\infty$. Only in the later stages is a_∞ approached asymptotically, as shown in Fig. 12.10.

Fig. 12.10. Transient thermoelastic variations of the radius of a circular area in sliding contact from its initial value a_0 to its steady-state value a_∞.

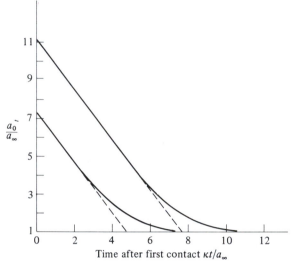

Time after first contact $\kappa t/a_\infty$

13

Rough surfaces

13.1   Real and apparent contact

It has been tacitly assumed so far in this book that the surfaces of contacting bodies are topographically smooth; that the actual surfaces follow precisely the gently curving nominal profiles discussed in Chapters 1 and 4. In consequence contact between them is continuous within the nominal contact area and absent outside it. In reality such circumstances are extremely rare. Mica can be cleaved along atomic planes to give an atomically smooth surface and two such surfaces have been used to obtain perfect contact under laboratory conditions. The asperities on the surface of very compliant solids such as soft rubber, if sufficiently small, may be squashed flat elastically by the contact pressure, so that perfect contact is obtained throughout the nominal contact area. In general, however, contact between solid surfaces is discontinuous and the *real* area of contact is a small fraction of the *nominal* contact area. Nor is it easy to flatten initially rough surfaces by plastic deformation of the asperities. For example the serrations produced by a lathe tool in the nominally flat ends of a ductile compression specimen will be crushed plastically by the hard flat platens of the testing machine. They will behave like plastic wedges (§6.2(c)) and deform plastically at a contact pressure $\approx 3Y$ where Y is the yield strength of the material. The specimen as a whole will yield in bulk at a nominal pressure of Y. Hence the maximum ratio of the real area of contact between the platen and the specimen to the nominal area is about $\frac{1}{3}$. Strain hardening of the crushed asperities will decrease this ratio further.

We are concerned in this chapter with the effect of surface roughness and discontinuous contact on the results of conventional contact theory which have been derived on the basis of smooth surface profiles in continuous contact.

Most real surfaces, for example those produced by grinding, are not regular: the heights and the wavelengths of the surface asperities vary in a random way.

A machined surface as produced by a lathe has a regular structure associated with the depth of cut and feed rate, but the heights of the ridges will still show some statistical variation. Most man-made surfaces such as those produced by grinding or machining have a pronounced 'lay', which may be modelled to a first approximation by one-dimensional roughness. It is not easy to produce a wholly isotropic roughness. The usual procedure for experimental purposes is to air-blast a metal surface with a cloud of fine particles, in the manner of shot-peening, which gives rise to a randomly cratered surface. Before discussing random rough surfaces, however, we shall consider the contact of regular wavy surfaces.

The simplest model of a rough surface is a regular wavy surface which has a sinusoidal profile. Provided that the amplitude Δ is small compared with the wavelength λ so that the deformation remains elastic, the contact of such a surface with an elastic half-space can be analysed by the methods of Chapters 2 and 3.

13.2 Contact of regular wavy surfaces

(a) One-dimensional wavy surface

We will start by considering an elastic half-space subjected to a sinusoidal surface traction

$$p = p^* \cos(2\pi x/\lambda) \tag{13.1}$$

which alternately pushes the surface down and pulls it up. The normal displacements of the surface under this traction can be found by substituting (13.1) into equation (2.25*b*), i.e.

$$\frac{\partial \bar{u}_z}{\partial x} = -\frac{2(1-\nu^2)}{\pi E} \int_{-\infty}^{\infty} \frac{p^* \cos(2\pi s/\lambda)}{x-s} \, ds$$

$$= -\frac{2(1-\nu^2)}{\pi E} p^* \int_{-\infty}^{\infty} \frac{\cos\{2\pi(x-\xi)/\lambda\}}{\xi} \, d\xi$$

Expanding the numerator and integrating gives

$$\frac{\partial \bar{u}_z}{\partial x} = -\frac{2(1-\nu^2)}{E} p^* \sin(2\pi x/\lambda) \tag{13.2}$$

or

$$\bar{u}_z = \frac{(1-\nu^2)\lambda}{\pi E} p^* \cos(2\pi x/\lambda) + \text{const.} \tag{13.3}$$

Not surprisingly the sinusoidal variation in traction produces a sinusoidal surface of the same wavelength.

The stresses within the solid may be found by superposition of the stresses under a line load (eq. (2.23)) or, more directly, by equations (2.6) from the stress function

$$\phi(x, z) = (p^*/\alpha^2)(1 + \alpha z)\, e^{-\alpha z} \cos \alpha x \tag{13.4}$$

where $\alpha = 2\pi/\lambda$. The maximum principal shear stress τ_1 occurs at a depth $z = \lambda/2\pi$ beneath the points of maximum traction $(x = n\pi)$. Its value is p^*/e.

If an elastic half-space with a flat surface is brought into contact with an elastic solid, whose nominally flat surface before loading has a one-dimensional wave of small amplitude Δ and wavelength λ, the gap between the surfaces may be expressed by† (see Fig. 13.1(*a*))

$$h(x) = \Delta\{1 - \cos(2\pi x/\lambda)\} \tag{13.5}$$

If the surfaces are now pressed into contact by a mean pressure $\bar{p}$ sufficient to compress the wave completely so that the surfaces are in continuous contact, the pressure distribution can be found from the above results. The elastic displacements of the surfaces are such that

$$(\bar{u}_z)_1 + (\bar{u}_z)_2 = \delta - h(x) \tag{13.6}$$

where δ is the approach of datum points in each body. Equation (13.3) shows that this equation is satisfied by a pressure distribution of the form

$$p(x) = \bar{p} + p^* \cos(2\pi x/\lambda) \tag{13.7}$$

where $p^* = \pi E^* \Delta/\lambda$, since the uniform pressure $\bar{p}$ produces a uniform displacement. For contact to be continuous the pressure must be positive everywhere so that $\bar{p} \geqslant p^*$ (see Fig. 13.1(*b*)).

If the mean pressure is less than p^* there will not be continuous contact between the two surfaces. They will make contact in parallel strips of width $2a$ located at the crests of the undulations and will separate in the troughs (see Fig. 13.1(*c*)). Westergaard (1939) has shown that a pressure distribution:

$$p(x) = \frac{2\bar{p}\,\cos(\pi x/\lambda)}{\sin^2(\pi a/\lambda)}\{\sin^2(\pi a/\lambda) - \sin^2(\pi x/\lambda)\}^{1/2} \tag{13.8}$$

acting on an elastic half-space produces normal surface displacements:

$$\bar{u}_z(x) = \frac{(1 - \nu^2)\bar{p}\lambda}{\pi E \sin^2 \psi_a} \cos 2\psi + C, \quad 0 \leqslant |x| \leqslant a \tag{13.9a}$$

---

† The same form of expression for the undeformed gap is obtained if *both* surfaces have parallel undulations of the same wavelength, even though the undulations are displaced in phase so that initial contact does not occur at the crests.

$$\bar{u}_z(x) = \frac{(1-\nu^2)\bar{p}\lambda}{\pi E \sin^2 \psi_a} \left[\cos 2\psi + 2 \sin \psi \, (\sin^2 \psi - \sin^2 \psi_a)^{1/2} \right.$$

$$\left. - 2 \sin^2 \psi_a \ln \left\{ \frac{\sin \psi + (\sin^2 \psi - \sin^2 \psi_a)^{1/2}}{\sin \psi_a} \right\} \right] + C,$$

$$a \leqslant |x| \leqslant \lambda/2 ,$$ (13.9b)

where $\bar{p}$ is the mean pressure, $\psi = \pi x/\lambda$, $\psi_a = \pi a/\lambda$, and C is a constant determined by the datum chosen for displacements. The displacements of (13.9a) satisfy the contact condition (13.6) for $|x| \leqslant a$ provided that

$$\bar{p} = (\pi E^* \Delta/\lambda) \sin^2 (\pi a/\lambda)$$ (13.10)

and the displacements outside the contact strip ($a \leqslant |x| \leqslant \lambda/2$) are such that the gap remains positive. Noting from equation (13.7) that $(\pi E^* \Delta/\lambda) = p^*$, equation (13.10) can be inverted to express the ratio of the 'real' to the 'apparent' area of contact, i.e.

$$2a/\lambda = (2/\pi) \sin^{-1}(\bar{p}/p^*)^{1/2}$$ (13.11)

This relationship is plotted in Fig. 13.2. When $\bar{p} \ll p^*$, then $2a \ll \lambda$, and the compression of the crest of each wave should be independent of the other waves,

Fig. 13.1. Contact of a one-dimensional wavy surface with an elastic half-space. (*a*) Unloaded ($\bar{p} = 0$), (*b*) Complete contact ($\bar{p} = p^*$), (*c*) Partial contact ($\bar{p} < p^*$).

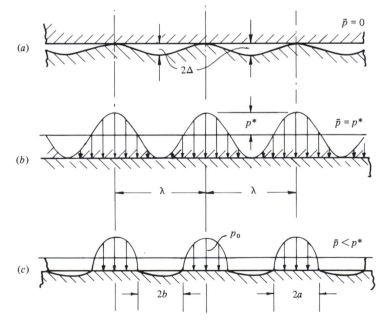

so that the Hertz theory might be applied. The load P carried by each crest is $\bar{p}\lambda$ and the curvature $1/R$ of each crest is $4\pi^2 \Delta/\lambda^2$. Substituting these values in the Hertz equation (4.43) for line contact gives

$$2a/\lambda = (2/\pi)(\bar{p}/p^*)^{1/2} \tag{13.12}$$

which is the limit of equation (13.11) for $\bar{p} \ll p^*$. It is shown dotted in Fig. 13.2.

At the other limit when $\bar{p} \to p^*$ only a small strip of width $2b$ ($\ll \lambda$) remains out of contact, where $b = \lambda/2 - a$. An asymptotic expression for b can be found by regarding the non-contact zone as a pressurised crack of length $2b$ in an infinite solid. The contact pressure in the non-contact zone is, of course, zero, but it can be thought of as the superposition of the pressure necessary to maintain the surfaces in contact, given by (13.7), and an equal negative pressure acting on the surface of the 'crack' ($a \leqslant x \leqslant \lambda - a$). Provided $b \ll \lambda/2$ the pressure within the crack may be written

$$p(x') \approx 2\pi^2 (x'/\lambda)^2 p^* - (p^* - \bar{p}) \tag{13.13}$$

where $x' = x - \lambda/2$. Since the interface has no strength it will open until the stress intensity factor at its ends falls to zero. The stress intensity factor at the ends of a pressurised crack of length $2b$ is given by (see Paris & Sim, 1965)

$$K_{\mathrm{I}} = (\pi b)^{-1/2} \int_{-b}^{b} p(x')\{(b + x')/(b - x')\}^{1/2}\, dx' \tag{13.14}$$

Fig. 13.2. Real area of contact of a one-dimensional wavy surface with an elastic half-space. Solid line – exact, eq. (13.11); broken line – asymptotic, eqs. (13.12) and (13.15).

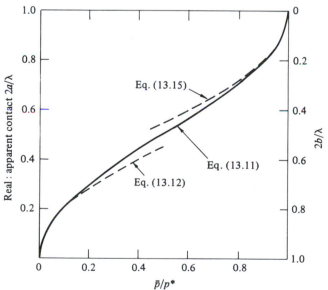

Substituting $p(x')$ from (13.13), integrating and equating K_I to zero give the length of the no-contact zone to be

$$2b/\lambda = (2/\pi)(1 - \bar{p}/p^*)^{1/2} \tag{13.15}$$

which is the limit of equation (13.11) as $\bar{p} \to p^*$. It is also plotted in Fig. 13.2, where it may be seen that the asymptotic equations (13.12) and (13.15) provide close bounds on the exact result. As we shall see below, in the case of a two-dimensional wavy surface, only the asymptotic results can be obtained in closed form.

(b) Two-dimensional waviness

The gap between a flat surface and one which has a regular orthogonal waviness can be expressed by

$$h(x, y) = \Delta_1 + \Delta_2 - \Delta_1 \cos(2\pi x/\lambda_1) - \Delta_2 \cos(2\pi y/\lambda_2) \tag{13.16}$$

The surfaces touch at the corners of a rectangular grid of mesh size $\lambda_1 \times \lambda_2$; the maximum gap, which coincides with a hollow in the surface, has a depth $2(\Delta_1 + \Delta_2)$ at the mid-point of the rectangle. After compression the elastic displacements, in the area over which the surfaces are in contact, are given by substituting (13.16) in (13.6). The pressure to make contact over the whole surface can now be found by superposition of the pressures necessary to compress each of the component waves taken separately. Thus by equation (13.7) we have

$$p(x, y) = \bar{p} + p_x^* \cos(2\pi x/\lambda_1) + p_y^* \cos(2\pi y/\lambda_2) \tag{13.17}$$

where $p_x^* = \pi E^* \Delta_1/\lambda_1$ and $p_y^* = \pi E^* \Delta_2/\lambda_2$. To maintain contact everywhere $\bar{p} \geqslant p_x^* + p_y^*$. When the mean pressure is less than this value the interface will comprise areas of contact and separation. At low pressures the contact area at a crest will be elliptical and given by Hertz theory. At high pressures the small area of separation will also be elliptical and can be found by modelling it as a pressurised crack. In between, the shape of the contact area will not be elliptical and a closed-form solution seems improbable.

For an isotropic wavy surface $\Delta_1 = \Delta_2 = \Delta$, $\lambda_1 = \lambda_2 = \lambda$ whereupon we can write equation (13.17) as

$$p(x, y) = \bar{p} + \tfrac{1}{2}p^* \{\cos(2\pi x/\lambda) + \cos(2\pi y/\lambda)\} \tag{13.18}$$

where $p^* = 2\pi E^* \Delta/\lambda$. The curvature of a peak in the surface $1/R = 4\pi^2 \Delta/\lambda^2$ and the load carried by each peak $P = \bar{p}\lambda^2$. Substituting these values in the Hertz equation (4.22) gives the ratio of the real (circular) area of contact to the nominal (square) area:

$$\frac{\pi a^2}{\lambda^2} = \pi \left(\frac{3}{8\pi} \frac{\bar{p}}{p^*}\right)^{2/3} \tag{13.19}$$

At the other extreme, when contact is nearly complete, the approximately circular region of no-contact is regarded as a 'penny-shaped crack' subjected to an internal pressure equal and opposite to that given by equation (13.18). As before the radius of the crack b is found from the condition that the stress intensity factor at the edge of the crack should be zero. In this way (see Johnson *et al.*, 1985) it is found that

$$\frac{\pi b^2}{\lambda^2} = \frac{3}{2\pi}\left(1 - \frac{\bar{p}}{p^*}\right) \tag{13.20}$$

Both asymptotic results, given by equations (13.19) and (13.20) are plotted in Fig. 13.3. In between, numerical solutions by Johnson *et al.* (1985) show how the ratio of the real to apparent contact area varies with contact pressure. The photographs of a rubber model in Fig. 13.4 illustrate the changing shape of the contact area.

(c) Plastic crushing of a serrated surface

If a regular wavy surface, compressed by a rigid flat die, yields before it is flattened elastically, the crests of the waves will be crushed plastically. It

Fig. 13.3. Real area of contact of a two-dimensional wavy surface with an elastic half-space. Broken line – asymptotic, eqs. (13.19) and (13.20); solid circle – numerical solutions; open circle – experimental from Fig. 13.4.

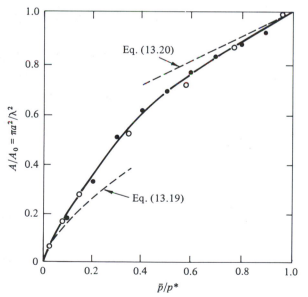

has then been observed that it is difficult to flatten the surface by plastic crushing of the asperities (see Greenwood & Rowe, 1965). This behaviour has been modelled by Childs (1973) by representing the rough surface by a regular array of wedge-shaped serrations, which is indented by a flat rigid punch whose width L is much greater than the pitch of the serrations, as shown in Fig. 13.5(a). When the punch is loaded by a normal force P the tips of the serrations are crushed, each with a contact width l. Taking the material of the serrated surface to be rigid-perfectly-plastic, the serrations will first crush according to the slip-line field of Fig. 6.8(b) from which the asperity pressure p can be calculated. With increasing load, this mode of deformation will continue until the deformation fields of adjacent serrations begin to overlap, i.e. when point C in Fig. 6.8(b) reaches the trough between two serrations. For a semi-wedge-angle $\alpha = 65°$ this point is reached when $l/\lambda = 0.36$. Further deformation is now constrained by the interference between adjacent serrations. Slip-line fields have been constructed by Childs (1973) for this situation which results in a sharp increase in asperity pressure as $l/\lambda \to 1.0$. This configuration may be visualised as a back extrusion, where the material displaced by the downward motion of the punch has to be extruded upwards through the small remaining

Fig. 13.4. Area of contact between a perspex flat and a rubber block with an isotropic wavy surface $\bar{p}/p^*$: (a) 0.024, (b) 0.080, (c) 0.139, (d) 0.345, (e) 0.550, (f) 0.759.

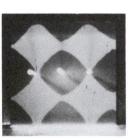
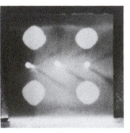
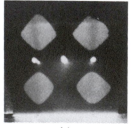

(a)   (b)   (c)

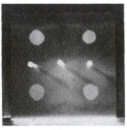

(d)   (e)   (f)

gap between the contact areas. The mean pressure on the punch $\bar{p}$ is given by:

$$\frac{\bar{p}}{2k} \equiv \frac{P}{2kL} \approx \frac{p}{2k}\frac{l}{\lambda} \qquad (13.21)$$

When $\bar{p}$ reaches the limiting pressure for plastic indentation of a flat punch $(5.14k)$, the punch will indent the block as a whole and no further deformation of the asperities will take place. In the example shown in Fig. 13.5(b) $(\alpha = 65°)$, this limit is reached when $l/\lambda = 0.81$. In practice the asperities strain-harden relative to the bulk so that the maximum value of l/λ is less than 0.81. Thus, under purely normal loading of an extended surface, it is not possible to crush the asperities flat by purely plastic deformation. We have seen that this arises

Fig. 13.5. Crushing a regular serrated plastic surface by a rigid flat punch $(\alpha = 65°)$. Bulk indentation by the punch will occur when $pl/\lambda = \bar{p} = P/L = 5.14k$, i.e. when $l/\lambda = 0.81$.

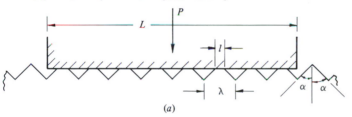

(a)

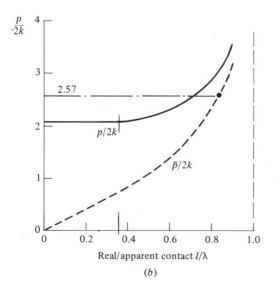

Real/apparent contact l/λ

(b)

from the constraint imposed by adjacent asperities. If, however, the block as a whole is extending plastically parallel to the surface of the punch (neglecting friction) this constraint is removed and the asperities are flattened by small punch pressures. This state of affairs is common at the surface of a die in a metal forming operation. It also occurs if the block in Fig. 13.5(a) is not as wide as the punch so that bulk plastic flow takes place when $\bar{p}$ exceeds $2k$. Finally, frictional shearing of the serrations by a tangential force applied to the block facilitates the growth of the real/apparent contact area l/λ. The mode of plastic deformation of the serrations is then similar to that of the wedge shown in Fig. 7.15.

13.3   Characteristics of random rough surfaces

We will now discuss briefly the topographical characteristics of random rough surfaces which are relevant to their behaviour when pressed into contact.

Surface texture is most commonly measured by a profilometer which draws a stylus over a sample length of the surface of the component and reproduces a magnified trace of the surface profile as shown in Fig. 13.6. Note that the trace is a much distorted image of the actual profile through using a larger magnification in the normal than in the tangential direction. Modern profilometers digitise the trace at a suitable sampling interval and couple the output to a computer in order to extract statistical information from the data. First, a datum or centre-line is established by finding the straight line (or circular arc in the case of round components) from which the mean square deviation is a minimum. This implies that the area of the trace above the datum is equal to that below it. The average roughness is now defined by

$$R_a \equiv \frac{1}{L} \int_0^L |z| \, dx \tag{13.22}$$

where $z(x)$ is the height of the surface above the datum and L is the sampling length. A less common but statistically more meaningful measure of average roughness is the 'root-mean-square' or standard deviation σ of the height of the

Fig. 13.6. Profilometer trace.

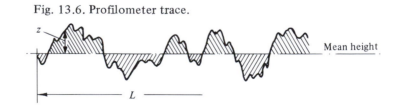

surface from the centre-line, i.e.

$$\sigma^2 = \frac{1}{L} \int_0^L z^2 \, dx \tag{13.23}$$

The relationship between σ and R_a depends, to some extent, on the nature of the surface; for a regular sinusoidal profile $\sigma = (\pi/2 \sqrt{2})R_a$, for a Gaussian random profile $\sigma = (\pi/2)^{1/2}R_a$.

The R_a value by itself gives no information about the shape of the surface profile, i.e. about the *distribution* of the deviations from the mean. The first attempt to do this was by the so-called bearing area curve (Abbott & Firestone, 1933). This curve expresses, as a function of height z, the fraction of the nominal area lying within the surface contour at elevation z. It would be obtained from a profile trace, such as Fig. 13.6, by drawing lines parallel to the datum at varying heights z and measuring the fraction of the length of the line at each height which lies within the profile (see Fig. 13.7). We note in passing that the 'bearing area curve' does not give the true bearing area when the rough surface is in contact with a smooth flat one. It implies that the material in the area of interpenetration vanishes, no account being taken of contact deformation. The true bearing or contact area will be discussed in §4.

An alternative approach to the bearing area curve is through elementary statistics. If we denote by $\phi(z)$ the probability that the height of a particular point in the surface will lie between z and $z + dz$, then the probability that the height of a point on the surface is greater than z is given by the cumulative probability function: $\Phi(z) = \int_z^\infty \phi(z') \, dz'$. This yields an S-shaped curve identical with the bearing area curve.

It has been found that many real surfaces, notably freshly ground surfaces, exhibit a height distribution which is close to the 'normal' or Gaussian probability

Fig. 13.7. Height distribution $\phi(z)$ and 'bearing area' curve given by the cumulative height distribution $\Phi(z)$.

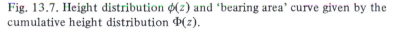

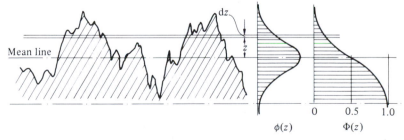

function:

$$\phi(z) = \sigma^{-1}(2\pi)^{-1/2} \exp\left(-z^2/2\sigma^2\right) \tag{13.24}$$

where σ is the standard (r.m.s.) deviation from the mean height. The cumulative probability

$$\Phi(z) = \tfrac{1}{2} - \frac{1}{(2\pi)^{1/2}} \int_0^{z/\sigma} \exp\left(-z'^2/2\sigma^2\right) d(z'/\sigma) \tag{13.25}$$

is to be found in any statistical tables. When plotted on normal probability graph paper, data which follow the normal or Gaussian distribution will fall on a straight line whose gradient gives a measure of the standard deviation σ, as shown by the ground surface in Fig. 13.8(*a*). It is convenient from a mathematical point of view to use the normal probability function in the analysis of randomly rough surfaces, but it must be kept in mind that few real surfaces are Gaussian. For example, a ground surface which is subsequently polished so that the tips of the higher asperities are removed (Fig. 13.8(*b*)) departs markedly from the straight line in the upper height range. A lathe-turned surface is far from random; its peaks are nearly all the same height and its troughs the same depth. It appears on probability paper as shown in Fig. 13.8(*c*).

So far we have discussed only variations in height of the surface; spatial variations must also be considered. There are several ways in which the spatial variation can be specified. We shall use the root-mean-square slope σ_m and r.m.s. curvature σ_κ defined as follows.†

A sample length L of the surface is traversed by a stylus profilometer and the height z is sampled at discrete intervals of length h. If z_{i-1}, z_i and z_{i+1} are three consecutive heights, the slope is defined by

$$m = (z_{i+1} - z_i)/h \tag{13.26}$$

and the curvature by

$$\kappa = (z_{i+1} - 2z_i + z_{i-1})/h^2 \tag{13.27}$$

The r.m.s. slope and curvature are then found from

$$\sigma_m^2 = (1/n) \sum_{i=1}^{i=n} m^2 \tag{13.28}$$

$$\sigma_\kappa^2 = (1/n) \sum_{i=1}^{i=n} \kappa^2 \tag{13.29}$$

where $n = L/h$ is the total number of heights sampled.

† For alternative specifications of a random rough surface in terms of the *auto-correlation function* or the *spectral density*, the interested reader is referred to the book: *Rough Surfaces*, Ed. T. R. Thomas (1982).

Fig. 13.8. Cumulative height distributions plotted on normal probability paper: solid circle – surface heights; cross – peak heights; triangle – summit heights. (*a*) Bead-blasted aluminium, (*b*) mild steel abraded and polished, (*c*) lathe-turned mild steel (Williamson, 1967–8).

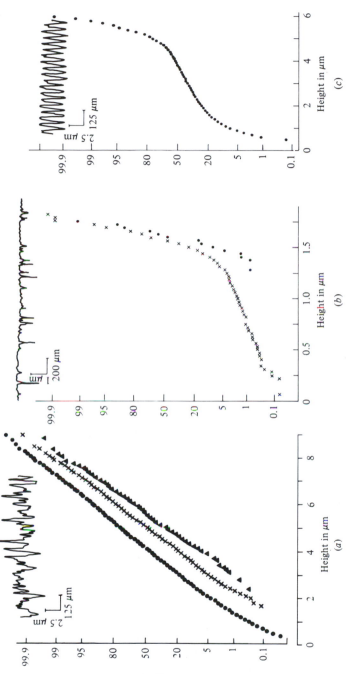

We should like to think of the parameters σ, σ_m and σ_κ as properties of the surface which they describe. Unfortunately their values in practice depend upon both the sample length L and the sampling interval h used in their measurement. If we think of a random rough surface as having a continuous spectrum of wavelengths, neither wavelengths which are longer than the sample length nor those which are shorter than the sampling interval will be recorded faithfully by a profilometer. A practical upper limit for the sample length is imposed by the size of the specimen and a lower limit to the meaningful sampling interval by the radius of the profilometer stylus. The mean square roughness σ is virtually independent of the sampling interval h provided that h is small compared with the sample length L. The parameters σ_m and σ_κ, however, are very sensitive to sampling interval: their values tend to increase without limit as h is made smaller, and shorter and shorter wavelengths are included. This uncomfortable fact has led to the concept of *functional filtering* whereby both the sample length and sampling interval are chosen to be appropriate to the particular application under consideration. We shall return to this point in §5.

When rough surfaces are pressed into contact they touch at the high spots of the two surfaces, which deform to bring more spots into contact. We shall see in the next section that, to quantify this behaviour, we need to know the standard deviation of the asperity heights σ_s, the mean curvature of their summits $\bar{\kappa}_s$, and the asperity density η_s, i.e. the number of asperities per unit area of the surface. These quantities have to be deduced from the information contained in a profilometer trace. It must be kept in mind that a maximum in the profilometer trace, referred to as a 'peak', does not necessarily correspond to a true maximum in the surface, referred to as a 'summit', since the trace is only a one-dimensional section of a two-dimensional surface. On the basis of random process theory, following the work of Longuet-Higgins (1957a & b), Nayak (1971) and Whitehouse & Phillips (1978, 1982), Greenwood (1984) has investigated the relationship between the summit properties of interest and the properties of a profilometer trace as influenced by sampling interval, with the following conclusions:

(i) For an isotropic surface having a Gaussian height distribution with standard deviation σ, the distribution of summit heights is very nearly Gaussian with a standard deviation

$$\sigma_s \approx \sigma \qquad (13.30)$$

The mean height of the summits lies between 0.5σ and 1.5σ above the mean level of the surface. The same result is true for peak heights in a profilometer trace, as shown by the data in Fig. 13.8(a), where the fact that the data for the peak heights and summit heights lie approximately parallel to those for the surface as a whole, shows that they have nearly the same standard deviation. A peak in the profilometer trace is identified when, of three adjacent sample heights, z_{i-1}, z_i and z_{i+1}, the middle one z_i is greater than both the outer two.

(ii) The mean summit curvature is of the same order as the root-mean-square curvature of the surface, i.e.,

$$\bar{\kappa}_s \approx \sigma_\kappa \qquad (13.31)$$

(iii) By identifying peaks in the profile trace as explained above, the number of peaks per unit length of trace η_p can be counted. If the wavy surface were regular, as discussed in §2(*b*), the number of summits per unit area η_s would be η_p^2. Nayak (1971) showed that, for a *random* isotropic surface, with a vanishingly small sampling interval, $\eta_s = 1.209\eta_p^2$. Over a wide range of finite sampling intervals Greenwood (1984) showed that

$$\eta_s \approx 1.8\eta_p^2 \qquad (13.32)$$

Although the sampling interval has only a second-order effect on the relationships between summit and profile properties expressed in equations (13.31) and (13.32), it must be emphasised that the profile properties themselves σ_κ and η_p are both very sensitive to the size of the sampling interval.

13.4   Contact of nominally flat rough surfaces

We have seen throughout this book that, in the frictionless contact of elastic solids, the contact stresses depend only upon the *relative* profile of their two surfaces, i.e. upon the shape of the gap between them before loading. The system may then be replaced, without loss of generality, by a flat, rigid surface in contact with a body having an effective modulus E^* and a profile which results in the same undeformed gap between the surfaces. We are concerned here with the contact of two nominally flat surfaces, which have r.m.s. roughnesses σ_1 and σ_2 respectively. However we shall consider the contact of a rigid flat plane with a deformable surface of equivalent roughness $\sigma = (\sigma_1^2 + \sigma_2^2)^{1/2}$.

The situation is illustrated in Fig. 13.9. We shall follow the analysis of Greenwood & Williamson (1966). The mean level of the surface is taken as

Fig. 13.9. Contact of a randomly rough surface with a smooth flat.

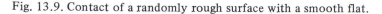

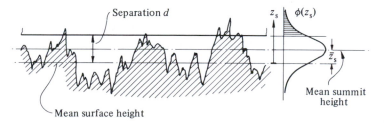

datum and the distance between the datum and the rigid flat is referred to as
the separation. Fig. 13.9 shows the peaks in a profile trace but, from the contact
point of view we are interested in the summits of the surface asperities. We
shall denote the summit heights by z_s, having a mean $\bar{z}_s$ and a distribution
function $\phi(z_s)$, which expresses the probability of finding a summit of height
z_s lying in the interval z_s to $z_s + dz_s$. If there are N summits in the nominal
surface area A_0, the number of summits in contact at a separation d is given by

$$n = N \int_d^\infty \phi(z_s)\, dz_s \tag{13.33}$$

For simplicity we shall follow Greenwood & Williamson (1966) and assume that
the asperity summits are spherical with a *constant* curvature κ_s. If a summit
height exceeds the separation it will be compressed by $\delta = z_s - d$ and make
contact with the flat in a small circular area of radius a. The ith summit, therefore, has a contact area

$$A_i = \pi a_i^2 = f(\delta_i) \tag{13.34}$$

and the force required to compress it may be written

$$P_i = g(\delta_i) \tag{13.35}$$

where the functions $f(\delta)$ and $g(\delta)$ depend upon the material properties of the
surfaces. If the deformation is entirely within the elastic limit, from the Hertz
equation (4.23),

$$f(\delta) = \pi\delta/\kappa_s \tag{13.36}$$

and

$$g(\delta) = (\tfrac{4}{3})E^*\kappa_s^{-1/2}\delta^{3/2} \tag{13.37}$$

For perfectly plastic compression of an asperity, if 'piling-up' or 'sinking in'
is neglected (see §6.3):

$$f(\delta) \approx 2\pi\delta/\kappa_s \tag{13.38}$$

and

$$g(\delta) \approx \bar{p}A \approx 6\pi Y\delta/\kappa_s \tag{13.39}$$

where Y is the yield strength of the softer surface. Halling & Nuri (1975) have
proposed alternative functions $f(\delta)$ and $g(\delta)$ appropriate to a material which
displays power-law strain hardening.

To find the total real area of contact A and the total nominal pressure
$\bar{p}\ (= P/A_0)$ we must sum equations (13.34) and (13.35) for all the asperities
in contact, i.e. those whose height z_s exceeds the separation d. Thus

$$A = N \int_d^\infty f(z_s - d)\phi(z_s)\, dz_s \tag{13.40}$$

and

$$\bar{p}A_0 \equiv P = N \int_d^\infty g(z_s - d)\phi(z_s)\, dz_s \tag{13.41}$$

Greenwood & Williamson (1966) have evaluated these integrals numerically for elastically deforming asperities (eq. (13.36) and (13.37)) and a Gaussian distribution of asperity heights.

The contact area (non-dimensionalised) is plotted against the non-dimensional load in Fig. 13.10 from which it may be seen that the contact area is approximately proportional to the load over three decades of load. The separation may be normalised by subtracting the mean summit height and dividing by the standard deviation, i.e. by putting $\zeta = (d - \bar{z}_s)/\sigma_s$. The theory is based on the assumption that the deformation of each asperity is independent of its neighbours. This will become increasingly in error when the real contact area is no longer small compared with the nominal area, at normalised separations less than 0.5, say. At the other end of the scale, the probability of contact becomes negligible if the normalised separation exceeds about 3.0.

The significant features of Greenwood & Williamson's results may be demonstrated without recourse to numerical analysis by using an exponential rather

Fig. 13.10. Real area of contact of a randomly rough elastic surface with a smooth flat: A – Gaussian distribution of asperity heights (Greenwood & Williamson, 1966); B – exponential distribution (eq. (13.46)).

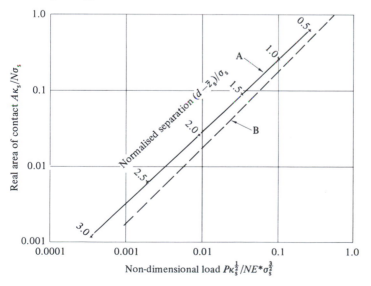

than a Gaussian distribution function, i.e. by putting

$$\phi(z_s) = (C/\sigma_s) \exp(-z_s/\sigma_s), \quad z_s > 0 \tag{13.42}$$

where C is an arbitrary constant. The number of asperities in contact, given by (13.33), then becomes

$$n = (CN/\sigma_s) \int_d^\infty \exp(-z_s/\sigma_s) \, dz_s$$

$$= CN \exp(-d/\sigma_s) \int_0^\infty \exp(-\delta/\sigma_s) \, d(\delta/\sigma_s)$$

$$= CN \exp(-d/\sigma_s) \tag{13.43}$$

Similarly equation (13.40) for the contact area becomes

$$A = (CN/\sigma_s) \int_d^\infty f(\delta) \exp(-z_s/\sigma_s) \, dz_s$$

$$= CN \exp(-d/\sigma_s) \int_0^\infty f(\delta/\sigma_s) \exp(-\delta/\sigma_s) \, d(\delta/\sigma_s)$$

$$\equiv n I_f \tag{13.44}$$

and equation (13.41) for the load becomes

$$P = CN \exp(-d/\sigma_s) \int_0^\infty g(\delta/\sigma_s) \exp(-\delta/\sigma_s) \, d(\delta/\sigma_s)$$

$$\equiv n I_g \tag{13.45}$$

The definite integrals in equations (13.44) and (13.45), denoted by I_f and I_g respectively, are constants independent of the separation d. Thus the real contact area A and the mean pressure $\bar{p}$ are both proportional to the number of asperities in contact n, and are hence proportional to each other irrespective of the mode of deformation of the asperities expressed by f and g. These results imply that, although the size of each particular contact spot grows as the separation is decreased, the number of spots brought into contact also increases at a rate such that the *average* size $\bar{a}$ remains constant.

If the deformation of the asperities is elastic $f(\delta/\sigma_s)$ and $g(\delta/\sigma_s)$ are given by equations (13.36) and (13.37) which, when substituted in (13.44) and (13.45), give $I_f = \pi\sigma_s/\kappa_s$ and $I_g = \pi^{1/2}E^*\sigma_s^{3/2}\kappa_s^{-1/2}$. Thus the ratio of the real to apparent contact area is given by

$$\frac{A}{A_0} = \frac{A}{P}\frac{P}{A_0} = \frac{I_f}{I_g}\bar{p} = \pi^{1/2}(\sigma_s\kappa_s)^{-1/2}(\bar{p}/E^*) \tag{13.46}$$

where $\bar{p}$ is the nominal contact pressure P/A_0. In this case, with an exponential probability function, the number of asperities in contact and the real area of contact are both exactly proportional to the load. The comparison with the more realistic Gaussian distribution is shown in Fig. 13.10.

It is instructive to compare the ratio of real to apparent area of contact for a random rough surface, given by equation (13.46), with that for a *regular* wavy surface. Now equation (13.19) for the real area of contact of a flat surface with a regular wavy surface of amplitude Δ may be written

$$A/A_0 = 0.762(\Delta\,\kappa_\mathrm{s})^{-1/2}(\bar{p}/E^*)^{2/3} \tag{13.47}$$

Recognising that the standard deviation σ of a random rough surface is a comparable quantity to the amplitude Δ of a regular wavy surface, we see that equations (13.46) and (13.47) involve the same dimensionless variables. Whereas the regular surface deforms such that the real area of contact grows as the (load)$^{2/3}$, the contact area with a randomly rough surface grows in direct proportion with the load. This conclusion is consistent with Amontons' law of friction. Frictional forces must be developed at the points of real contact and we would expect, therefore, that the total force of friction would be proportional to the real area of contact, which we have seen is in direct proportion to the load. Further experimental support for the conclusions presented above is provided by measurements of thermal and electrical conductance between conducting bodies across a rough interface. The conductance of a single circular contact area is given by $2Ka$, where K is the bulk conductivity of the solids. The total conductance of the interface therefore is $2K\Sigma a = 2Kn\bar{a}$. Now we have seen that the mean contact size $\bar{a}$ remains approximately constant while the number of contact spots n increases in direct proportion to the load, thereby ensuring that the total conductance increases in proportion to the load. The law of friction demands that the total area (Σa^2) increases in proportion to the nominal contact pressure (load); the conductance experiments demand that Σa increases in direct proportion to load. The only simple way in which both these conditions can be fulfilled simultaneously is by the mean contact *size $\bar{a}$* remaining constant and the *number* of real contact spots n increasing in proportion to the load.

A further consequence of the area of real contact being proportional to the load is that the real mean contact pressure is nearly constant. For an exponential probability function and elastically deforming asperities,

$$\bar{p}_\mathrm{r} = \frac{P}{A} = \frac{I_g}{I_f} = 0.56E^*(\sigma_\mathrm{s}\kappa_\mathrm{s})^{1/2} \tag{13.48}$$

For a Gaussian height distribution $\bar{p}_\mathrm{r}$ varies from 0.3 to $0.4E^*(\sigma_\mathrm{s}\kappa_\mathrm{s})^{1/2}$ over the relevant range of loading. Since each asperity in the model is spherical the onset of plastic yield is given by equation (6.9), i.e. when $\bar{p} = 1.1Y \approx 0.39H$, where

H is the hardness of the material and Y its yield stress. Thus the average contact pressure will be sufficient to cause yield if $\bar{p}_r \geqslant 0.39H$, that is if

$$\psi \equiv (E^*/H)(\sigma_s \kappa_s)^{1/2} \geqslant C \tag{13.49}$$

where C is a constant which depends somewhat on the height distribution but has a value close to unity The non-dimensional parameter ψ is known as the 'plasticity index'. It describes the deformation properties of a rough surface. If its value is appreciably less than unity the deformation of the asperities when in contact with a flat surface will be entirely elastic; if the value exceeds unity the deformation will be predominantly plastic.

The theory outlined above assumes that the asperities have a constant curvature κ_s whereas, in reality, the summit curvatures will have a random variation. As an approximation we can use the mean summit curvature $\bar{\kappa}_s$ in equations (13.44)–(13.49) which, as we saw in the last section (eq. (13.31)), is approximately equal to the r.m.s. curvature of the surface found from a profilometer trace. This procedure is not strictly correct because the summit curvature is not independent of the summit height, but it has been justified by Onions & Archard (1973).

An alternative definition of the plasticity index has been proposed by Mikic (1974). We saw in §6.3 that the extent of plastic deformation during the indentation of an elastic-plastic surface by a rigid wedge or cone was governed by the non-dimensional parameter $(E^* \tan \beta/Y)$ where β is the angle of inclination of the face of the wedge or cone to the surface of the solid and Y is the yield stress. Thus, when two rough surfaces are in contact we might expect the degree of plastic deformation of the asperities to be proportional to the *slope* of the asperities. Remembering that yield stress is proportional to hardness, Mikic proposed a plasticity index defined by

$$\psi = E^* \sigma_m/H \tag{13.50}$$

where σ_m is the r.m.s. slope of the surface which is obtained directly from a profile trace. This definition avoids the difficulty of two statistical quantities which are not independent, but does not escape the dependence of σ_m on the sampling interval used to measure it.

13.5 Elastic contact of rough curved surfaces

We come now to the main question posed in this chapter: how are the elastic contact stresses and deformation between curved surfaces in contact, which form the main subject of this book, influenced by surface roughness? The qualitative behaviour is clear from what has been said already. There are two scales of size in the problem: (i) the bulk (nominal) contact dimensions and elastic compression which would be calculated by the Hertz theory for the 'smooth' mean profiles of the two surfaces and (ii) the height and spatial distribution of the asperities. For the situation to be amenable to quantitative analysis these two scales of size should be very different. In other words, there should be many asperities lying within the nominal contact area. When the two bodies are pressed together true contact occurs only at the tips of the asperities, which are compressed in the manner discussed in §4. At any point in the nominal contact area the nominal pressure increases with overall load and the real contact area increases in proportion; the average real contact pressure remains constant at a value given by equation (13.48) for elastically deforming asperities. Points of real contact with the tips of the higher asperities will be found outside the nominal contact area, just as a rough seabed results in a ragged coastline with fjords and off-shore islands. The asperities act like a compliant layer on the surface of the body, so that contact is extended over a larger area than it would be if the surfaces were smooth and, in consequence, the contact pressure for a given load will be reduced. Quantitative analysis of these effects, using the Greenwood & Williamson model of a rough surface (spherically tipped elastic asperities of constant curvature), has been applied to the point contact of spheres by Greenwood & Tripp (1967) and Mikic (1974) and to the line contact of cylinders by Lo (1969). We shall consider the axi-symmetric case which can be simplified to the contact of a smooth sphere of radius R with a nominally flat rough surface having a standard distribution of summit heights σ_s, where R and σ_s are related to the radii and roughnesses of the two surfaces by: $1/R = 1/R_1 + 1/R_2$ and $\sigma_s^2 = \sigma_{s1}^2 + \sigma_{s2}^2$.

Referring to Fig. 13.11, a datum is taken at the mean level of the rough surface. The profile of the undeformed sphere relative to the datum is given by

$$y = y_0 - r^2/2R$$

At any radius the combined normal displacement of both surfaces is made up of a bulk displacement w_b and an asperity displacement w_a. The 'separation' d between the two surfaces contains only the bulk deformation, i.e.

$$d(r) = w_b(r) - y(r) = -y_0 + (r^2/2R) + w_b(r) \tag{13.51}$$

The asperity displacement $w_a = z_s - d$, where z_s is the height of the asperity summit above the datum. If now we assume that the asperities deform elastically, the function $g(w_a)$ is given by equation (13.37) with δ replaced by w_a. Then, by substitution in equation (13.41), the effective pressure at radius r is found to be

$$p(r) = (4\eta_s E^*/3\kappa_s^{1/2}) \int_d^\infty \{z_s - d(r)\}^{3/2} \phi(z_s) \, dz_s \tag{13.52}$$

where η_s is the asperity density N/A_0. The bulk compression w_b, which is contained in the expression for separation (13.51), is related to the effective pressure $p(r)$ by the equations for the axi-symmetric deformation of an elastic half-space presented in §3.8. In the notation of the present section equation (3.98a) for the normal displacement of an axi-symmetric distribution of pressure $p(r)$ can be written:

$$w_b(r) = \frac{4}{\pi E^*} \int_0^a \frac{t}{t+r} p(t) K(k) \, dt \tag{13.53}$$

where $\mathbf{K}$ is the complete elliptic integral of the first kind with argument $k = 2(rt)^{1/2}/(r + t)$. Equations (13.51), (13.52) and (13.53) have been solved by Greenwood & Tripp (1967) for a Gaussian distribution of asperity heights,

Fig. 13.11. Contact of a smooth elastic sphere with a nominally flat randomly rough surface: solid line — effective pressure distribution $p(r)$; broken line – Hertz pressure (smooth surfaces). Effective radius a^* defined by eq. (13.56).

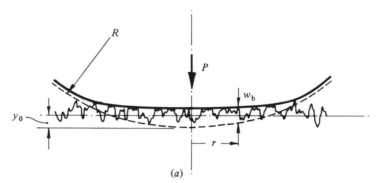

(a)

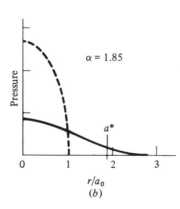

$\alpha = 1.85$

a^*

Pressure

r/a_0
(b)

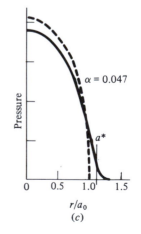

$\alpha = 0.047$

a^*

Pressure

r/a_0
(c)

using an iterative numerical technique to find the effective pressure distribution $p(r)$.†

Effective pressure distributions, normalised by the maximum pressure p_0 and contact radius a_0 for smooth surfaces under the same load P, are plotted in Fig. 13.11(b) and (c). As expected, the effect of surface roughness is to reduce the maximum contact pressure $p(0)$ and to spread the load over an area of greater radius. The solution to equations (13.51), (13.52) and (13.53) depends upon two independent non-dimensional parameters. The first, which we shall denote by α, can be expressed variously by

$$\alpha \equiv \frac{\sigma_s}{\delta_0} = \frac{\sigma_s R}{a_0^2} = \sigma_s \left(\frac{16RE^{*2}}{9P^2} \right)^{1/3} \tag{13.54}$$

where δ_0 is the bulk compression and a_0 is the contact radius for smooth surfaces under the load P, i.e. given by the Hertz theory.‡ The second parameter is defined by Greenwood & Tripp as

$$\mu = \tfrac{8}{3}\eta_s\sigma_s(2R/\kappa_s)^{1/2} \tag{13.55}$$

which depends on the topography of the surfaces but not upon the load. The ratio of maximum effective pressure with a rough surface $p(0)$ to the maximum pressure with a smooth surface p_0 (given by Hertz) is plotted against α in Fig. 13.12 for two values of μ which bracket a wide range of practical rough surfaces. It is clear from Fig. 13.11(b) and (c) that, with a rough surface, the effective pressure falls asymptotically to zero. The contact area, therefore, is not precisely defined. One possibility is to define the 'contact' radius as the radius at which the effective pressure falls to some arbitrarily chosen small fraction of the maximum pressure. Greenwood & Tripp arbitrarily define an effective 'contact' radius a^* by

$$a^* = \frac{3\pi \int_0^\infty rp(r)\,\mathrm{d}r}{4\int_0^\infty p(r)\,\mathrm{d}r} \tag{13.56}$$

Its value is indicated in the examples shown in Fig. 13.11(b) and (c). With this definition theoretical values of a^*/a are plotted against α for $\mu = 4$ and $\mu = 17$ in Fig. 13.13, where they are compared with experimental measurements of the

---

† If the elastic foundation model is used in place of equation (13.53) to find the bulk displacement w_b, together with an exponential probability of asperity heights, a solution can be obtained in closed form as shown by Johnson (1975).

‡ Greenwood & Tripp (1967) use the non-dimensional parameter $T = (8/3\sqrt{2})\alpha^{-3/2}$.

contact size. In reality the contact area has a ragged edge (see Fig. 13.14) which makes its measurement subject to uncertainty. The rather arbitrary definition of $a*$ is therefore not of serious consequence.

It is clear from Figs. 13.12 and 13.13 that the effect of surface roughness on the contact pressure and contact area is governed primarily by the parameter α; the parameter μ has a secondary effect. Further, we can conclude that the Hertz theory for smooth surfaces can be used with only a few per cent error provided the parameter α is less than about 0.05, i.e. provided the combined roughness of the two surfaces σ_s is less than about 5% of the bulk elastic compression δ_0.

By influencing the pressure distribution and the contact area, surface roughness influences the magnitude and position of the maximum shear stress in the solid and hence the load at which bulk yield will take place. For a smooth (Hertzian) contact the maximum shear stress has the value $0.31p_0$ at a depth $z = 0.48a$. Greenwood & Tripp find that with the pressure distributions appropriate to a rough surface, as shown in Fig. 13.11, the maximum shear stress

Fig. 13.12. Influence of surface roughness on the maximum contact pressure $p(0)$ compared with the maximum Hertz pressure p_0.

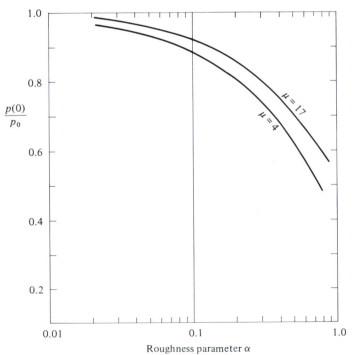

$\approx 0.29 p(0)$ at a depth $z \approx 0.35 a^*$. However, since $p(0)$ decreases significantly with increasing roughness (Fig. 13.12) and a^* increases (Fig. 13.13), the maximum shear stress is reduced and occurs at a greater depth than with smooth surfaces.

We now return to the question of 'functional filtering': how should the sample length and sampling interval be chosen in order to obtain appropriate values for the parameters α and μ? The sample length presents no great difficulty. Wavelengths in the surface profile which greatly exceed the nominal contact diameter are not going to influence the contact deformation appreciably, so that we should take $L < 4a_0$. Since σ is found in practice to vary approximately as $L^{1/2}$, the precise choice of L is not critical. We then use Greenwood's result (13.30) and put the standard deviation of the summit heights σ_s equal to the r.m.s. height of the surface σ measured from the profilometer trace. If the parameter α only is required no further considerations are necessary since σ is not sensitive to sampling interval. If the parameter μ is also required then values for the asperity density η_s and the asperity summit curvature $\bar{\kappa}_s$ must be determined. They are given by equations (13.31) and (13.32) in terms of the peak density η_p and the r.m.s. curvature σ_κ found from the profile trace, but both these latter quantities are strongly dependent on sampling interval. Intuitively we would expect there to be a scale of roughness below which the asperities would be immediately destroyed by plastic deformation when the surfaces were brought into contact and would not contribute significantly to the contact pressure:

Fig. 13.13. Influence of the surface roughness on the effective contact radius a^* compared with the Hertz radius a_0. Experiments: circle – $\mu = 4$; triangle – $\mu = 5$; square – $\mu = 15$.

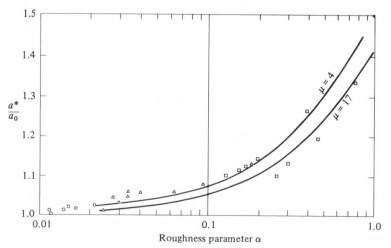

it would then be reasonable not to reduce the sampling interval below this size. However, it has proved difficult to quantify this cut-off point. At the present time most profilometers arbitrarily use a sampling interval of $\sim 10\ \mu m$.

Tangential forces

When two bodies having rough surfaces are in contact the influence of the surface roughness upon their tangential compliance is also of interest. In rolling contact surface roughness might be expected to affect the creep coefficients.

An experimental study (O'Connor & Johnson, 1963) of a hard steel smooth sphere, pressed into contact with a rough flat by a constant normal force, showed

Fig. 13.14. Contact of a smooth steel ball, diameter 25.4 mm, with a rough steel flat: (*a*) $\sigma = 0.19\ \mu m$, $P = 10$ kg, $\alpha = 0.043$; (*b*) $\sigma = 0.54\ \mu m$, $P = 60$ kg, $\alpha = 0.043$; (*c*) $\sigma = 0.54\ \mu m$, $P = 4$ kg, $\alpha = 0.22$; (*d*) $\sigma = 2.4\ \mu m$, $P = 40$ kg, $\alpha = 0.22$.

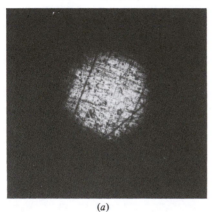

(*a*)

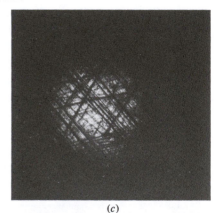

(*c*)

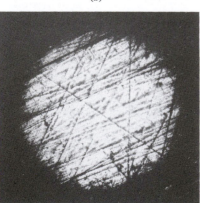

(*b*)

(*d*)

that the compliance under the action of a superimposed tangential force, given by equation (7.42), was affected very little by the roughness of the surface. This was the case from the start with a rough ground hard steel surface in which the asperity deformation was predominantly elastic. With a scratched surface of soft steel, in which the asperity deformation was fully plastic, the tangential compliance was somewhat greater on first load than that given by equation (7.42) for smooth elastic surfaces, but was close to equation (7.42) on subsequent loadings. The small effect of roughness on tangential compliance can be explained as follows. First, the tangential elastic compliance of an individual asperity is comparable with its normal compliance as demonstrated by equations (7.43) and (7.44), provided that the ratio of the tangential to the normal traction is relatively small compared with the coefficient of limiting friction. In the central region of the contact area, as Fig. 7.7 shows, the tangential traction is a minimum while the normal pressure is a maximum. The real contact area in this region will be high and, in consequence, the compliance of the asperities will be small. Since the tangential traction is also small in that region the contribution of the asperity deformation to the bulk compliance is negligible. At the edges of the contact when the tangential traction is larger and the normal pressure smaller, some micro-slip will take place in the same manner as described in §7.2(d).

In rolling contact, the creep ratio ξ is determined by the strains in the region of no slip at the leading edge of the contact area. In this region the tangential traction is less than the normal pressure so the same argument applies, though with less force, to explain the small influence of surface roughness on creep in rolling contact.

The parameter α, which was used as a measure of the effect of surface roughness on static contact under a purely normal load should also apply to static and rolling contacts under the action of tangential loads. However the condition that $\alpha < 0.05$ for the effect of surface roughness to be negligible in normal contacts is likely to be somewhat conservative when applied to tangential forces.

APPENDIX 1

Cauchy Principal Values of some useful integrals

(1) $\displaystyle\int_b^a \frac{ds}{x-s} = \ln\left(\frac{x-b}{a-x}\right), \quad b \leqslant x \leqslant a$

(2) $\displaystyle\int_{-1}^{+1} \frac{1-S^2}{X-S}\, dS = 2X + (1-X^2)\ln\left(\frac{1+X}{1-X}\right), \quad -1 \leqslant X \leqslant +1$

(3) $\displaystyle\int_{-1}^{+1} \frac{S(1-S^2)^{1/2}}{X-S}\, dS = \pi X^2 - \pi/2, \quad -1 \leqslant X \leqslant +1$

(4) $\displaystyle\int_{-1}^{+1} \frac{(1-S^2)^{1/2}}{X-S}\, dS = \pi X, \quad -1 \leqslant X \leqslant +1$

(5) $\displaystyle\int_{-1}^{+1} \frac{\text{sign}\,(S)(1-X^2)^{1/2}\, dS}{X-S} = 2(1-X^2)^{1/2}\ln\left\{\frac{|X|}{1+(1-X^2)}\right\},$

$$-1 \leqslant X \leqslant +1$$

(6) $\displaystyle\int_{-1}^{+1} \frac{|S|(1-X^2)^{1/2}\, dS}{X-S} = 2X + 2X(1-X^2)\ln\left\{\frac{|X|}{1+(1+X^2)^{1/2}}\right\},$

$$-1 \leqslant X \leqslant +1$$

APPENDIX 2

Geometry of smooth non-conforming surfaces in contact

The profile of each body close to the origin O can be expressed:

$$z_1 = (1/2R_1')x_1^2 + (1/2R_1'')y_1^2$$

and

$$z_2 = -\{(1/2R_2')x_2^2 + (1/2R_2'')y_2^2\}$$

where the directions of the axes for each body are chosen to coincide with the principal curvatures of that body. In general the two sets of axes may be inclined to each other at an arbitrary angle θ, as shown in Fig. A2.1(a).

We now transform the coordinates to a common set of axes (x, y) inclined at α to x_1 and β to x_2 as shown. The gap between the surfaces can then be written

$$h = z_1 - z_2 = Ax^2 + By^2 + Cxy$$

Fig. A 2.1

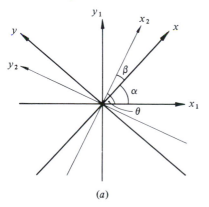

(a)

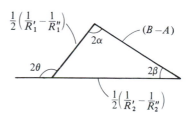

(b)

where

$$C = \tfrac{1}{2}\left(\frac{1}{R_2'} - \frac{1}{R_2''}\right)\sin 2\beta - \tfrac{1}{2}\left(\frac{1}{R_1'} - \frac{1}{R_1''}\right)\sin 2\alpha$$

The condition that C should vanish, so that

$$h = Ax^2 + By^2$$

is satisfied by the triangle shown in Fig. A2.1(b), with the result:

$$A - B = \tfrac{1}{2}\left(\frac{1}{R_1'} - \frac{1}{R_1''}\right)\cos 2\alpha + \tfrac{1}{2}\left(\frac{1}{R_2'} - \frac{1}{R_2''}\right)\cos 2\beta$$

$$|A - B| = \tfrac{1}{2}\left\{\left(\frac{1}{R_1'} - \frac{1}{R_1''}\right)^2 + \left(\frac{1}{R_2'} - \frac{1}{R_2''}\right)^2\right.$$

$$\left. + 2\left(\frac{1}{R_1'} - \frac{1}{R_1''}\right)\left(\frac{1}{R_2'} - \frac{1}{R_2''}\right)\cos 2\theta\right\}^{1/2}$$

Finally

$$A + B = \tfrac{1}{2}\left(\frac{1}{R_1'} + \frac{1}{R_1''} + \frac{1}{R_2'} + \frac{1}{R_2''}\right)$$

from which the values of A $(= 1/2R')$ and B $(= 1/2R'')$ can be found. N.B. Concave curvatures are *negative*.

APPENDIX 3

Summary of Hertz elastic contact stress formulae

$$E^* \equiv \left(\frac{1 - \nu_1^2}{E_1} + \frac{1 - \nu_2^2}{E_2} \right)^{-1}$$

$$R \equiv (1/R_1 + 1/R_2)^{-1}$$

(a) *Line contacts* (load P per unit length)

Semi-contact-width:

$$a = \left(\frac{4PR}{\pi E^*} \right)^{1/2}$$

Max. contact pressure:

$$p_0 = \frac{2P}{\pi a} = \left(\frac{PE^*}{\pi R} \right)^{1/2}$$

Max. shear stress:

$$\tau_1 = 0.30 p_0 \text{ at } x = 0, \quad z = 0.78a$$

(b) *Circular point contacts* (load P)

Radius of contact circle:

$$a = \left(\frac{3PR}{4E^*} \right)^{1/3}$$

Max. contact pressure:

$$p_0 = \left(\frac{3P}{2\pi a^2} \right) = \left(\frac{6PE^{*2}}{\pi^3 R^2} \right)^{1/3}$$

Approach of distant points:

$$\delta = \frac{a^2}{R} = \left(\frac{9}{16} \frac{P^2}{RE^{*2}} \right)^{1/3}$$

Max. shear stress:

$\tau_1 = 0.31 p_0$ at $r = 0$, $\quad z = 0.48a$

Max. tensile stress:

$\sigma_r = \frac{1}{3}(1 - 2\nu)p_0$ at $r = a$, $\quad z = 0$

(c) *Elliptical point contacts* (load P)

a = major semi-axis; b = minor semi-axis; $c = (ab)^{1/2}$; R' and R'' are major and minor *relative* radii of curvature (see Appendix 2); equivalent radius of curvature $R_e = (R'R'')^{1/2}$

$$a/b \approx (R'/R'')^{2/3}$$

$$c = (ab)^{1/2} = \left(\frac{3PR_e}{4E^*}\right)^{1/3} F_1(R'/R'')$$

Max. contact pressure:

$$p_0 = \frac{3P}{2\pi ab} = \left(\frac{6PE^{*2}}{\pi^3 R_e^2}\right)^{1/3} [F_1(R'/R'')]^{-2/3}$$

Approach of distant points:

$$\delta = \left(\frac{9P^2}{16R_e E^{*2}}\right)^{1/3} F_2(R'/R'')$$

The functions $F_1(R'/R'')$ and $F_2(R'/R'')$ are plotted in Fig. 4.4 (p. 97). To a first approximation they may be taken to be unity.

For values of maximum shear stress τ_1, see Table 4.1 (p. 99).

APPENDIX 4

Subsurface stresses in line contact

Tractions: $p = p_0\{1 - (x/a)^2\}^{1/2}$; $q = q_0\{1 - (x/a)^2\}^{1/2}$

$-(\sigma_x)_p/p_0$ and $-(\tau_{zx})_q/q_0$

					$\pm x/a$			
z/a	0	0.2	0.4	0.6	0.8	1.0	1.5	2.0
0	1.000	0.980	0.917	0.800	0.600	0	0	0
0.2	0.659	0.642	0.591	0.507	0.402	0.329	0.124	0.060
0.4	0.426	0.416	0.391	0.357	0.330	0.316	0.197	0.109
0.6	0.275	0.272	0.267	0.265	0.270	0.276	0.221	0.142
0.8	0.180	0.182	0.188	0.200	0.217	0.232	0.218	0.160
1.0	0.121	0.125	0.135	0.153	0.173	0.192	0.201	0.165
1.5	0.051	0.054	0.065	0.081	0.099	0.118	0.148	0.148
2.0	0.025	0.027	0.034	0.045	0.059	0.073	0.103	0.117

$-(\sigma_z)_p/p_0$

					$\pm x/a$			
z/a	0	0.2	0.4	0.6	0.8	1.0	1.5	2.0
0	1.000	0.980	0.917	0.800	0.600	0	0	0
0.2	0.981	0.959	0.892	0.767	0.549	0.212	0.006	0.001
0.4	0.928	0.906	0.834	0.705	0.509	0.281	0.034	0.007
0.6	0.857	0.834	0.765	0.648	0.490	0.320	0.074	0.020
0.8	0.781	0.760	0.699	0.600	0.474	0.342	0.114	0.038
1.0	0.707	0.690	0.638	0.557	0.457	0.352	0.148	0.059
1.5	0.555	0.544	0.514	0.468	0.410	0.346	0.202	0.107
2.0	0.447	0.441	0.424	0.396	0.361	0.322	0.221	0.140

$\mp(\tau_{zx})_p/p_0$ and $\mp(\sigma_z)_q/q_0$

z/a	0	0.2	0.4	0.6	0.8	1.0	1.5	2.0
					±x/a			
0	0	0	0	0	0	0	0	0
0.2	0	0.038	0.080	0.131	0.192	0.192	0.025	0.007
0.4	0	0.064	0.130	0.195	0.242	0.230	0.076	0.027
0.6	0	0.075	0.147	0.209	0.245	0.238	0.119	0.051
0.8	0	0.075	0.145	0.200	0.231	0.231	0.147	0.075
1.0	0	0.070	0.133	0.182	0.211	0.217	0.161	0.095
1.5	0	0.050	0.096	0.134	0.160	0.173	0.162	0.121
2.0	0	0.035	0.068	0.096	0.118	0.133	0.142	0.124

$\mp(\sigma_x)_q/q_0$

z/a	0	0.2	0.4	0.6	0.8	1.0	1.5	2.0
					±x/a			
0	0	0.400	0.800	1.200	1.600	2.000	0.764	0.536
0.2	0	0.282	0.550	0.782	0.934	0.958	0.712	0.521
0.4	0	0.185	0.354	0.489	0.577	0.625	0.596	0.481
0.6	0	0.117	0.223	0.309	0.375	0.426	0.473	0.425
0.8	0	0.073	0.141	0.200	0.251	0.297	0.368	0.365
1.0	0	0.046	0.090	0.132	0.172	0.210	0.285	0.307
1.5	0	0.016	0.033	0.052	0.073	0.095	0.151	0.192
2.0	0	0.007	0.014	0.023	0.034	0.046	0.083	0.118

APPENDIX 5

Linear creep coefficients as defined in equations (8.41), (8.42) and (8.43) ($\nu = 0.3$)

	Ellipticity	$-C_{11}$	$-C_{22}$	$C_{23} = -C_{32}$	$-C_{33}$
$\dfrac{a}{b}$ (i)	0.2	3.56	2.65	0.62	4.68
	0.4	3.72	2.91	0.83	2.42
	0.6	3.90	3.17	1.04	1.74
	0.8	4.08	3.46	1.27	1.38
	1.0	4.29	3.73	1.50	1.18
$\dfrac{b}{a}$	0.8	4.58	4.11	1.82	1.01
	0.6	4.99	4.65	2.29	0.86
	0.4	5.80	5.59	3.27	0.70
	0.2	7.93	8.45	6.76	0.57
	$1.0^{(ii)}$	5.16	4.32	$\left\{\begin{matrix} 2.04 \\ -1.44 \end{matrix}\right\}$	1.84
	$All^{(iii)}$	3.52	2.47	–	–

Note: (i) a is the semi-axis of the ellipse of contact in the rolling direction.
(ii) Approximate values from equations (8.34), (8.36), (8.38) and (8.39).
(iii) Strip theory, equation (8.50).

From Kalker (1967a).

References and author index

Numbers in square brackets refer to pages on which the reference is cited.

Abbott, E. J. & Firestone, F. A. (1933). Specifying surface quality. *Mechanical Engineering (ASME)* **55**, 569, [407].

Abramian, B. L., Arutiunian, N. Kh. & Babloian, A. A. (1966). On symmetric pressure of a circular stamp on an elastic half-space in the presence of adhesion. *PMM* **30**, 143, [79].

Ahmadi, N., Keer, L. M. & Mura, T. (1983). Non-Hertzian contact stress analysis – normal and sliding contact. *International Journal of Solids and Structures*, **19**, 357, [134].

Akyuz, F. A. & Merwin, J. E. (1968). Solution of nonlinear problems of elastoplasticity by finite element method. *AIAA Journal*, **6**, 1825, [172].

Alblas, J. B. & Kuipers, M. (1970). On the two-dimensional problem of a cylindrical stamp pressed into a thin elastic layer. *Acta Mechanica*, **9**, 292, [138].

Aleksandrov, V. M. (1968). Asymptotic methods in contact problems. *PMM*, **32**, 691, [141].

Aleksandrov, V. M. (1969). Asymptotic solution of the contact problem for a thin layer. *PMM* **33**, 49, [141].

Alexander, J. M. (1955). A slip-line field for the hot rolling process. *Proceedings, Institution of Mechanical Engineers*, **169**, 1021, [323, 324, 326].

Alexander, J. M. (1972). On the theory of rolling. *Proceedings, Royal Society*, A326, 535, [322].

Andersson, T., Fredriksson, B. & Persson, B. G. A. (1980). The boundary element method applied to 2-dimensional contact problems. In *New Developments in Boundary Element Methods*, ed. Brebbia. Southampton: CML Publishers, [152].

Archard, J. F. & Cowking, E. W. (1956). Elastohydrodynamic lubrication at point contacts, Symposium on elastohydrodynamic lubrication, *Proceedings, Institution of Mechanical Engineers*, **180**, Pt 3B, 47, [339].

Arnold, R. N., Bycroft, G. N. & Warburton, G. B. (1955). Forced vibrations of a body on an infinite elastic solid. *Trans. ASME. Series E, Journal of Applied Mechanics*, **22**, 391, [348, 349].

Arutiunian, N. Kh. (1959). The plane contact problem in the theory of creep. *PMM* **23**, 1283, [197, 198].

Barber, J. R. (1969). Thermoelastic instabilities in the sliding of conforming solids. *Proceedings, Royal Society*, A312, 381, [391].

Barber, J. R. (1971a). Solution of heated punch problem by point source methods. *International Journal of Engineering Sciences*, **9**, 1165, [381].

Barber, J. R. (1971b). Effect of thermal distortion on constriction resistance. *International Journal of Heat and Mass Transfer*, **14**, 751, [382, 390].

Barber, J. R. (1972). Distortion of the semi-infinite solid due to transient surface heating. *International Journal of Mechanical Sciences*, **14**, 377, [395].

Barber, J. R. (1973). Indentation of a semi-infinite solid by a hot sphere. *International Journal of Mechanical Sciences*, 15, 813, [385].

Barber, J. R. (1975). Thermoelastic contact problems. In *The Mechanics of Contact between Deformable Bodies*, Eds. de Pater & Kalker. Delft: University Press, [380].

Barber, J. R. (1976). Some thermoelastic contact problems involving frictional heating. *Quarterly Journal of Mechanics and Applied Mathematics*, 29, 2, [393, 395].

Barber, J. R. (1978). Contact problems involving a cooled punch. *Journal of Elasticity*, 8, 409, [387, 389, 390].

Barber, J. R. (1980a). Some implications of Dundurs' Theorem for thermoelastic contact and crack problems. *Journal of Mechanical Engineering Sciences*, 22, 229, [381].

Barber, J. R. (1980b). The transient thermoelastic contact of a sphere sliding on a plane. *Wear*, 59, 21, [396].

Barber, J. R. (1982). Indentation of an elastic half-space by a cooled flat punch. *Quarterly Journal of Mechanics and Applied Mathematics*, 35, 141, [390].

Barber, J. R. & Martin-Moran, C. J. (1982). Green's functions for transient thermoelastic contact problems for the half-plane. *Wear*, 79, 11, [395].

Barovich, D., Kingsley, S. C. & Ku, T. C. (1964). Stresses on a thin strip or slab with different elastic properties from that of the substrate. *International Journal of Engineering Sciences*, 2, 253, [140].

Beale, E. M. L. (1959). On quadratic programming. *Naval Research and Logistics Quarterly*, 6, [152].

Bedding, R. J. & Willis, J. R. (1973). Dynamic indentation of an elastic half-space. *Journal of Elasticity*, 3, 289, [359].

Beeching, R. & Nicholls, W. (1948). Theoretical discussion of pitting failures in gears. *Proceedings, Institution of Mechanical Engineers*, 158, 317, [104].

Belajev, N. M. (1917). *Bulletin of Institution of Ways and Communications*, St Petersburg, [66].

Bentall, R. H. & Johnson, K. L. (1967). Slip in the rolling contact of two dissimilar elastic rollers. *International Journal of Mechanical Sciences*, 9, 389, [150, 243, 248, 249, 314].

Bentall, R. H. & Johnson, K. L. (1968). An elastic strip in plane rolling contact. *International Journal of Mechanical Sciences*, 10, 637, [138, 152, 314].

Bhasin, Y. P., Oxley, P. L. B. & Roth, R. N. (1980). An experimentally-determined slip-line field for plane-strain wedge indentation of a strain-hardening material. *Journal of the Mechanics and Physics of Solids*, 28, 149, [165].

Bishop, J. F. W. (1953). Complete solutions to problems of deformation of a plastic-rigid material. *Journal of Mechanics and Physics of Solids*, 2, 43, [170].

Bishop, R. F., Hill, R. & Mott, N. F. (1945). The theory of indentation and hardness tests. *Proceedings, Physics Society*, 57, 147, [172].

Bland, D. R. & Ford, H. (1948). The calculation of roll force and torque in cold strip rolling. *Proceedings, Institution of Mechanical Engineers*, 159, 144, [327].

Bogy, D. B. (1971). Two edge-bonded elastic wedges of different materials and wedge angles. *Trans. ASME, Series E, Journal of Applied Mechanics*, 38, 377, [109].

Boley, B. A. & Weiner, J. H. (1960). *Theory of Thermal Stresses*. New York: Wiley, [380].

Boussinesq, J. (1885). *Application des Potentials à l'etude de l'équilibre et du mouvement des solides élastiques*. Paris: Gauthier-Villars, [45, 108].

Bowden, F. P. & Tabor, D. (1951) Vol. I, (1964) Vol. II. *Friction and Lubrication of Solids*. London: Oxford University Press, [204, 234].

Brewe, D. E. & Hamrock, B. J. (1977). Simplified solution for elliptical-contact deformation between two elastic solids. *Journal of Lubrication Technology, Trans. ASME, Series F*, 99, 485, [98].

Bryant, M. D. & Keer, L. M. (1982). Rough contact between elastically and geometrically similar curved bodies. *Trans. ASME, Series E, Journal of Applied Mechanics*, 49, 345, [75, 210].

Bufler, H. (1959). Zur Theorie der rollenden Reibung. *Ing. Arch.*, 27, 137, [207, 247, 248].

Burton, R. A. (1980). Thermal deformation in frictionally heated contact. *Wear*, 59, 1, [391].

Burton, R. A. & Nerlikar, V. (1975). Large disturbance solutions for thermoelastic deformed surfaces. *Journal of Lubrication Technology, Trans. ASME, Series F*, 97, 539, [393].

Burton, R. A., Nerlikar, V. & Kilaparta, S. R. (1973). Thermoelastic instability in a seal-like configuration. *Wear*, 24, 177, [395].

Calladine, C. R. & Greenwood, J. A. (1978). Line and point loads on a non-homogeneous incompressible elastic half-space. *Quarterly Journal of Mechanics and Applied Mathematics*, 31, 507, [135].

Calvit, H. H. (1967). Numerical solution of impact of a rigid sphere on a linear visocelastic half-space and comparison with experiment. *International Journal of Solids and Structures*, 3, 951, [369].

Cameron, A. (1966). *Principles of Lubrication.* London: Longmans, [329].

Carslaw, H. S. & Jaeger, J. C. (1959). *Conduction of Heat in Solids*, 2nd Ed., Oxford: Clarendon, [375, 376].

Carter, F. W. (1926). On the action of a locomotive driving wheel. *Proceedings, Royal Society*, A112, 151, [252, 255, 262].

Cattaneo, C. (1938). Sul contatto di due corpi elastici: distribuzione locale degli sforzi. *Rendiconti dell' Accademia nazionale dei Lincei*, 27, Ser. 6, 342, 434, 474, [214].

Cerruti, V. (1882). Roma, Acc. Lincei, *Mem. fis. mat.* [45].

Challen, J. M. & Oxley, P. L. B. (1979). Different regimes of friction and wear using asperity, deformation models. *Wear*, 53, 229, [240].

Chartet, A. (1947). Propriétés générales des contacts de roulement. *Comptes. rend. Acad. Sci.*, 225, 986, [252].

Cheng, H. (1970). A numerical solution of the elastohydrodynamic film thickness in elliptical contact. *Trans. ASME, Series F*, 92, 155, [339].

Childs, T. H. C. (1970). The sliding of rigid cones over metals in high adhesion conditions. *International Journal of Mechanical Sciences*, 12, 393, [241].

Childs, T. H. C. (1973). The persistence of asperities in indentation experiments. *Wear*, 25, 3, [404].

Clark, S. K. (Ed.) (1971). *Mechanics of Pneumatic Tyres.* Washington, D.C.: National Bureau of Standards, Monograph 122, [283].

Cocks, M. (1962). Interaction of sliding metal surfaces. *Journal of Applies Physics*, 33, 2152, [240].

Cole, J. D. & Huth, J. H. (1958). Stresses produced in a half-plane by moving loads. *Trans. ASME, Journal of Applied Mechanics*, 25, 433, [370, 372].

Collins, I. F. (1978). On the rolling of a rigid cylinder on rigid/perfectly plastic half-space. *J. Mech. Appliquée*, 2, 431, [299].

Collins, I. F. (1980). Geometrically self-similar deformations of a plastic wedge. *International Journal of Mechanical Sciences*, 22, 735, [236, 237].

Comninou, M. (1976). Stress singularities at a sharp edge in contact problems with friction. *ZAMP*, 27, 493, [109].

Comninou, M. & Dundurs, J. (1979). On the Barber boundary conditions for the thermoelastic contact. *Trans. ASME, Series E, Journal of Applied Mechanics*, 46, 849, [389].

Comninou, M., Dundurs, J. & Barber, J. R. (1981). Planar Hertz contact with heat conduction. *Trans. ASME, Series E, Journal of Applied Mechanics*, 48, 549, [390].

Conway, H. D. & Engel, P. A. (1969). Contact stresses in slabs due to round rough indenters. *International Journal of Mechanical Sciences*, 11, 709, [138, 141].

Conway, H. D., Vogel, S. M., Farnham, K. A. & So, S. (1966). Normal and shearing contact stresses in indented strips and slabs. *International Journal of Engineering Sciences*, 4, 343, [138].

Cooper, D. H. (1969). Hertzian contact-stress deformation coefficients. *Trans. ASME, Series E, Journal of Applied Mechanics*, **36**, 296, [98].

Craggs, J. W. & Roberts, A. M. (1967). On the motion of a heavy cylinder over the surface of an elastic solid. *Trans. ASME, Series E, Journal of Applied Mechanics*, **34**, 207, [372, 373].

Crane, F. A. A. & Alexander, J. M. (1968). Slip-line fields and deformation in hot rolling of strips. *Journal of the Institute of Metals*, **96**, 289, [324].

Crook, A. W. (1952). A study of some impacts between metal bodies by a piezoelectric method. *Proceedings, Royal Society*, **A212**, 377, [360, 364].

Crook, A. W. (1957). Simulated gear-tooth contacts: some experiments on their lubrication and subsurface deformation. *Proceedings, Institution of Mechanical Engineers*, **171**, 187, [292].

Crook, A. W. (1961). Elastohydrodynamic lubrication of rollers, *Nature*, **190**, 1182, [335].

Crook, A. W. (1963). Lubrication of rollers, Pt IV, *Phil. Trans. Royal Society*, A No. 1056, **255**, 281, [338].

Davies, R. M. (1948). A critical study of the Hopkinson pressure bar. *Phil. Trans. Royal Society*, **A240**, 375, [360].

Davies, R. M. (1949). The determination of static and dynamic yield stresses using a steel ball. *Proceedings, Royal Society*, **A197**, 416, [157].

de Pater, A. D. (1964). On the reciprocal pressure between two elastic bodies in contact. In *Rolling Contact Phenomena*, ed. Bidwell. New York: Elsevier, [49].

Den Hartog, J. P. (1956). *Mechanical Vibrations*, 4th Ed., Chapter 8, §8. London: McGraw-Hill, [350].

Denton, B. K. & Crane, F. A. A. (1972). Roll load and torque in the hot rolling of steel strip. *Journal of Iron and Steel Institute*, **210**, 606, [326].

Deresiewicz, H. (1954). Contact of elastic spheres under an oscillating torsional couple. *Trans. ASME, Journal of Applied Mechanics*, **21**, 52, [233].

Deresiewicz, H. (1957). Oblique contact of non-spherical bodies. *Trans. ASME, Journal of Applied Mechanics*, **24**, 623, [220].

Deresiewicz, H. (1958). Mechanics of granular materials. *Advances in Applied Mechanics*, **5**, 233, [231].

Deresiewicz, H. (1968). A note on Hertz impact. *Acta Mechanica*, **6**, 110, [353].

Dewhurst, P., Collins, I. F. & Johnson, W. (1973). A class of slip-line field solutions for the hot rolling of strip. *Journal of Mechanical Engineering Sciences*, **15**, 439, [324, 325].

Dow, T. A. & Burton, R. A. (1972). Thermoelastic instability in the absence of wear. *Wear*, **19**, 315, [395].

Dowson, D. & Higginson, G. R. (1977). *Elastohydrodynamic Lubrication*, 2nd Ed. Oxford: Pergamon, [333].

Dowson, D., Higginson, G. R. & Whitaker, A. V. (1962). Elastohydrodynamic lubrication: A survey of isothermal solutions. *Journal of Mechanical Engineering Science*, **4**, 121, [334].

Drutowski, R. C. (1959). Energy losses of balls rolling on plates. *Trans. ASME, Series D*, **81**, 233, [311].

Dugdale, D. S. (1953). Wedge indentation experiments with cold worked metals. *Journal of Mechanics and Physics of Solids*, **2**, 14, [171].

Dugdale, D. S. (1954). Cone indentation experiments. *Journal of Mechanics and Physics of Solids*, **2**, 265, [171].

Dumas, G. & Baronet, C. N. (1971). Elasto-plastic indentation of a half-space by a long rigid cylinder. *International Journal of Mechanical Sciences*, **13**, 519, [172].

Dundurs, J. (1974). Distortion of a body caused by thermal expansion. *Mech. Res. Comm.*, **1**, 121, [381].

Dundurs, J. (1975). Properties of elastic bodies in contact. In *Mechanics of Contact between Deformable Bodies*, ed. de Pater & Kalker. Delft: University Press, [142].

Dundurs, J. & Lee, M.-S. (1972). Stress concentration at a sharp edge in contact problems. *Journal of Elasticity*, 2, 109, [109].

Duvaut, G. & Lions, J.-L. (1972). *Les inéquations en mécanique et en physique.* Paris: Dunod, [151, 264].

Dyson, A. (1965). Approximate calculations of Hertzian compressive stresses and contact dimensions. *Journal of Mechanical Engineering Sciences*, 7, 224, [98].

Dyson, A., Naylor, H. & Wilson, A. R. (1956). Measurement of oil film thickness in elasto-hydrodynamic contacts. Symposium on EHL, *Proceedings, Institution of Mechanical Engineers*, 180, Pt. 3B, 119, [339].

Eason, G. (1965). The stresses produced in a semi-infinite solid by a moving surface force. *International Journal of Engineering Sciences*, 2, 581, [370].

Eason, G. & Shield, R. T. (1960). The plastic indentation of a semi-infinite solid by a perfectly rough punch. *ZAMP*, 11, 33, [169, 177].

Edwards, C. M. & Halling, J. (1968a). An analysis of the plastic interaction of surface asperities. *Journal of Mechanical Engineering Sciences*, 10, 101, [240].

Edwards, C. M. & Halling, J. (1968b). Experimental study of plastic interaction of model surface asperities during sliding. *Journal of Mechanical Engineering Sciences*, 10, 121, [240].

Eldridge, K. R. & Tabor, D. (1955). The mechanism of rolling friction. I: The plastic range. *Proceedings, Royal Society*, A229, 181, [291].

Essenburg, F. (1962). On surface constraints in plate problems. *Trans. ASME, Series E, Journal of Applied Mechanics*, 29, 340, [144].

Fessler, H. & Ollerton, E. (1957). Contact stresses in toroids under radial loads. *British Journal of Applied Physics*, 8, 387, [67, 99].

Fichera, G. (1964). Problemi elastostatici con vincoli unilaterale: il problema di Signorini con ambigue condizioni al contorno. *Mem. Accad. Naz. Lincei*, Series 8, 7, 91, [151].

Flamant (1892). *Compt. Rendus*, 114, 1465, Paris, [15].

Flom, D. G. & Bueche, A. M. (1959). Theory of rolling friction for spheres. *Journal of Applied Physics*, 30, 1725, [305].

Föppl, L. (1947). *Die Strange Lösung die Rollende Reibung.* München, [252].

Follansbee, P. S. & Sinclair, G. B. (1984). Quasi-static normal indentation of an elastic-plastic half-space by a rigid sphere. *International Journal of Solids and Structures*, 20, 81, [172, 176, 177, 183].

Ford, H. & Alexander, J. M. (1963). *Advanced Mechanics of Materials.* London: Longmans, [157, 166, 172].

Foss, F. E. & Brumfield, R. C. (1922). Some measurements of the shape of Brinell ball indentation. *Proceedings, ASTM*, 22, 312, [179, 182].

Francis, H. A. (1976). Phenomenological analysis of plastic spherical indentation. *Trans. ASME, Series H*, 98, 272, [177].

Frank, F. (1965). Grundlagen zur Berechnung der Seiten führung Skennlinien von Reifen. *Kaut. Gummi*, 8, 515, [283].

Fredriksson, B. (1976). On elastostatic contact problems with friction: a finite element analysis. Doctoral dissertation, University of Linköping, Sweden, [152].

Fromm, H. (1927). Berechnung des Schlupfes beim Rollen deformierbaren Scheiben, *ZAMP*, 7, [252, 281].

Fromm, H. (1943). Lilienthal Gesellschaft Report, 169. For English summary see Hadekel (1952) and Clark (1971), [281].

Galin, L. A. (1945). Imprint of stamp in the presence of friction and adhesion. *PMM*, 9 (5), 413, [40].

Galin, L. A. (1953). *Contact Problems in the Theory of Elasticity.* Moscow. (English translation by H. Moss, North Carolina State College, Department of Mathematics, 1961), [29, 49, 135].

Garg, V. K., Anand, S. C. & Hodge, P. G. (1974). Elastic-plastic analysis of a wheel rolling on a rigid track. *International Journal of Solids and Structures*, **10**, 945, [290].

Gdoutos, E. E. & Theocaris, P. S. (1975). Stress concentrations at the apex of a plane indenter acting on an elastic half-plane. *Journal of Applied Mechanics, Trans ASME, Series E*, **42**, 688, [109].

Gladwell, G. M. L. (1968). The calculation of mechanical impedances relating to an indenter vibrating on the surface of a semi-infinite elastic body. *Journal of Sound and Vibration*, **8**, 215, [348, 349].

Gladwell, G. M. L. (1976). On some unbounded contact problems in plane elasticity theory. *Trans. ASME, Series E, Journal of Applied Mechanics*, **43**, 263, [142].

Gladwell, G. M. L. (1980). *Contact Problems in the Classical Theory of Elasticity.* Alphen aan den Rijn: Sijthoff and Noordhoff, [39, 49, 68, 78, 135, 137, 141].

Gladwell, G. M. L. & England, A. H. (1975). Contact problems for the spherical shell. *Proceedings 12th Annual Meeting of Society of Engineering Sciences.* Austin: University of Texas, [144].

Goldsmith, W. (1960). *Impact.* London: Arnold, [342, 360, 363].

Goodman, L. E. (1960). A review of progress in analysis of interfacial slip damping. *Proceedings of ASME Colloquium on Structural Damping*, ed. Ruzicka. New York: Pergamon, [230].

Goodman, L. E. (1962). Contact stress analysis of normally loaded rough spheres. *Trans. ASME, Series E, Journal of Applied Mechanics*, **29**, 515, [119, 121].

Goodman, L. E. & Brown, C. B. (1962). Energy dissipation in contact friction: constant normal and cyclic tangential loading. *Trans. ASME, Series E, Journal of Applied Mechanics*, **29**, 17, [227].

Goodman, L. E. & Keer, L. M. (1965). The contact stress problem for an elastic sphere indenting an elastic cavity. *International Journal of Solids and Structures*, **1**, 407, [117].

Graff, K. F. (1975). *Wave Motion in Elastic Solids.* Oxford: Clarendon, [340].

Graham, G. A. C. (1967). The contact problem in the linear theory of viscoelasticity when the contact area has any number of maxima and minima. *International Journal of Engineering Sciences*, **5**, 495, [193].

Graham, G. A. C. (1973). A contribution to the Hertz theory of impact. *International Journal of Engineering Sciences*, **11**, 409, [353].

Gray, G. G. & Johnson, K. L. (1972). The dynamic response of elastic bodies in rolling contact to random roughness of their surfaces. *Journal of Sound and Vibration*, **22**, 323, [350].

Green, A. E. (1949). On Boussinesq's problem and penny-shaped cracks. *Proceedings, Cambridge Philosophical Society*, **45**, 251, [210].

Green, A. E. & Zerna, W. (1954). *Theoretical Elasticity.* Oxford: Clarendon, [49, 135].

Green, A. P. (1954). The plastic yielding of metal junctions due to combined shear and pressure. *Journal of Mechanics and Physics of Solids*, **2**, 197, [240].

Greenwood, J. A. (1972). An extension of the Grubin theory of elastohydrodynamic lubrication. *J. Phys. D (App. Phys.)*, **5**, 2195, [336, 337, 338].

Greenwood, J. A. (1984). A unified theory of surface roughness. *Proceedings, Royal Society*, **A393**, 133, [410, 411].

Greenwood, J. A. & Johnson, K. L. (1981). The mechanics of adhesion of viscoelastic solids. *Philosophical Magazine*, **43**, 697, [128].

Greenwood, J. A., Minshall, J. & Tabor, D. (1961). Hysteresis losses in rolling and sliding friction. *Proceedings, Royal Society*, **A259**, 480, [285, 286].

Greenwood, J. A. & Rowe, G. W. (1965). Deformation of surface asperities during bulk plastic flow. *Journal of Applied Physics*, **36**, 667, [404].

Greenwood, J. A. & Tabor, D. (1955). Deformation properties of friction junctions. *Proceedings, Physical Society, Series B*, **68**, 609, [240].

Greenwood, J. A. & Tabor, D. (1958). The friction of hard sliders on lubricated rubber: the importance of deformation losses. *Proceedings, Physical Society*, 71, 989, [286].

Greenwood, J. A. & Tripp, J. H. (1967). The elastic contact of rough spheres. *Trans. ASME, Series E, Journal of Applied Mechanics*, 34, 153, [417, 418, 419, 420].

Greenwood, J. A. & Williamson, J. B. P. (1966). Contact of nominally flat surfaces. *Proceedings, Royal Society*, A295, 300, [411, 412, 413].

Grubin, A. N. (1949). Central Sc. Res. Inst. for Tech. and Mech. Eng., Book No. 30, Moscow (DSIR Translation No. 337) (see Cameron, 1966), [335, 336, 337].

Grunsweig, J., Longman, I. M. & Petsch, N. J. (1954). Calculations and measurements of wedge indentation. *Journal of Mechanics and Physics of Solids*, 2, 81, [164].

Gupta, P. K. & Walowit, J. A. (1974). Contact stresses between an elastic cylinder and a layered solid. *Journal of Lubrication Technology, Trans. ASME, Series F*, 94, 251, [140].

Gupta, P. K., Walowit, J. A. & Finkin, E. F. (1973). Stress distributions in plane strain layered elastic solids subjected to arbitrary boundary loading. *Journal of Lubrication Technology, Trans. ASME, Series F*, 93, 427, [140].

Haddow, J. B. (1967). On a plane strain wedge indentation paradox. *International Journal of Mechanical Sciences*, 9, 159, [165].

Hadekel, R. (1952). *The Mechanical Characteristics of Pneumatic Tyres.* Ministry of Supply, London, S & T Memo 10/52, [281].

Haines, D. J. (1964–5). Contact stresses in flat elliptical contact surfaces which support radial and shearing forces during rolling. *Proceedings, Institution of Mechanical Engineers*, 179 (Part 3), [265].

Haines, D. J. & Ollerton, E. (1963). Contact stress distributions on elliptical contact surfaces subjected to radial and tangential forces. *Proceedings, Institution of Mechanical Engineers*, 177, 95, [261, 265].

Halling, J. (1959). Effect of deformation of the surface texture on rolling resistance. *British Journal of Applied Physics*, 10, 172, [311].

Halling, J. & Al-Qishtaini, M. A. (1967–8). Some factors affecting the contact conditions and slip between a rolling ball and its track. *Proceedings, Institution of Mechanical Engineers*, 182 (Part I), 757, [268].

Halling, J. & Nuri, K. A. (1975). Contact of rough surfaces of working-hardening materials. In *The Mechanics of Contact between Deformable Bodies*, Eds. de Pater & Kalker. Delft: University Press, [412].

Hamilton, G. M. (1963). Plastic flow in rollers loaded above the yield point. *Proceedings, Institution of Mechanical Engineers*, 177, 667, [292, 310].

Hamilton, G. M. (1983). Explicit equations for the stresses beneath a sliding spherical contact. *Proceedings, Institution of Mechanical Engineers*, 197C, 53, [75, 210].

Hamilton, G. M. & Goodman, L. E. (1966). The stress field created by a circular sliding contact. *Trans. ASME, Journal of Applied Mechanics*, 33, 371, [75, 210].

Hamilton, G. M. & Moore, S. L. (1971). Deformation and pressure in an elastohydrodynamic contact, *Proceedings, Royal Society*, A322, 313, [335].

Hamrock, B. J. & Dowson, D. (1977). Isothermal elastohydrodynamic lubrication of point contacts. *Trans. ASME, Series F*, 99, 264, [339].

Hardy, C., Baronet, C. N. & Tordion, G. V. (1971). Elastoplastic indentation of a half-space by a rigid sphere. *Journal of Numerical Methods in Engineering*, 3, 451, [172, 173, 176, 183].

Harris, T. A. (1966). *Rolling Bearing Analysis.* London, New York and Sidney: Wiley, [129].

Heathcote, H. L. (1921). The ball bearing: In the making, under test and on service. *Proceedings of the Institute of Automobile Engineers*, 15, 569, [269, 271].

Heinrich, G. & Desoyer, K. (1967). Rollreibung mit axialem Schub. *Ing. Arch.*, 36, 48, [256].

Herrebrugh, K. (1968). Solving the incompressible and isothermal problem in elastohydro-dynamic lubrication through an integral equation. *Journal of Lubrication Technology, Trans. ASME, Series F*, **90**, 262, [332, 335].

Hertz, H. (1882a). Über die Berührung fester elastischer Körper (On the contact of elastic solids). *J. reine und angewandte Mathematik*, **92**, 156–171. (For English translation see *Miscellaneous Papers by H. Hertz*, Eds. Jones and Schott, London: Macmillan, 1896.) [90, 243].

Hertz, H. (1882b). Über die Berührung fester elastische Körper and über die Harte (On the contact of rigid elastic solids and on hardness). *Verhandlungen des Vereins zur Beförderung des Gewerbefleisses*, Leipzig, Nov. 1882. (For English translation see *Miscellaneous Papers by H. Hertz*, Eds. Jones and Schott, London: Macmillan, 1896.) [90, 156].

Hetenyi, M. & McDonald, P. H. (1958). Contact stresses under combined pressure and twist. *Trans. ASME, Journal of Applied Mechanics*, **25**, 396, [82, 233].

Hill, R. (1950a). *Theory of Plasticity*. Oxford: University Press, [157, 165, 172, 174, 175].

Hill, R. (1950b). A theoretical investigation of the effect of specimen size in the measurement of hardness. *Philosophical Magazine*, **41**, 745, [171].

Hill, R., Lee, E. H. & Tupper, S. J. (1947). Theory of wedge indentation of ductile metals. *Proceedings, Royal Society*, **A188**, 273, [162].

Hirst, W. & Howse, M. G. J. W. (1969). The indentation of materials by wedges. *Proceedings, Royal Society*, **A311**, 429, [183].

Hobbs, A. E. W. (1967). *A Survey of Creep*. British Rail Dept. DYN 52, Derby, [268].

Huber, M. T. (1904). Zur Theorie der Berührung fester elastische Körper. *Ann. der Phys.*, **14**, 153, [62].

Hunter, S. C. (1956–7). Energy absorbed by elastic waves during impact. *Journal of Mechanics and Physics of Solids*, **5**, 162, [354].

Hunter, S. C. (1960). The Hertz problem for a rigid spherical indenter and a viscoelastic half-space. *Journal of Mechanics and Physics of Solids*, **8**, 219, [196, 369].

Hunter, S. C. (1961). The rolling contact of a rigid cylinder with a viscoelastic half-space. *Trans. ASME, Series E, Journal of Applied Mechanics*, **28**, 611, [305, 306].

Hutchings, I. M., Macmillan, N. H. & Rickerby, D. G. (1981). Further studies of the oblique impact of a hard sphere against a ductile solid. *International Journal of Mechanical Sciences*, **23**, 639, [365].

Ishlinsky, A. J. (1944). The axi-symmetrical problem in plasticity and the Brinell test. *PMM*, **8**, 201, [170].

Jaeger, J. C. (1942). Moving sources of heat and temperature of sliding contacts. *Proceedings, Royal Society, NSW*, **56**, 203, [378].

Johnson, K. L. (1955). Surface interaction between elastically loaded bodies under tangential forces. *Proceedings, Royal Society*, **A230**, 531, [75, 220].

Johnson, K. L. (1958a). The effect of a tangential contact force upon the rolling motion of an elastic sphere on a plane. *Trans. ASME, Journal of Applied Mechanics*, **25**, 339, [258, 260, 261, 266].

Johnson, K. L. (1958b). The effect of spin upon the rolling motion of an elastic sphere on a plane. *Trans ASME, Journal of Applied Mechanics*, **25**, 332, [258, 266].

Johnson, K. L. (1959). The influence of elastic deformation upon the motion of a ball rolling between two surfaces. *Proceedings, Institution of Mechanical Engineers*, **173**, 795, [266].

Johnson, K. L. (1961). Energy dissipation at spherical surfaces in contact transmitting oscillating forces. *Journal of Mechanical Engineering Science*, **3**, 362, [228, 229, 231].

Johnson, K. L. (1962a). Tangential tractions and microslip in rolling contact. In *Rolling Contact Phenomena*, Ed. Bidwell, p. 6. New York: Elsevier, [260].

Johnson, K. L. (1962b). A shakedown limit in rolling contact. *Proceedings, 4th US National Congress of Applied Mechanics, Berkeley, ASME*, [288].

Johnson, K. L. (1968*a*). Deformation of a plastic wedge by a rigid flat die under the action of a tangential force. *Journal of Mechanics and Physics of Solids*, 16, 395, [234].

Johnson, K. L. (1968*b*). An experimental determination of the contact stresses between plastically deformed cylinders and spheres. In *Engineering Plasticity*, Eds. Heyman & Leckie. Cambridge: University Press, [183].

Johnson, K. L. (1970*a*). The correlation of indentation experiments. *Journal of Mechanics and Physics of Solids*, 18, 115, [172, 175].

Johnson, K. L. (1970*b*). Regimes of elastohydrodynamic lubrication. *Journal of Mechanical Engineering Science*, 12, 9, [335].

Johnson, K. L. (1975). Non-Hertzian contact of elastic spheres. In *The Mechanics of Contact between Deformable Bodies*, Eds. de Pater & Kalker, p. 26. Delft: University Press, [419].

Johnson, K. L. (1976). Adhesion at the contact of solids. *Theoretical and Applied Mechanics*, Proc. 4th IUTAM Congress, Ed. Koiter, p. 133. Amsterdam: North-Holland, [126].

Johnson, K. L. & Bentall, R. H. (1969). The onset of yield in the cold rolling of thin strips. *Journal of Mechanics and Physics of Solids*, 17, 253, [319].

Johnson, K. L. & Bentall, R. H. (1977). A numerical method for finding elastic contact pressures. Cambridge University Engineering Department Report C-MECH/TR14, [56].

Johnson, K. L., Greenwood, J. A. & Higginson, J. G. (1985). The contact of elastic wavy surfaces. *International Journal of Mechanical Sciences*, 27, [403].

Johnson, K. L. & Jefferis, J. A. (1963). Plastic flow and residual stresses in rolling and sliding contact. *Proc. Institution of Mechanical Engineers Symposium on Rolling Contact Fatigue*, London, p. 50, [207, 290].

Johnson, K. L., Kendall, K. & Roberts, A. D. (1971). Surface energy and the contact of elastic solids. *Proceedings, Royal Society*, A324, 301, [126].

Johnson, K. L. & O'Connor, J. J. (1964). The mechanics of fretting. *Proc. Institution of Mechanical Engineers, Applied Mechanics Convention*, Newcastle, 178, Part 3J, 7, [230].

Johnson, K. L., O'Connor, J. J. & Woodward, A. C. (1973). The effect of indenter elasticity on the Hertzian fracture of brittle materials. *Proceedings, Royal Society*, A334, 95, [124].

Johnson, K. L. & Tevaarwerk, J. L. (1977). Shear behaviour of elastohydrodynamic oil films. *Proceedings, Royal Society*, A352, 215, [339].

Johnson, K. L. & White, I. C. (1974). Rolling resistance measurements at high loads. *International Journal of Mechanical Sciences*, 16, 939, [301, 310].

Johnson, W. (1972). *Impact Strength of Materials.* London: Arnold, [342, 360, 361, 366].

Johnson, W., Mahtab, F. U. & Haddow, J. B. (1964). Indentation of a semi-infinite block by a wedge of comparable hardness. *International Journal of Mechanical Sciences*, 6, 329, [168].

Johnson, W., Travis, F. W. & Loh, S. Y. (1968). High speed cratering in wax and plasticine. *International Journal of Mechanical Sciences*, 10, 593, [367].

Kalker, J. J. (1964). The transmission of a force and couple between two elastically similar rolling spheres. *Proc. Kon. Ned. Akad. van Wetenschappen*, B67, 135, [258].

Kalker, J. J. (1967*a*). On the rolling contact of two elastic bodies in the presence of dry friction. Doctoral Dissertation, Technical University Delft (Nad. Drukkerig Bedrigf NV – Leiden), [76, 256, 258, 259, 260, 263, 265, 266, 431].

Kalker, J. J. (1967*b*). A strip theory for rolling with slip and spin. *Proc. Kon. Ned. Akad. van Wetenschappen*, B70, 10, [261, 263].

Kalker, J. J. (1969). *Transient phenomena in two elastically similar rolling cylinders in the presence of dry friction.* Lab. v. Tech. Mech., Technical University Delft, Report No. 11, [264, 271, 272].

Kalker, J. J. (1970). Transient phenomena in two elastic cylinders rolling over each other with dry friction. *Trans AMSE, Series E, Journal of Applied Mechanics*, 37, 677, [271, 272].

Kalker, J. J. (1971a). A minimum principle for the law of dry friction, Pt I. *Trans. ASME, Series E, Journal of Applied Mechanics*, 38, 875, [256, 272].

Kalker, J. J. (1971b). A minimum principle for the law of dry friction, Pt. 2 Application to non-steadily rolling elastic cylinders. *Trans ASME, Series E, Journal of Applied Mechanics*, 38, 881, [271, 272].

Kalker, J. J. (1973). Simplified theory of rolling contact. Delft Progress Report, Series C, 1, 1, [276].

Kalker, J. J. (1977). Variational principles of contact elastostatics. *Journal Inst. Math. & Appl.*, 20, 199, [151].

Kalker, J. J. (1978). Numerical contact elastostatics. In *Contact Problems and Load Transfer in Mechanical Assemblages*, Proc. Euromech Coll. No. 110, Linköping, Sweden, [151].

Kalker, J. J. (1979). The computation of 3-D rolling contact with dry friction. *International Journal of Numerical Methods in Engineering*, 14, 1293, [264].

Kalker, J. J. & van Randen, Y. (1972). A minimum principle for frictionless elastic contact with application to non-Hertzian half-space contact problems. *Journal of Engineering Mathematics*, 6, 193, [56, 152].

von Kármán, T. (1925). On the theory of rolling. *Z. angew Math. Mech.* 5, 139, [321].

Keer, L. M., Dundurs, J. & Tsai, K. C. (1972). Problems involving receding contact between a layer and a half-space. *Trans ASME, Series E, Journal of Applied Mechanics*, 39, 1115, [142].

Khadem, R. & O'Connor, J. J. (1969a). Adhesive or frictionless compression of an elastic rectangle between two identical elastic half-spaces. *International Journal of Engineering Sciences*, 7, 153, [111].

Khadem, R. & O'Connor, J. J. (1969b). Axial compression of an elastic circular cylinder in contact with two identical elastic half-spaces. *International Journal of Engineering Sciences*, 7, 785, [111].

Kolsky, H. (1953). *Stress Waves in Solids*. Oxford: University Press, [340].

Kornhauser, M. (1964). *Structural Effects of Impact*. Sutton, Surrey: Spartan, [367].

von Kunert, K. (1961). Spannungsverteilung im Halbraum bei elliptische Flächenpressungs-verteilung über einer rechteckigen Druckfläche. *Forschung a. d. Gebeite des Ingenieur-wesens*, 27, 165, [56, 134].

Kuznetsov, A. I. (1962). Penetration of rigid dies into a half-space with power-law strain hardening and nonlinear creep. *PMM*, 26, 717, [198, 199].

Lamb, H. (1904). On the propagation of tremors over the surface of an elastic solid. *Phil. Trans. Royal Society*, A203, 1, [345].

Lazan, B. J. (1968). *Damping of Materials and Members in Structural Mechanics*. Oxford: Pergamon, [181].

Lee, C. H., Masaki, S. & Kobayashi, S. (1972). Analysis of ball indentation. *International Journal of Mechanical Sciences*, 14, 417, [172].

Lee, E. H. & Radok, J. R. M. (1960). The contact problem for viscoelastic bodies. *Trans. ASME, Series E, Journal of Applied Mechanics*, 27, 438, [188, 189, 193].

Lee, E. H. & Rogers, T. G. (1963). Solution of viscoelastic stress analysis problems using measured creep or relaxation functions. *Trans ASME, Series E, Journal of Applied Mechanics*, 30, 127, [186].

Lo, C. C. (1969). Elastic contact of rough cylinders. *International Journal of Mechanical Sciences*, 11, 105, [417].

Lockett, F. J. (1963). Indentation of a rigid-plastic material by a conical indenter. *Journal of Mechanics and Physics of Solids*, 11, 345, [168, 169].

Longuet-Higgins, M. S. (1957a). The statistical properties of an isotropic random surface. *Phil. Trans. Royal Society*, **A249**, 321, [410].

Longuet-Higgins, M. S. (1957b). The statistical analysis of a random moving surface. *Phil. Trans. Royal Society*, **A250**, 157, [410].

Love, A. E. H. (1929). Stress produced in a semi-infinite solid by pressure on part of the boundary. *Phil. Trans. Royal Society*, **A228**, 377, [54, 59].

Love, A. E. H. (1939). Boussinesq's problem for a rigid cone. *Quarterly Journal of Mathematics* (Oxford series), **10**, 161, [114].

Love, A. E. H. (1952). *A Treatise on the Mathematical Theory of Elasticity*, 4th Edn. Cambridge: University Press, [45, 47, 354].

Lubkin, J. L. (1951). Torsion of elastic spheres in contact. *Trans. ASME, Series E, Journal of Applied Mechanics*, **18**, 183, [233].

Ludema, K. C. & Tabor, D. (1966). The friction and viscoelastic properties of polymeric solids. *Wear*, **9**, 329, [311].

Lundberg, G. (1939). Elastische Berührung zweier Helbräume. *Forschung a. d. Gebeite des Ingenieurwesens*, **10**, 201, [134].

Lundberg, G. & Sjövall, H. (1958). *Stress and Deformation in Elastic Solids*. Pub. No. 4, Inst. Th. of Elast., Chalmers University of Technology, Göteborg, Sweden, [66, 99].

Lur'e, A. I. (1964). *Three-dimensional Problems of the Theory of Elasticity*. English translation by J. R. M. Radok, Interscience, [49, 63].

Lutz, O. (1955 et seq.). Grundsätzliches über stufenlos verstellbare Wälzgetriebe. *Konstruktion*; 7, 330; 9, 169 (1957); **10**, 425 (1958), [260].

Mandel, J. (1967). Résistance au roulement d'un cylindre indéformable sur un massif parfaitement plastique. In *Le Frottement & l'Usure*, p. 25. Paris: GAMI, [296, 297, 301].

Marsh, D. M. (1964). Plastic flow in glass. *Proceedings, Royal Society*, **A279**, 420, [172, 176].

Marshall, E. A. (1968). Rolling contact with plastic deformation. *Journal of Mechanics and Physics of Solids*, **16**, 243, [301].

Matthews, J. R. (1980). Indentation hardness and hot pressing. *Acta Met.*, **28**, 311, [176, 199].

Matthewson, M. J. (1981). Axi-symmetric contact on thin compliant coatings. *Journal of Mechanics and Physics of Solids*, **29**, 89, [141].

Maugis, D. & Barquins, M. (1978). Fracture mechanics and the adherence of viscoelastic bodies. *Journal of Physics D (Applied Physics)*, **11**, 1989, [128].

Maugis, D., Barquins, M. & Courtel, R. (1976). Griffith's crack and adhesion of elastic bodies. *Métaux, Corrosion, Industries*, **51**, 1, [128].

Maw, N., Barber, J. R. & Fawcett, J. N. (1976). The oblique impact of elastic spheres. *Wear*, **38**, 101, [356].

Maw, N., Barber, J. R. & Fawcett, J. N. (1981). The role of tangential compliance in oblique impact. *Journal of Lubrication Technology, Trans. ASME, Series F*, **103**, 74, [356, 357].

May, W. D., Morris, E. L. & Atack, D. (1959). Rolling friction of a hard cylinder over a viscoelastic material. *Journal of Applied Physics*, **30**, 1713, [303].

McCormick, J. A. (1978). *A Numerical Solution for a Generalised Elliptical Contact of Layered Elastic Solids*, MTI Report No. 78 TR 52, Mech. Tech. Inc., Latham, New York, [141].

McEwen, E. (1949). Stresses in elastic cylinders in contact along a generatrix. *Philosophical Magazine*, **40**, 454, [102, 205].

Meijers, P. (1968). The contact problem of a rigid cylinder on an elastic layer. *Applied Science Research*, **18**, 353, [138, 140, 141].

Merritt, H. E. (1935). Worm gear performance. *Proceedings, Institution of Mechanical Engineers*, **129**, 127, [8, 10].

Merwin, J. E. & Johnson, K. L. (1963). An analysis of plastic deformation in rolling contact. *Proceedings, Institution of Mechanical Engineers*, **177**, 676, [292, 294, 309].

Mikhlin, S. G. (1948). *Singular Integral Equations.* Leningrad (English translation by
A. H. Armstrong, Pergamon, 1957), [29, 30].

Mikic, B. B. (1974). Thermal contact conductance: theoretical considerations. *International
Journal of Heat and Mass Transfer*, **17**, 205, [416, 417].

Miller, G. F. & Pursey, H. (1954). Field and radiation impedance of mechanical radiators.
Proceedings, Royal Society, **A223**, 521, [345, 346, 348, 359].

Miller, G. F. & Pursey, H. (1955). Partition of energy between elastic waves. *Proceedings,
Royal Society*, **A233**, 55, [347].

Mindlin, R. D. (1949). Compliance of elastic bodies in contact. *Trans. ASME, Series E,
Journal of Applied Mechanics*, **16**, 259, [74, 82, 214, 220].

Mindlin, R. D. (1954). Mechanics of Granular Media, *Proc. 2nd US National Congress on
Applied Mechanics*, p. 13. New York: ASME, [231].

Mindlin, R. D. & Deresiewicz, H. (1953). Elastic spheres in contact under varying oblique
forces. *Trans. ASME, Series E, Journal of Applied Mechanics*, **20**, 327, [221, 227].

Mindlin, R. D., Mason, W. P., Osmer, J. F. & Deresiewicz, H. (1952). Effects of an oscillating tangential force on the contact surfaces of elastic spheres. *Proc. 1st US National
Congress of Applied Mechanics*, p. 203. New York: ASME, [227].

Mok, C. H. & Duffy, J. (1965). The dynamic stress–strain relation as determined from
impact tests. *International Journal of Mechanical Sciences*, **7**, 355, [363].

Morland, L. W. (1967a). Exact solutions for rolling contact between viscoelastic cylinders.
Quarterly Journal of Mechanics and Applied Mathematics, **20**, 73, [306].

Morland, L. W. (1967b). Rolling contact between dissimilar viscoelastic cylinders.
Quarterly Journal of Applied Mathematics, **25**, 363, [306].

Morton, W. B. & Close, L. J. (1922). Notes on Hertz' theory of contact problems.
Philosophical Magazine, **43**, 320, [62].

Mossakovski, V. I. (1954). The fundamental general problem of the theory of elasticity,
etc. *PMM*, **18**, 187 (in Russian), [74, 80, 119].

Mossakovski, V. I. (1963). Compression of elastic bodies under conditions of adhesion.
PMM, **27**, 418, [119, 123].

Mulhearn, T. O. (1959). Deformation of metals by Vickers-type pyramidal indenters.
Journal of Mechanics and Physics of Solids, **7**, 85, [172].

Murch, L. E. & Wilson, W. R. D. (1975). A thermal elastohydrodynamic inlet zone
analysis. *Journal of Lubrication Technology, Trans. ASME, Series F*, **97**, 212, [339].

Muskhelishvili, N. I. (1946). *Singular Integral Equations.* Moscow. (English translation
by J. R. M. Radok, Noordhoff, 1953.) [29].

Muskhelishvili, N. I. (1949). *Some Basic Problems of the Mathematical Theory of
Elasticity*, 3rd Edn., Moscow (English translation by J. R. M. Radok, Noordhoff, 1953.)
[29, 39].

Nadai, A. I. (1963). *Theory of Flow and Fracture of Solids, Vol. II*, p. 221. New York,
Toronto and London: McGraw-Hill, [37].

Nayak, L. & Johnson, K. L. (1979). Pressure between elastic bodies having a slender area
of contact and arbitrary profiles. *International Journal of Mechanical Sciences*, **21**,
237, [134].

Nayak, P. R. (1971). Random process model of rough surfaces. *Journal of Lubrication
Technology, Trans. ASME, Series F*, **93**, 398, [410, 411].

Noble, B. & Spence, D. A. (1971). *Formulation of Two-dimensional and Axi-symmetric
Boundary Value Problems.* University of Wisconsin Math. Res. Centre Report TR1089,
[49, 78].

Norbury, A. L. & Samuel, T. (1928). The recovery and sinking-in or piling-up of material
in the Brinell test. *Journal Iron and Steel Institute*, **117**, 673, [200].

O'Connor, J. J. & Johnson, K. L. (1963). The role of surface asperities in transmitting
tangential forces between metals. *Wear*, **6**, 118, [422].

Olle₁ ₗon, E. & Haines, D. J. (1963). Contact stress distributions on elliptical contact surfaces subjected to radial and tangential forces. *Proceedings Institution of Mechanical Engineers*, 177, 95, [204, 260].

Onions, R. A. & Archard, J. F. (1973). The contact of surfaces having a random structure. *Journal of Physics D*, 6, 289, [416].

Pacejka, H. B. (1981). The tyre as a vehicle component. In Clark (1971), [282, 283].

Pacejka, H. B. & Dorgham, M. A. (1983). *Tyre Mechanics and its Impact on Vehicle Dynamics*. St Helier, Jersey: Int. Association for Vehicle Design, Interscience Enterprises, [283].

Panek, C. & Dundurs, J. (1979). Thermoelastic contact between bodies with wavy surfaces. *Trans. ASME, Series E, Journal of Applied Mechanics*, 46, 854, [390].

Pao, Y. C., Wu, T.-S. & Chiu, Y. P. (1971). Bounds on the maximum contact stress of an indented layer. *Trans. ASME, Series E, Journal of Applied Mechanics*, 38, 608, [138].

Paris, P. C. & Sih, G. C. (1965). Stress analysis of cracks. In *Fracture Toughness Testing and its Applications*, ASTM STP 381, 30, [401].

Paul, B. & Hashemi, J. (1981). Contact pressure on closely conforming elastic bodies. *Trans. ASME, Series E, Journal of Applied Mechanics*, 48, 543, [150].

Pekeris, C. L. (1955). The seismic surface pulse. *Proc. Nat. Acad. Sci. USA*, 41, 469, [344].

Persson, A. (1964). *On the Stress Distribution of Cylindrical Elastic Bodies in Contact*, Dissertation, Chalmers Tekniska Hogskola, Göteborg, [117, 118].

Petryk, H. (1983). A slip-line field analysis of the rolling contact problem at high loads. *International Journal of Mechanical Sciences*, 25, 265, [299].

Pinnock, P. R., Ward, I. M. & Wolfe, J. M. (1966). The compression of anisotropic fibre monofilaments II. *Proceedings, Royal Society*, A291, 267 (see also, Hadley, D. W., Ward, I. M. & Ward, J. (1965), Proc. Roy. Soc. A285, 275), [135].

Pomeroy, R. J. & Johnson, K. L. (1969). Residual stresses in rolling contact. *Institution of Mechanical Engineers, Journal of Strain Analysis*, 4, 208, [295].

Ponter, A. R. S. (1976). A general shakedown theorem for inelastic materials. *Proc. 3rd Int. Conf. on Struct. Mech. in Reactor Tech.*, Section L, Imperial College, London, [292].

Ponter, A. R. S., Hearle, A. D. & Johnson, K. L. (1985). Application of the kinematical shakedown theorem to rolling and sliding contact. *Journal of Mechanics and Physics of Solids*, 33, [292].

Popov. G. Ia. (1962). The contact problem of the theory of elasticity for the case of a circular contact area. *PMM*, 26, 152, [63].

Poritsky, H. (1950). Stresses and deflections of cylindrical bodies in contact. *Trans. ASME, Series E, Journal of Applied Mechanics*, 17, 191, [104, 205, 215, 252].

Puttick, K. E., Smith, L. S. A. & Miller, L. E. (1977). Stress and strain fields round indentations in polymethylmethacrylate. *Journal of Physics D: Applied Physics*, 10, 617, [178].

Radok, J. R. M. (1957). Viscoelastic stress analysis. *Q. App. Math.*, 15, 198, [187].

Ratwani, M. & Erdogan, F. (1973). On the plane contact problem for a frictionless elastic layer. *International Journal of Solids and Structures*, 43, 921, [142].

Reynolds, O. (1875). On rolling friction. *Phil. Trans. Royal Society*, 166, 155, [243, 247].

Richart, F. E., Woods, R. D. & Hall, J. R. (1970). *Vibration of Soils and Foundations*. Englewood Cliffs, New Jersey: Prentice-hall, [349].

Richmond, O., Morrison, H. L. & Devenpeck, M. L. (1974). Sphere indentation with application to the Brinell hardness test. *International Journal of Mechanical Sciences*, 16, 75, [170, 176, 200].

Rickerby, D. G. & Macmillan, N. H. (1980). On the oblique impact of a rigid sphere on a rigid-plastic solid. *International Journal of Mechanical Sciences*, 22, 491, [365].

Roark, R. J. (1965). *Formulas for Stress and Strain*, 4th Edition. New York, St Louis, San Francisco, Toronto, London, Sidney: McGraw-Hill, [129].

Robertson, I. A. (1966). Forced vertical vibration of a rigid circular disc on a semi-infinite solid. *Proc. Cambridge Philosophical Society*, **62**, 547, [348].

Rostovtzev, N. A. (1953). Complex stress functions in the axi-symmetric contact problem of the theory of elasticity. *PMM*, **17**, 611, [49].

Routh, E. J. (1908). *Analytical Statics*, Vol. III. Cambridge: University Press, [64].

Rydholm, G. (1981). *On Inequalities and Shakedown in Contact Problems*. Dissertation No. 61, Mech. Eng., University of Linköping, [292].

Sackfield, A. & Hills, D. A. (1983a). Some useful results in the classical Hertz contact problem. *Journal of Strain Analysis*, **18**, 101, [67, 104].

Sackfield, A. & Hills, D. A. (1983b). Some useful results in the tangentially loaded Hertz contact problem. *Journal of Strain Analysis*, **18**, 107, [75, 210].

Sackfield, A. & Hills, D. A. (1983c). A note on the Hertz contact problem: Correlation of standard formulae. *Journal of Strain Analysis*, **18**, 195, [205, 210].

Samuels, L. E. & Mulhearn, T. O. (1956). The deformed zone associated with indentation hardness impressions. *Journal of Mechanics and Physics of Solids*, **5**, 125, [172].

von Schlippe, B. & Dietrich, R. (1941). Das Flattern eines bepneuten Rades. *Ber. Lilienthal Ges.*, **140**, 35, [281, 282].

Segal, V. M. (1971). Plastic contact in the motion of a rough cylinder over a perfectly plastic half-space. *Igv. AN SSSR. Mekh. Tverdogo Tela*, **6** (3), 184, [301].

Shail, R. (1978). Lamé polynomial solutions to some elliptic crack and punch problems. *International Journal of Engineering Science*, **16**, 551, [68].

Shield, R. T. (1955). On plastic flow of metals under conditions of axial symmetry. *Proc. Royal Society*, **A233**, 267, [168, 171, 177].

Sims, R. D. (1954). Calculation of roll force and torque in hot rolling mills. *Proceedings, Institution of Mechanical Engineers*, **168**, 191, [323].

Skalski, K. (1979). Contact problem analysis of an elastoplastic body. Prace Naukowe Mechanica, z. 67, Warsaw Polytechnic, [172].

Smith, J. O. & Liu, C. K. (1953). Stresses due to tangential and normal loads on an elastic solid. *Trans. ASME, Series E, Journal of Applied Mechanics*, **20**, 157, [205].

Sneddon, I. N. (1946). Boussinesq's problem for a flat-ended cylinder. *Proc. Cambridge Philosophical Society*, **42**, 29, [60].

Sneddon, I. N. (1948). Boussinesq's problem for a rigid cone. *Proc. Cambridge Philosophical Society*, **44**, 492, [114].

Sneddon, I. N. (1951). *Fourier Transforms*. New York, Toronto, London: McGraw-Hill, [49, 78, 79, 137, 140, 312].

Söhngen, H. (1954). Zur Theorie der endlichen Hilbert Transformation. *Math. Zeitschrift*, **60**, 31, [30].

Sokolovskii, V. V. (1969). *Theory of Plasticity*. Moscow, [197].

Spence, D. A. (1968). Self-similar solutions to adhesive contact problems with incremental loading. *Proceedings, Royal Society*, **A305**, 55, [74, 78, 79, 80, 114, 119, 123, 251].

Spence, D. A. (1973). An eigenvalue problem for elastic contact with finite friction. *Proc. Cambridge Philosophical Society*, **73**, 249, [37, 40].

Spence, D. A. (1975). The Hertz contact problem with finite friction. *Journal of Elasticity*, **5**, 297, [80, 121, 123].

Steuermann, E. (1939). On Hertz theory of local deformation of compressed bodies. *Comptes Rendus (Doklady) de l'Académie des Sciences de l'URSS*, **25**, 359, [63, 114, 115, 116, 144].

Stilwell, N. A. & Tabor, D. (1961). Elastic recovery of conical indentations. *Proc. Physical Society*, **78**, 169, [183].

Svec, O. J. & Gladwell, G. M. L. (1971). An explicit Boussinesq solution for a polygonal distribution of pressure over a triangular region. *Journal of Elasticity*, **1**, 167, [56].

Symonds, P. S. (1951). Shakedown in continuous media. *Trans. ASME, Series E, Journal of Applied Mechanics*, **18**, 85, [288].

Tabor, D. (1948). A simple theory of static and dynamic hardness. *Proceedings, Royal Society*, **A192**, 247, [181, 364].

Tabor, D. (1951). *Hardness of Metals.* Oxford: University Press, [176, 199].

Tabor, D. (1955). The mechanism of rolling friction: the elastic range. *Proceedings, Royal Society*, **A229**, 198, [285].

Tabor, D. (1959). Junction growth in metallic friction. *Proceedings, Royal Society*, **A251**, 378, [235].

Tabor, D. (1975). Interaction between surfaces: adhesion and friction. In *Surface Physics of Materials*, Vol. II, Chap. 10, Ed. Blakely. New York, San Francisco and London: Academic Press, [125].

Thomas, H. R. & Hoersch, V. A. (1930). *Stress Due to the Pressure of One Elastic Solid Upon Another.* University of Illinois, Engineering Experimental Station, Bulletin No. 212, [66, 99].

Thomas, T. R. (Ed.) (1982). *Rough Surfaces.* London: Longman, [408].

Thompson, J. C. & Robinson, A. R. (1977). An exact solution for the superseismic stage of dynamic contact between a punch and an elastic body. *Trans. ASME, Series E, Journal of Applied Mechanics*, **44**, 583, [355].

Timoshenko, S. & Goodier, J. N. (1951). *Theory of Elasticity*, 3rd Edn. New York, London *et al.*: McGraw-Hill, [12, 16, 25, 51, 52, 76, 130, 143, 340].

Ting, T. C. T. (1966). The contact stress between a rigid indenter and a viscoelastic half-space. *Trans. ASME, Series E, Journal of Applied Mechanics*, **33**, 845, [193, 194, 368].

Ting, T. C. T. (1968). Contact problems in the linear theory of viscoelasticity. *Trans. ASME, Series E, Journal of Applied Mechanics*, **35**, 248, [193].

Tsai, K. C., Dundurs, J. & Keer, L. M. (1974). Elastic layer pressed against a half-space. *Trans. ASME, Series E, Journal of Applied Mechanics*, **41**, 707, [142].

Tsai, Y. M. (1968). A note on surface waves produced by Hertzian impact. *Journal of the Mechanics and Physics of solids*, **16**, 133, [358, 359].

Tsai, Y. M. (1971). Dynamic contact stresses produced by the impact of an axisymmetrical projectile on an elastic half-space. *International Journal of Solids and Structures*, **7**, 543, [358, 359].

Tsai, Y. M. & Kolsky, H. (1967). A study of the fractures produced in glass blocks by impact. *Journal of the Mechanics and Physics of Solids*, **15**, 263, [358].

Turner, J. R. (1979). The frictional unloading problem on linear elastic half-space. *Journal of Institute of Mathematics and its Applications*, **24**, 439, [80, 125].

Turner, J. R. (1980). Contact on a transversely isotropic half-space, or between two transversely isotropic bodies. *International Journal of Solids and Structures*, **16**, 409, [135].

Tyler, J. C., Burton, R. A. & Ku, P. M. (1963). Contact fatigue under an oscillatory normal load. *Trans. ASLE*, **6**, 255, [229].

Ufliand, Ia. S. (1967). *Integral Transforms in the Theory of Elasticity*, 2nd Ed., Nauka, Leningrad. English trans. in *Survey of articles on Appl. of Integral Transforms in the Th. of Elasticity.* North Carolina State Univ., Appl. Math. Res. Grp. File No. PSR-24/6, 1965.

Updike, D. P. & Kalnins, A. (1970). Axisymmetric behaviour of an elastic spherical shell compressed between rigid plates. *Trans. ASME, Series E, Journal of Applied Mechanics*, **37**, 635, [144].

Updike, D. P. & Kalnins, A. (1972). Contact pressure between an elastic spherical shell compressed between rigid plates. *Trans. ASME, Series E, Journal of Applied Mechanics*, **39**, 1110, [144].

Venkatraman, B. (1964). Creep in the presence of a concentrated force at a point. *Conference on Thermal Loading and Creep*, London, Proceedings, Institution of Mechanical Engineers, **178**, Pt. 3L, 111, [197].

Vermeulen, P. J. & Johnson, K. L. (1964). Contact of non-spherical elastic bodies transmitting tangential forces. *Trans. ASME, Series E, Journal of Applied Mechanics*, **31**, 338, [75, 260].

Wannop, G. L. & Archard, J. F. (1973). Elastic hysteresis and a catastrophic wear mechanism for polymers. *Proceedings, Institution of Mechanical Engineers*, **187**, 615, [308].

Ward, I. M. (1971). *Mechanical Properties of Solid Polymers*. New York: Wiley-Interscience, [310].

Wernitz, W. (1958). *Walz-Bohrreibung*. Braunschweig: F. Vieweg & Sohn, [260, 266].

Westergaard, H. M. (1939). Bearing pressures and cracks. *Trans. ASME, Journal of Applied Mechanics*, **6**, 49, [399].

Whitehouse, D. J. & Phillips, M. J. (1978). Discrete properties of random surfaces. *Phil. Trans. Royal Society*, **A290**, 267, [410].

Whitehouse, D. J. & Phillips, M. J. (1982). Two-dimensional discrete properties of random surfaces. *Phil. Trans. Royal Society*, **A305**, 441, [410].

Williams, W. E. (1961). A solution to the steady state thermoelastic equations. *ZAMP*, **12**, 452, [380].

Williamson, J. B. P. (1967–8). The microtopography of surfaces, in *Properties and metrology of surfaces, Proc. Inst. Mech. Eng.* **182** Pt 3K, 21, [409].

Willis, J. R. (1966). Hertzian contact of anisotropic bodies. *Journal of the Mechanics and Physics of Solids*, **14**, 163, [135].

Wilsea, M., Johnson, K. L. & Ashby, M. F. (1975). Indentation of foamed plastics. *International Journal of Mechanical Sciences*, **17**, 457, [178].

Wolfe, P. (1959). The simplex method for quadratic programming. *Econometrica*, **27**, 382, [152].

Wu, T.-S. & Plunkett, R. (1965). On contact problems of thin circular rings. *Trans. ASME, Series E, Journal of Applied Mechanics*, **32**, 11, [144].

Wymer, D. G. & Cameron, A. (1974). Elastohydrodynamic lubrication of a line contact. *Proceedings, Institution of Mechanical Engineers*, **188**, 211, [339].

Yang, W. H. (1966). The contact problem for viscoelastic bodies. *Trans. ASME, Series E, Journal of Applied Mechanics*, **33**, 395, [189, 193].

Zukas, J. A. & Nicholas, T. (1982). *Impact Dynamics*. New York: Wiley, [361].

Subject index